Simulation and Modeling of Nanomaterials

Simulation and Modeling of Nanomaterials

Editor

Vladimir S. Bystrov

MDPI • Basel • Beijing • Wuhan • Barcelona • Belgrade • Manchester • Tokyo • Cluj • Tianjin

Editor
Vladimir S. Bystrov
Institute of Mathematical
Problems of Biology
RAS—The Branch of Keldysh
Institute of Applied
Mathematics of Russian
Academy of Sciences (IMPB
RAS—Branch of KIAM RAS)
Russia

Editorial Office
MDPI
St. Alban-Anlage 66
4052 Basel, Switzerland

This is a reprint of articles from the Special Issue published online in the open access journal *Nanomaterials* (ISSN 2079-4991) (available at: https://www.mdpi.com/journal/nanomaterials/special_issues/simul_model_nano).

For citation purposes, cite each article independently as indicated on the article page online and as indicated below:

LastName, A.A.; LastName, B.B.; LastName, C.C. Article Title. *Journal Name* **Year**, *Volume Number*, Page Range.

ISBN 978-3-0365-4739-8 (Hbk)
ISBN 978-3-0365-4740-4 (PDF)

Contents

Preface to "Simulation and Modeling of Nanomaterials"

Fast progress in nanosciences and materials sciences is closely related to the development and application of computer methods in this field. The development of modern computer modeling and contemporary computational methods of simulation and structure and property calculations of nanomaterials leads to the inevitable development and rise of the works on computer simulation and modeling in investigations of many various types of new nanomaterials, which allows us to reduce the cost of their design and significantly increase the efficiency of the creation of such new and very necessary nanomaterials, which is extremely important. At present, computational modeling and simulation have become the leading approach when creating new materials with predefined properties. These approaches allow us to select the optimal parameters for the nanomaterials themselves (and predict their physical properties, characteristics, and behavior under the different conditions) as well as determine the parameters of the technologies for their practical manufacture.

For this Special Issue, we seek computer modeling, molecular modeling, and numerical studies of nanomaterials carried out through various contemporary methods using molecular mechanics (MM), quantum chemical calculations (QM), including semi-empirical approaches and density functional theory (DFT), partly with a molecular dynamics (MD) approach, and using multiprocessor clusters. The high-level software, such as Quantum ESPRESSO, VASP, etc., is in use for these computational studies.

Vladimir S. Bystrov
Editor

nanomaterials

MDPI

Review

Polarization Switching in 2D Nanoscale Ferroelectrics: Computer Simulation and Experimental Data Analysis

Ekaterina Paramonova [1], Vladimir Bystrov [1,*], Xiangjian Meng [2], Hong Shen [2], Jianlu Wang [2] and Vladimir Fridkin [3]

1 Institute of Mathematical Problems of Biology, Keldysh Institute of Applied Mathematics, RAS, Moscow 142290, Russia; ekatp@yandex.ru
2 National Lab. Infrared Physics, Shanghai Institute of Technical Physics, CAS, Shanghai 200083, China; xjmeng@mail.sitp.ac.cn (X.M.); hongshen@mail.sitp.ac.cn (H.S.); jlwang@mail.sitp.ac.cn (J.W.)
3 Federal Center of Photonics and Crystallography RAS, Shubnikov Institute of Crystallography RAS, Moscow 117333, Russia; fridkinv@gmail.com
* Correspondence: vsbys@mail.ru

Received: 22 August 2020; Accepted: 12 September 2020; Published: 15 September 2020

Abstract: The polarization switching kinetics of nanosized ferroelectric crystals and the transition between homogeneous and domain switching in nanoscale ferroelectric films are considered. Homogeneous switching according to the Ginzburg-Landau-Devonshire (LGD) theory is possible only in two-dimensional (2D) ferroelectrics. The main condition for the applicability of the LGD theory in such systems is its homogeneity along the polarization switching direction. A review is given of the experimental results for two-dimensional (2D) films of a ferroelectric polymer, nanosized barium titanate nanofilms, and hafnium oxide-based films. For ultrathin 2D ferroelectric polymer films, the results are confirmed by first-principle calculations. Fitting of the transition region from homogeneous to domain switching by sigmoidal Boltzmann functions was carried out. Boltzmann function fitting data enabled us to correctly estimate the region sizes of the homogeneous switching in which the LGD theory is valid. These sizes contain several lattice constants or monolayers of a nanosized ferroelectrics.

Keywords: LGD theory; polarization; nanoscale ferroelectrics; kinetics; homogeneous switching; computer simulation; fitting

1. Introduction

Studies of polarization switching in ultrathin (nanoscale) polymer ferroelectric films of polyvinylidene fluoride-trifluoroethylene (P(VDF-TrFE)), obtained experimentally by the Langmuir-Blodgett (LB) method [1–7], have shown that for nanosized (within the film thickness, when their sizes are less than or equal to the critical sizes required for the formation of a domain core), homogeneous non-domain switching of polarization is observed [1,6]. This occurs in accordance with the Landau theory of phase transitions [8], developed for ferroelectrics by Ginzburg and Devonshire (Ginzburg-Landau-Devonshire (LGD) theory) [8–10]. In such homogeneous media, the kinetics of the process and the time of polarization switching are well described by the Landau-Khalatnikov equation [11–13], in the approximation of continuous homogeneous media.

This was shown experimentally [13–16] and theoretically (in first-principle calculations [17], including with molecular dynamics (MD) approaches [18] and using quantum–mechanical semi-empirical methods [17–20]). The main condition for the applicability of the LGD theory in such systems is its homogeneity along the polarization switching direction.

It is clear that it is necessary to apply the numerical estimates obtained by the LGD theory with respect to specific and real ferroelectric samples (of any composition and geometry) with extreme caution, and it is necessary to take into account possible limitations and the approximate nature of the results obtained. Nevertheless, under certain conditions and for some structures, the estimates give quite reasonable values consistent with the experimental values, as we will see from the analysis of the results obtained for a number of different systems carried out in this article.

In this paper we do not deal with domain switching and consider only homogeneous switching and its transition to the domain one. But domain switching works well for the sizes when domain nuclei can already form—all these domain switching processes are remarkably described in the Tagantsev et al. monograph [21].

Naturally, many issues of domain formation and switching in various real systems remain quite complex and require careful analysis in each specific case. Note, in this case, that the phenomenological LGD theory itself does not answer all the questions and does not consider the mechanisms of polarization switching themselves (especially at the microscopic level). This is a continual theory and it describes the thermodynamics of changes in polarization during a phase transition, which is determined by the potential barrier between the polar and nonpolar phases in the LGD theory. This theory also considers the kinetics of changes in polarization in a homogeneous continuous medium according to the nonlinear Ladnau-Khalatnikov equations—again in the approximation of a homogeneous infinite medium. All this must be taken into account in applications to various real structures.

It is known that the LGD theory does not describe the switching of conventional bulk ferroelectrics, since it predicts the magnitude of the coercive field, which is 2 to 3 times higher than the experimental one. The large coercive fields predicted by the LGD theory came to be called *intrinsic* (or *proper*), and their experimental values are *extrinsic* (or *improper*).

Domain discovery helped resolve this inconsistency. As it turned out, such bulk ferroelectrics inevitably split into domains [21]. This division of the polar crystal into domains reduces the free energy of the crystal. In this case, the minimum of the free energy or thermodynamic potential of the crystal below the Curie point in the polar ferroelectric phase is achieved if the crystal is divided into domains.

The appearance of nanoscale ferroelectrics, namely, polymer ferroelectric films, the thickness of which reached the minimum possible values of one mono-molecular layer (~0.5 nm) [1–7], created a completely new and different situation.

The question of domain formation does not arise, since such layers are much smaller than the size of a possible domain (~10 nm [21]), but at the same time all these layers (as one layer and also all layers as a whole, if their formation is created from several layers) created significant spontaneous polarization in their polar phase. In this case, even the values of the coercive fields turned out to be much larger than those in bulk ferroelectrics and they were close to the values of the LGD theory. It became possible to talk about the applicability of the LGD theory here.

Thus, here and further in this article, we will not talk about the formation of domain nuclei and the growth of domains (and their role in polarization switching), but we are only talking about non-domain homogeneous polarization switching, under the conditions of a continuous homogeneous medium, when its size (thickness) does not exceed the dimensions of the formation of domains and one can speak of one continuous medium in the direction of polarization.

We also note that for some time it was still believed that such uniform switching is only possible in polymer ferroelectrics, and cannot be observed in other perovskite ferroelectrics and similar crystals. Nevertheless, since the discovery of homogeneous switching in ferroelectric polymers, it has become evident that all nanoscale ferroelectric films and crystals whose thickness is less than or comparable to the size of a domain nucleus can have a homogeneous nature of polarization switching. This has recently been proved by the example of ferroelectric films of a classical ferroelectric–barium titanate [1,3,22]—in their nanoscale form. In this work, a comparative analysis of such homogeneous polarization switching in some different nanoscale ferroelectrics is carried out.

2. Methods and Models

Before we move on to modeling and discussion of results obtained, we should make a short introduction into the LGD theory and its estimations and predictions. As it was written above, the values of the intrinsic coercive field E_C are determined by the potential barrier between the polar and nonpolar phases as described by the polarization expansion coefficients of thermodynamic potential (or free energy) known from the LGD theory [4,5]:

$$\Phi = \Phi_0 + \frac{\alpha}{2}P^2 + \frac{\beta}{4}P^4 + \frac{\gamma}{6}P^6 - EP \tag{1}$$

where Φ is free energy, P is spontaneous polarization, E is an external field, and the coercive field E_C is expressed as:

$$E_C = \frac{P_0}{\chi_0}f(t) \tag{2}$$

where $P_0 = P(T = T_0) = \sqrt{(-\beta/\gamma)}$, $\chi_0 = \chi(T = T_0) = \gamma/2\beta^2$, $f(t)$ is the function:

$$f(t) = \frac{3}{25}\sqrt{\frac{3}{5}}\left(1 - \frac{25}{24}t\right) \tag{3}$$

and the reduced temperature is:

$$t = \frac{4\alpha\gamma}{\beta^2} = \frac{4\gamma}{\varepsilon_0 C\beta^2}(T - T_0) \tag{4}$$

Here, χ_0 is the ferroelectric contribution to the dielectric susceptibility, and α, β, γ are coefficients of the expansion of free energy in even degrees of polarization, known from the LGD theory. For estimates, it can be assumed that approximately $E_C \sim P/\chi_0 \sim P/\varepsilon\varepsilon_0$, where ε is the relative permittivity and ε_0 is the dielectric constant of the vacuum.

The polarization switching kinetics of two-dimensional (2D) polymer ferroelectrics was described by the Landau–Khalatnikov Equation [11] and its solution for first-order phase transitions in two-dimenstional ferroelectrics was considered in [12,13]:

$$\xi\frac{dP}{dt} = \frac{-\partial\Phi}{\partial P} = -\alpha P - \beta P^3 - \gamma P^5 + E \tag{5}$$

where ξ is the damping coefficient. In the general case, the gradient term can be taken into account. An investigation of the solution of this equation showed that in the vicinity of the coercive field E_C, the switching time sharply increases, and its reciprocal can be expressed as [12]:

$$\tau^{-1} \approx \frac{1}{\tau_0}\left(\frac{E}{E_C} - 1\right)^{\frac{1}{2}} \tag{6}$$

where $\tau_0 \cong 6.3\beta\gamma\xi/\beta^2$, otherwise:

$$\tau^{-2} \approx \frac{1}{\tau_0^2}\left(\frac{E}{E_C} - 1\right) \tag{7}$$

In this case, E_C is the proper (or intrinsic) coercive field of the ferroelectrics. This relation (7) shows the linear behavior of τ^{-2} along E in the vicinity of E_C (for $E > E_C$). This relation is more suitable for comparison with experimentally measured data and is used in [14,15,18]. This relationship turned out to be convenient for analyzing the results of theoretical calculations when modeling the polarization switching processes in polymer ferroelectrics by molecular dynamics (MD) methods [18,19].

The discovery of two-dimensional ferroelectrics [1,2] led to a new stage of polarization switching development, using the study of switching of these ultrathin single-crystal films. A start was made by the development of a new method for growing single-crystal ferroelectric polymer films of polyvinylidene fluoride-trifluoroethylene (P(VDF-TrFE)) [1–7].

To grow the ferroelectric films of this polymer, the Langmuir–Blodgett (LB) method [1–7,14–16,22–24] was used, based on the transfer of polymer chains (or mono-layers-ML) from the surface of the water to a substrate that carries an electrode. Figure 1a–d shows polymer chains and cells in the polar (ferroelectric) and nonpolar (paraelectric) phases (these images were built using the HyperChem 8.0 tool (Hypercube Inc., Gainesville, FL, USA) [20]); Figure 1e–g shows a transport scheme of LB layers and 1 ML observed in a scanning tunneling microscope. The Langmuir ferroelectric films obtained by this method in 1998 were thinner than any films that were previously obtained. The thickness of one monolayer (1 ML) was 0.5 nm; that is significantly less than the size of the critical domain nucleus known from the literature [21]. The thickness of Langmuir polymer films (two-dimensional ferroelectrics) was controlled by ellipsometry and atomic force spectroscopy, and two-dimensional ferroelectrics 0.5–1.0 nm thick were first obtained.

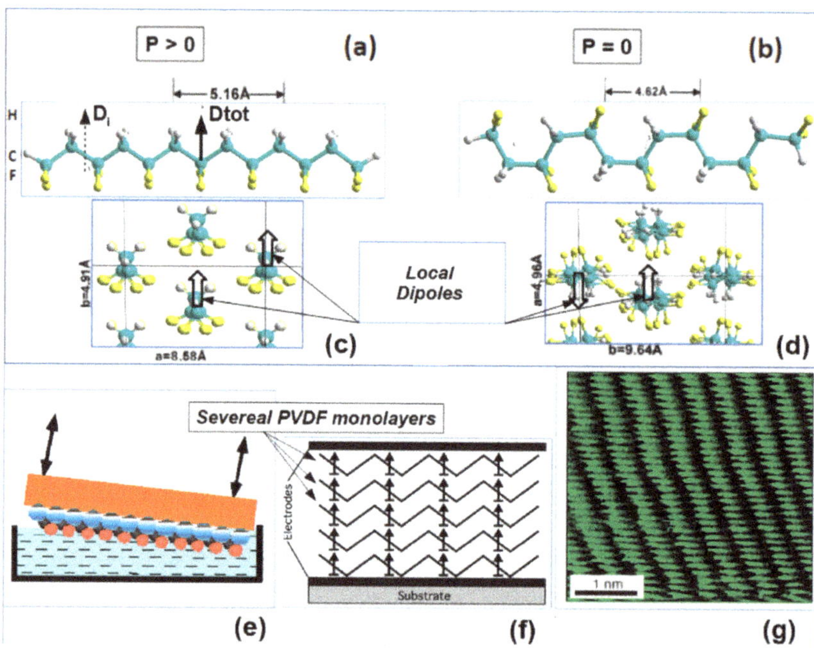

Figure 1. Ferroelectric polymer polyvinylidene fluoride (PVDF): (**a**,**c**) PVDF in polar trans conformation and total polarization $P > 0$; (**b**,**d**) PVDF in a nonpolar gauche conformation with total polarization $P = 0$. Reproduced with permission from [23]; AIP Publishing, 2012. (**e**) The formation of the PVDF Langmuir-Blodgett (LB) film on the surface of the water. Reproduced with permission from [4]; IOPscience, 2000. (**f**) Transferring several mono-layers (ML) of LB PVDF film onto a substrate with an electrode. Reproduced with permission from [24]; IEEE, 2005. (**g**) An image of the 1 M LB film of polyvinylidene fluoride-trifluoroethylene P(VDF-TrFE) by scanning tunneling microscopy [1,2]. Reproduced with permission from [2]; Springer Nature, 1998.

Experimental study [14,15] and computational simulation [17–19] of the polarization switching, carried out on these ultra-thin polymer ferroelectrics have confirmed the relationships (6) and (7). The MD approach calculations [18,19] were carried out using the HyperChem software [20], using various modes and semi-empirical methods (PM3, etc.) for correct quantum-chemical calculations at each step of the MD simulation run process. As an example of the MD run calculations, simple 2 PVDF chains affected by an external electric filed *E* are presented (Figure 2).

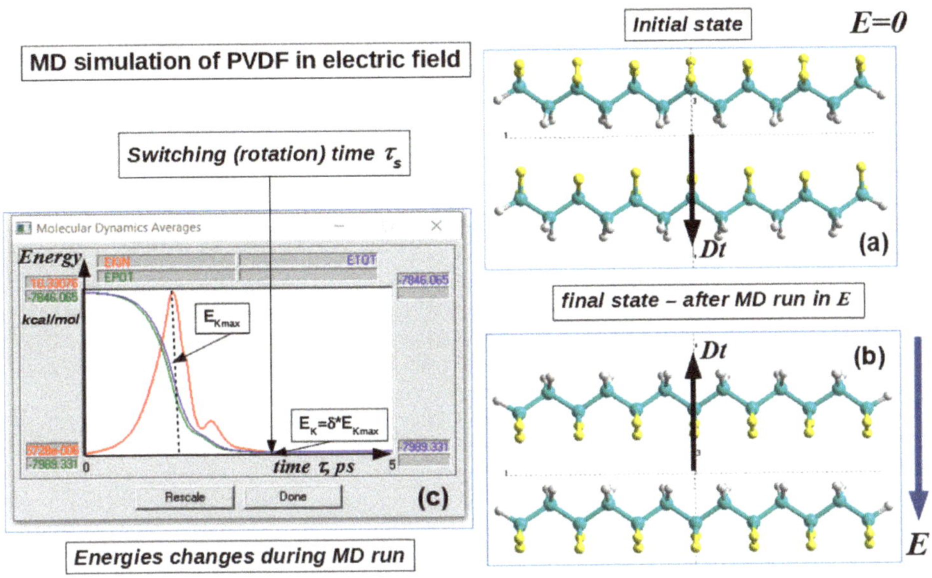

Figure 2. Scheme of the MD run process for a 2 PVDF-6 chains model with PM3 (in restricted Hartree-Fock approximation–RHF) calculations at each MD run step: (**a**) initial state; (**b**) final state after the MD run with dipole moment D orientation turned (switched) in the opposite direction in the electric field E; (**c**) changes of the MD average energies trajectory over time during the MD run (in ps) and the time of the switching τ_S for this PVDF system (red–kinetic energy, blue–total energy, green–potential energy). Reproduced with permission from [18]; Elsevier, 2014.

The final time, for example, for the case of the two-chain model rotation (switching time τ_S) was estimated from these MD energy trajectories (see Figure 2c) using criteria [18]: $\delta = E_K/E_{K\max} < 10^{-3}$, where E_K is the kinetic energy at the final point and $E_{K\max}$ is the kinetic energy at the maximum E_{KIN} point of the chain rotation (as shown in Figure 2c). As a result, this corresponds to reaching the rest point of the rotating chain and its new position from another opposite orientation of the total dipole ***Dt*** vector and polarization vector. With a similar precision we calculated the values of the coercive field, obtained from a hysteresis loop, corresponding to the switching of the PVDF chain polarization into the opposite direction (or the same rotation of the PVDF chain). These data were further used in the calculations of the thickness dependence of the coercive field (see below in Section 3.1).

3. Results and Discussion

3.1. Polymer Ferroelectrics

As a result, this MD approach allows us to determine switching (or rotation) time τ_S (Figure 2c) for any PVDF system as the applied electric field ***E*** varies. Using this approach, the switching time $\tau_S = \tau$ (Figure 3b) was calculated for various polymer ferroelectric models [18,19] and the results are compared with experimental data [14,15] (Figure 3a).

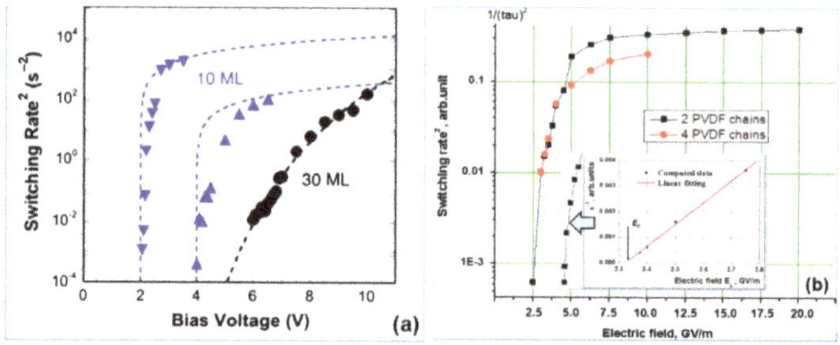

Figure 3. Switching time τs for polymer ferroelectrics presented as τ^{-2} along E in a logarithmic scale: (**a**) experimental data [14,15] for thin LB of P(VDF-TrFE) film of 10 ML (5 nm) with critical behavior of τ^{-2} for $E \rightarrow E_C$ in comparison with thick 30 ML (15 nm) films with exponential low (Reproduced with permission from [14]; AIP Publishing, 2011); (**b**) MD simulation run data [18,19] for PVDF models showing the critical behavior of τ^{-2} with a change in E, which qualitatively coincides with the experimentally observed behavior; the inset shows the linear behavior of τ^{-2} as $E \rightarrow E_C$ and determines the critical value of E_C ~ 2.3 GV/m (Reproduced with permission from [18]; Elsevier, 2014).

The critical behavior of τ^{-2} for $E \rightarrow E_C$ for the P(VDF-TrFE) LB films was shown in [14,15] (see Figure 3a). Figure 3a presents the data measured for a film of 10 ML (5 nm). Dotted lines show theoretical points corresponding to Equation (7), and triangles show experimental data. For a thicker film of 30 ML (15 nm), the circles indicate the experimental values of τ^{-2}, and the dotted line indicates the exponential dependence. Figure 3b shows the calculated data using the MD run for $\tau^{-2}(E)$ behavior in a similar logarithmic scale, which qualitatively coincides with the experimentally established character of the behavior [14,15] (Figure 3a). The inset in Figure 3b shows a linear dependence of Equation (7) $\tau^{-2}(E)$ on E near $E_{C,,}$ enabling interpolation to the intersection point with the horizontal axis, which determines the value of the coercive field E_C ~ 2.3–2.4 GV/m for 2–4 chain models (ML monolayers).

Taking into account that the field E is external and that for thin polymer layers representing monomolecular layers, the dielectric constant is ε ~ 2.4 [17,25–29] (while ε = 5 and greater is only for thick films), we obtain the limiting maximum value E_{CMAX} ~ $E_C/\varepsilon \cong 1$ GV/m, which is a proper coercive field and corresponds to many known experimental data as well as the LGD theory [1–7,16,29].

A similar study of the P(VDF-TrFE) LB films of different thicknesses was carried out in [16] (Figure 4). It was shown that for small thicknesses of 2–6 nm, the coercive field E_C is proper and practically unchanged, and in the region of thicknesses greater than 8 nm, a transition region arises. It was show that for thicknesses greater than 10–12 nm, the proper coercive field E_C becomes improper and is determined by the domain mechanism.

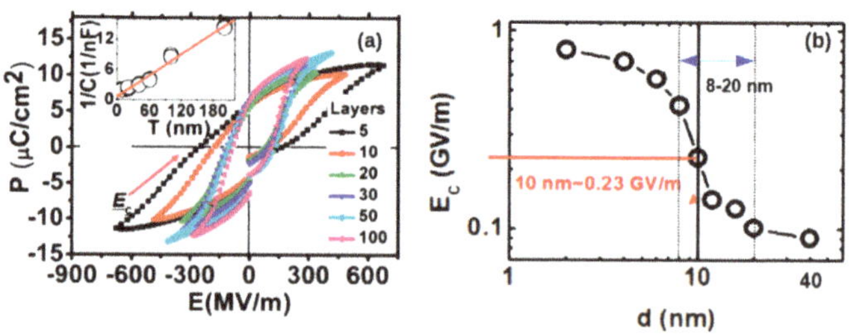

Figure 4. Hysteresis loops of the LB PVDF film [16]: (**a**) with different numbers of monolayers: 5, 10, 20, 30, 50 and 100 (the inset shows the linearity of the reciprocal capacity depending on the thickness); (**b**) E_C as a function of the LB thickness of the PVDF film (with the transition region at 8–20 nm). Reproduced with permission from [16]; AIP Publishing, 2014.

Another method for obtaining a coercive field is by calculation of hysteresis loops $P(E)$ [27,28]. Calculations performed for different numbers of chains and film thicknesses using both these methods showed that the dependence of the obtained coercive field [18,19,28] is in good agreement with the experimentally established dependence of the coercive field E_C on the film thickness [16] (Figure 5).

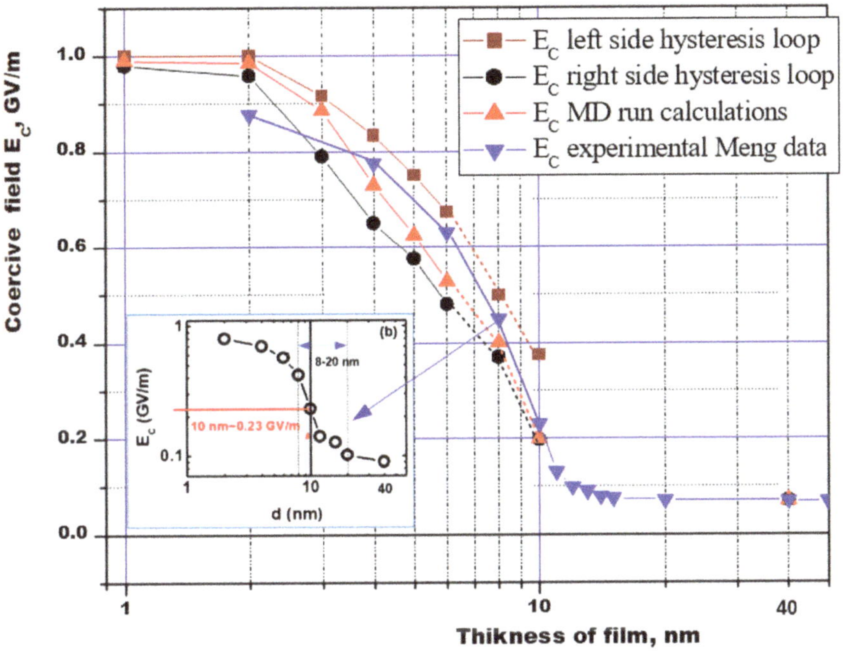

Figure 5. The dependence of the coercive field of the ferroelectric polymer PVDF on the film thickness, according to the results of calculations by different methods (from hysteresis loops and MD runs) in comparison with experimental data [16] (inset (b) is from Figure 4), taking into account the dielectric constant of an ultrathin molecular film $\varepsilon \approx 2.4$. (It is our new original recalculated data, which continues from our preliminary calculations and estimations [18,19,27,28]).

This dependence can be conditionally divided into 3 regions: the region of purely homogeneous LGD switching (up to 8 nm); the transition region (8–12 nm), where a kind of domain precursor is noted; and the region above 12–16 nm and further, where the domain switching mechanism predominates (the coercive field remains almost unchanged and is kept at a low level of ~0.07–0.05 GV/m) [16,18,19,27,28]. Thus, these switching calculations of ultrathin polymer ferroelectrics confirm that two-dimensional ferroelectrics can consist, in principle, of several monolayers or several unit cells. These data are also in good agreement with the results for thin $BaTiO_3$ films (see in Figure 6 below), in which homogeneous switching was recently found at the scale level up 10 nm, and with a further increase in thickness they already correspond to thick films [1–4,22].

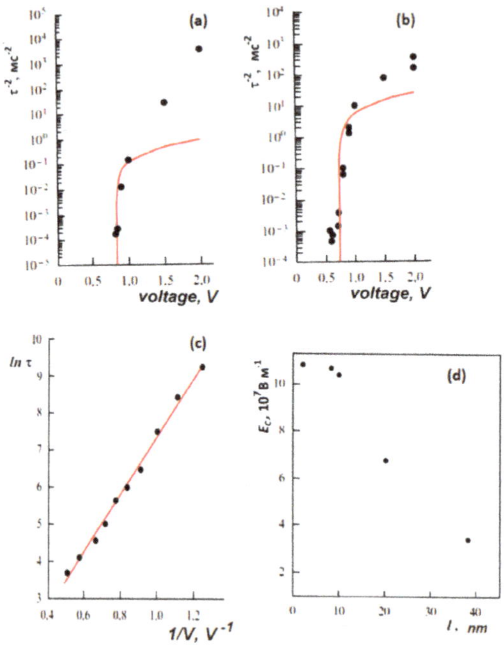

Figure 6. The dependence of the switching time τ on the applied voltage for 8-nm-thick $BaTiO_3$ film [1,3,22]: (**a**) in the capacitor and (**b**) in the probe mode in an atomic force microscope (lines τ^{-2} (V) on (**a**) and (**b**) correspond to the calculations according to formula (7)); (**c**) for a $BaTiO_3$ crystal, the line corresponds to $\ln[\tau(V^{-1})]$ and is obtained relying on the piezo-response mode in an atomic force microscope; (**d**) the dependence of the coercive field on the thickness of the $BaTiO_3$ film. Reproduced with permission from [22]; Elsevier, 2013.

3.2. Barium Titanate

The switching kinetics in these ultrathin single-crystal $BaTiO_3$ epitaxial laser films of 2–8 and 40 nm thickness synthesized on a $SrRuO_3/SrTiO_3$ substrate were studied in a $SrRuO_3$–ITO capacitor using an atomic force microscope. A detailed description of the technique is presented in [30]. The distinction between proper and improper behavior is shown in Figure 5, which shows the dependence of the switching time τ on the applied voltage V for a film of 8 nm thickness in a capacitor (Figure 6a), and under a tip of an atomic force microscope (Figure 6b). For comparison, the same dependence for a bulk $BaTiO_3$ single crystal of 1 mm thickness is shown in Figure 6c.

The dependence τ^{-2}(V) for a barium titanate film in a capacitor and in an atomic force microscope is shown to be consistent with formula (2) (Figure 6a,b). These experimental results agree well with relation (7). On the contrary, the switching kinetics of the bulk crystal (Figure 6c) is nearly exponential; the switching has a domain nature. Figure 6d shows the dependence of the coercive field on the film thickness of barium titanate in the range from 2 to 40 nm. It is seen that in the range from 2 to 10 nm, the coercive field is intrinsic and proper ($E_C \sim 0.12$ GV/m) and weakly depends on the thickness, in accordance with the LGD theory. For thicknesses of 40 nm (or more), the coercive field sharply decreases, which corresponds to the transition to domain switching. Correspondingly, films with a thickness of 2 and 8 nm exhibit their intrinsic homogeneous LGD switching (5)–(7), and thicker films exhibit a domain exponential dependence. These data are also in good agreement with the above results for LB polymer films [16] (Figures 4 and 5).

Thus, the obtained and experimentally observed dependencies of the switching time on the thickness of the barium titanate film (as well as the polymer films mentioned and described above) fully correspond to the Landau-Khalatnikov kinetics, and give the values of the coercive field E_C, which exactly follows from the phenomenological LGD theory.

At the same time, it should be noted that the surface field caused by the surface charge, in principle, affects the coercive field in thin two-dimensional and nanoscale ferroelectric fields [31]. However, as it was shown in paper [32], the switching kinetics of these two-dimensional and nanoscale films exactly follow the Landau-Khalatnikov kinetics. In addition, this has been verified in experiments [14,15,22] and demonstrated in this work for ferroelectric polymer films and perovskite ferroelectric films.

Moreover, it should be reiterated, that it is Landau-Khalatnikov kinetics that give the exact eigenvalue for the values of the coercive field E_C for nanosized ferroelectrics, which follows from the phenomenological LGD theory.

3.3. Nanosized Materials Based on Hafnium Oxide

In the past few years, there has also been interest in nanoscale films based on hafnium oxide [33–36], including those doped with silicon and several types of their solid solutions, such as for example, $Hf_{0.5}Zr_{0.5}O_2$ (HZO) [34,35]. Ferroelectricity was found in them and the corresponding hysteresis loops (Figure 7) suggest the proper intrinsic nature of the coercive field (~0.1–0.2 GV/m) here and, apparently, the homogeneous nature of the polarization switching. Though these samples are fully blended HZO solid solutions [36], there are some doubts as to whether they are fully homogeneous and fully monocrystallic, since they can consist of several phases. Therefore, it is too early to draw conclusions about homogeneous switching and its kinetics; these studies are ongoing.

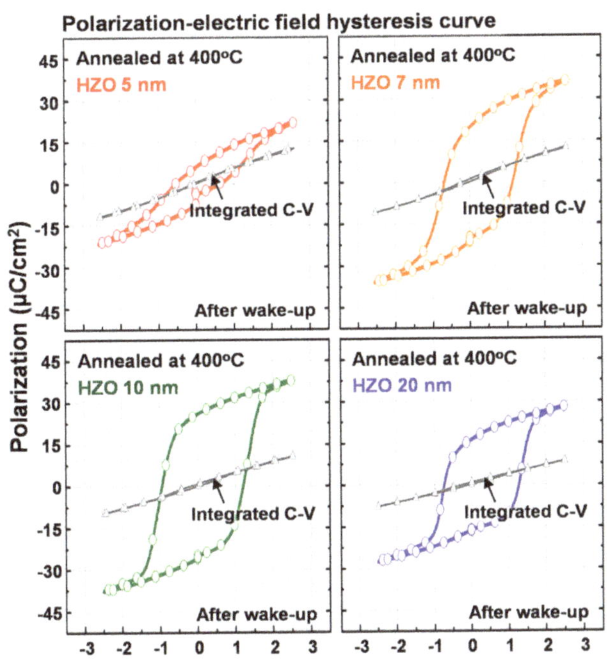

Figure 7. Polarization-electric field hysteresis and integrated capacitance voltage hysteresis curves of 5-, 7-, 10-, and 20-nm-thick HZO-based metal–insulator–metal (MIM) capacitors after wake-up field cycling (Reproduced with permission from [35]; AIP Publishing, 2018).

The dependence of the coercive field on the thickness of the ultrathin HZO films (4–20 nm) is also observed (some data are given in review [36]). However, there are no reliable and detailed measurements yet. Nevertheless, the study of these nanoscale films is of great interest and here we also tried to estimate whether their possible parameters match with homogeneous polarization switching. For example, one can estimate the data of [35] (Figure 7); at least approximately, they correspond to the dependence of the coercive field EC on the films' thickness x in such HZO-based films.

3.4. Boltzmann Function Fitting Data

Relying on the fact that the character of the transition region has a pronounced sigmoidal type for all the above-considered samples, we carried out fitting of various available experimental data in accordance with the formula of the sigmoidal Boltzmann function (in the form used by the OriginLab software, OriginLab Corp., Northampton, MA, USA) [37]:

$$E_C = \frac{A_1 - A_2}{1 + \exp\left(\frac{x - x_0}{dx}\right)} + A_2 \tag{8}$$

where A_1 and A_2 are parameters of the maximum and minimum function values (corresponding to the values of the coercive field E_C—a proper (intrinsic) and improper (extrinsic), respectively, x_0 is the average thickness of the films or samples corresponding to the middle of the transition region, and dx is the effective half-width of the transition region (see in Table 1).

Table 1. Fitted parameters of the Boltzmann function for various nanoscale ferroelectrics in comparison with modeling and calculations using MD runs and hysteresis loop data.

		Parameters of Boltzmann Function Fitting				
		A_1, GV/m	A_2, GV/m	dx, nm	x_0, nm	Accuracy
Modeling	MD [18,19]	1.28 ± 0.10 1.26 ± 0.10 1.22 ± 0.14	0.067 ± 0.023 0.067 ± 0.024 0.075 ± 0.038	3.12 ± 0.37 3.09 ± 0.38 2.67 ± 0.51	5.10 ± 0.69 5.21 ± 0.61 5.02 ± 0.77	0.00052 0.00056 0.00152
	MD run (av.)	1.25 ± 0.20	0.070 ± 0.051	2.96 ± 0.74	5.11 ± 1.19	
	Left HL *)	1.23 ± 0.07	0.072 ± 0.019	3.42 ± 0.36	6.30 ± 0.51	0.00035
	Right HL *)	1.51 ± 0.38	0.068 ± 0.038	3.29 ± 0.75	2.99 ± 1.82	0.00148
Experiments	PVDF [16]	0.89 ± 0.03	0.062 ± 0.009	1.71 ± 0.14	7.55 ± 0.19	0.00039
	$BaTiO_3$ [3,22]	0.11 ± 0.00005	0.035 ± 0.00004	3.83 ± 0.017	19.13 ± 0.012	1.668×10^{-9}
	HZO [30]	0.13 ± 0.04	0.055 ± 0.001	5.77 ± 1.66	4.61 ± 5.52	1.399×10^{-6}

*) Data obtained from calculations of the hysteresis loop (HL) [27,28]: Left HL is based on the E_C corresponding to the left side of HL, Right HL is based on the E_C that was obtained from the right side of HL.

The results obtained generally show that Equation (8) describes rather well how a coercive field changes as the thickness of all the samples varies. Moreover, for PVDF, the values obtained from the experiment and those obtained by MD modeling and hysteresis loops (especially the left side) calculations are close (see in Figure 5).

These values of the coercive field turn out to be the highest of all the experimental samples and indicate the existence of an intrinsic coercive field Ec up to sizes of at least 5–8 nm (which corresponds to 10–16 ML, where 1 ML = 0.5 nm [1–7]). That is, in this case, the existence of two-dimensional ferroelectrics is obvious up to ~10–16 PVDF ML (or up to ~8 nm).

At the same time, it turns out to be rather unexpected that for $BaTiO_3$ the inflection point of the sigmoid (8) here occurs at x_0 ~ 19–20 nm, that is, here the region of existence of a homogeneous ferroelectric in this perovskite crystal structure is almost 3 times greater than that of a polymer PVDF film. For HZO, the order of the width of the region is ~5 nm, which is rather close to that of PVDF, but with a significantly larger half-width dx ~ 6 nm, meaning there exists "smearing" of the transition region in a twice wider range. This, of course, is due to the insufficient number of accurate experimental data. It should be noted that the maximum value of Ec for barium titanate and HZO-based films turns out to be significantly lower than PVDF (almost by an order of magnitude). This, in principle, is not surprising and would be expected.

Thus, the proposed approach shows satisfactory fitting data and can be extended to the analysis of other similar data. In addition, it can be used to obtain significant parameters for two-dimensional ferroelectrics and homogeneous polarization switching in them.

4. Conclusions

The study of nanoscale ferroelectrics became possible for the first time when polymer ferroelectric films were synthesized by the Langmuir-Blodgett method [1–7]. This led to the discovery of two-dimensional ferroelectrics [2]. Soon, for classical perovskite ferroelectrics such as barium titanate, it became possible to create, by laser epitaxy, films several tens of nm thick [22]. The study of nanoscale perovskite films of barium titanate showed that they can also be two-dimensional ferroelectrics [1–4]. It can be assumed that other nanoscale single crystals can be two-dimensional and homogeneous. Recently, such nanoscale films have been obtained on the basis of hafnium oxide [33–36].

In this case, the phenomenological LGD theory describes well the switching kinetics, only if the medium is homogeneous and the size is slightly less or of the order of the critical size of the domain nucleus formation, which is the case for two-dimensional ferroelectrics. Thus, by two-dimensional ferroelectric we mean a nanoscale crystal, which in the direction of its switching can be considered to be homogeneous. It can consist, in principle, of several unit cells (or monolayers). Molecular modeling and quantum-mechanical calculations of polymer ferroelectrics using hysteresis loops and molecular dynamics methods were in good agreement and the presence of a transition region from homogeneous switching to domain switching of several nm (6–20 nm) length, thereby convincingly confirming that two-dimensional ferroelectrics can have several monolayers and cells. Similar results were obtained for nanoscale barium titanate and hafnium oxide.

The sigmoidal nature (described by the Boltzmann function (8)) of the transition region turned out to be common here. For various nanoscale ferroelectrics, it is possible to approximate quite effectively and thereby determine the important parameters of the transition region from the maximal value A1 (corresponding to the proper intrinsic coercive field E_C for nanoscale films) to the minimal A2 value (corresponding to the known improper coercive field in thick films and bulk crystals) with half-width dx and middle point x_0. These parameters, in general, determine the region of homogeneous switching in such nanosized ferroelectrics.

Therefore, one should distinguish between the intrinsic coercive field E_C for nanoscale homogeneous ferroelectric films and the previously known improper coercive field E_C in thick films and crystals, associated with the domain mechanism. The intrinsic field is several orders of magnitude greater than the experimental (improper) one.

In fact, in nanoscale ferroelectric films there is a competition between these two switching mechanisms: homogeneous and domain. As a result, when the film thickness increases, the domain mechanism should prevail. Thus, this parameter x_0 actually shows and estimates the size of the possible region of validity of the existence and applicability of the LGD theory for homogeneous polarization switching in the nanoscale ferroelectrics.

Author Contributions: E.P. and V.B. wrote the manuscript. V.F. supervised and supported this study and critically revised the manuscript. X.M., H.S. and J.W. contributed with the experiments and data analysis. All authors have read and agreed to the published version of the manuscript.

Funding: This research was funded by the Russian Foundation for Basic Research, RFBR grant # 20-51-53014_a.

Acknowledgments: This work was supported by the Russian Foundation for Basic Research (RFBR grants # 19-01-00519_a and # 20-51-53014_a) and was also completed as a part of the state assignment of the Ministry of Education and Science of the Russian Federation (No. 0017-2019-0009 and 01201373458 in Keldysh Institute of Applied Mathematics, RAS). Xiang-Jian Meng expresses his gratitude to the National Natural Science Foundation of China (NNSFC) for grant # 61574151 and Hong Shen expresses his gratitude for grant # 62011530043.

Conflicts of Interest: The authors declare no conflict of interest.

References

1. Fridkin, V.M.; Ducharme, S. *Ferroelectricity at the Nanoscale. Basic and Applications*, 1st ed.; Springer: New York, NY, USA, 2014.
2. Bune, A.V.; Fridkin, V.M.; Ducharme, S.; Blinov, L.M.; Palto, S.P.; Sorokin, A.V.; Yudin, S.G.; Zlatkin, A. Two-dimensional ferroelectric films. *Nature* **1998**, *391*, 874–877. [CrossRef]
3. Fridkin, V.M.; Ducharme, S. Ferroelectricity at the nanoscale. *Phys. Usp.* **2014**, *57*, 597–603. [CrossRef]
4. Blinov, L.M.; Fridkin, V.M.; Palto, S.P.; Bune, A.V.; Dowben, P.A.; Ducharme, S. Two-dimensional ferroelectrics. *Phys. Usp.* **2000**, *43*, 243–257. [CrossRef]
5. Palto, S.; Blinov, L.; Bune, A.; Dubovik, E.; Fridkin, V.; Petukhova, N.; Verkhovskaya, K.; Yudin, S. Ferroelectric Langmuir-Blodgett films. *Ferroelectr. Lett.* **1995**, *19*, 65–68. [CrossRef]
6. Bune, A.; Ducharme, S.; Fridkin, V.; Blinov, L.; Palto, S.; Petukhova, N.; Yudin, S. Novel switching phenomena in ferroelectric Langmuir–Blodgett films. *Appl. Phys. Lett.* **1995**, *67*, 3975. [CrossRef]
7. Palto, S.; Blinov, L.; Bune, A.; Dubovik, E.; Fridkin, V.; Petukhova, N.; Verkhovskaya, K.; Yudin, S. Ferroelectric Langmuir-Blodgett films. *Ferroelectrics* **1996**, *184*, 127–129. [CrossRef]
8. Landau, L.D. On the theory of phase transitions. *Phys. Z. Sowjet.* **1937**, *11*, 545. [CrossRef]
9. Ginzburg, V.L. On the *dielectric properties* of ferroelectric (seignetteelectric) crystals and *barium titanate*. *Zh. Eksp. Teor. Fiz.* **1945**, *15*, 739–750.
10. Ginzburg, V.L. On polarization and piezoelectric effect of barium titanate near the point of ferroelectric transition. *Zh. Eksp. Teor. Fiz.* **1949**, *19*, 36–41.
11. Landau, L.D.; Khalatnikov, I.M. On the anomalous absorption of sound near a second-order phase transition point. *Dokl. Akad. Nauk SSSR* **1954**, *96*, 469.
12. Vizdrik, G.; Ducharme, S.; Fridkin, V.M.; Yudin, S.G. Kinetic of ferroelectric switching in ultrathin films. *Phys. Rev. B* **2003**, *68*, 094113. [CrossRef]
13. Ievlev, A.; Verkhovskaya, K.; Fridkin, V. Landau-Khalatnikov switching kinetics in the ferroelectric copolymers nanostructures. *Ferroelectr. Lett.* **2006**, *33*, 147–152. [CrossRef]
14. Gaynutdinov, R.V.; Mitko, S.; Yudin, S.G.; Fridkin, V.M.; Ducharme, S. Polarization switching at the nanoscale in ferroelectric copolymer thin films. *Appl. Phys. Lett.* **2011**, *99*, 142904. [CrossRef]
15. Gaynutdinov, R.; Yudin, S.; Ducharme, S.; Fridkin, V. Homogeneous switching in ultrathin ferroelectric films. *J. Phys. Condens. Matter* **2012**, *24*, 015902. [CrossRef] [PubMed]
16. Wang, J.L.; Liu, B.L.; Zhao, X.L.; Tian, B.B.; Zou, Y.H.; Sun, S.; Shen, H.; Sun, J.L.; Meng, X.J.; Chu, J.H. Transition of the polarization switching from extrinsic to intrinsic in the ultrathin polyvinylidene fluoride homopolymer films. *Appl. Phys. Lett.* **2014**, *104*, 182907. [CrossRef]
17. Bystrov, V.S.; Bystrova, N.K.; Paramonova, E.V.; Vizdrik, G.; Sapronova, A.V.; Kuehn, M.; Kliem, H.; Kholkin, A.L. First principle calculations of molecular polarization switching in P(VDF–TrFE) ferroelectric thin Langmuir–Blodgett films. *J. Phys. Condens. Matter* **2007**, *19*, 456210. [CrossRef]
18. Bystrov, V.S. Molecular modeling and molecular dynamic simulation of the polarization switching phenomena in the ferroelectric polymers PVDF at the nanoscale. *Phys. B Condens. Matter* **2014**, *432*, 21–25. [CrossRef]
19. Paramonova, E.V.; Filippov, S.V.; Gevorkyan, V.E.; Avakyan, L.A.; Meng, X.J.; Tian, B.B.; Wang, J.L.; Bystrov, V.S. Polarization switching in ultrathin polyvinylidene fluoride homopolymer ferroelectric films. *Ferroelectrics* **2017**, *509*, 143–157. [CrossRef]
20. HyperChem (TM), Tools for Molecular Modeling; HyperChem 8.0, Professional Edition. Available online: http://www.hyper.com/?tabidD360 (accessed on 27 July 2020).
21. Tagantsev, A.K.; Cross, L.E.; Fousek, J. *Domains in Ferroelectric Crystals and Thin Films*; Springer: New York, NY, USA, 2010.
22. Gaynutdinov, R.; Minnekaev, M.; Mitko, S.; Tolstikhina, A.; Zenkevich, A.; Ducharme, S.; Fridkin, V. Polarization switching kinetics in ultrathin ferroelectric barium titanate film. *Phys. B Condens. Matter* **2013**, *424*, 8–12. [CrossRef]
23. Bystrov, V.S.; Paramonova, E.V.; Dekhtyar, Y.; Pullar, R.C.; Katashev, A.; Polyaka, N.; Bystrova, A.V.; Sapronova, A.V.; Fridlin, V.M.; Kliem, H.; et al. Polarization of poly(vinylidene fluoride) and poly(vinylidene

fluoride-trifluoroethylene) thin films revealed by emission spectroscopy with computational simulation during phase transition. *J. Appl. Phys.* **2012**, *111*, 104113. [CrossRef]

24. Ducharme, S.; Reece, T.J.; Othon, C.M.; Rannow, R.K. Ferroelectric polymer Langmuir-Blodgett films for nonvolatile memory applications. *IEEE Trans. Device Mater. Reliab.* **2005**, *5*, 720–735. [CrossRef]
25. Bystrov, V.S.; Paramonova, E.V.; Bdikin, I.K.; Bystrova, A.V.; Pullar, R.C.; Kholkin, A.L. Molecular modelling of the piezoelectric effect in the ferroelectric polymer poly(vinylidene fluoride) (PVDF). *J. Mol. Model.* **2013**, *19*, 3591–3602. [CrossRef]
26. Yamada, K.; Saiki, A.; Shingubara, S.; Takahagi, T. Study of a dielectric constant due to electronic polarization using a semi-empirical molecular orbital method I. *Jpn. J. Appl. Phys.* **2001**, *40*, 4829–4836. [CrossRef]
27. Gevorkyan, V.E.; Paramonova, E.V.; Avakyan, L.A.; Bystrov, V.S. Computer modeling and Molecular dynamics of polarization switching in the ferroelectric films PVDF and P(VDF-TrFE) on nanoscale. *Math. Biol. Bioinform.* **2015**, *10*, 131–153. [CrossRef]
28. Bystrov, V.S.; Paramonova, E.V.; Bystrova, A.V.; Gevorkyan, V.E.; Meng, X.J.; Tian, B.B.; Wang, J.L.; Avakyan, L.A. Analysis of the computational and experimental studies of the polarization switching in the PVDF and P(VDF-TrFE) ferroelectric films at the nanoscale. *Math. Biol. Bioinform.* **2015**, *10*, 372–386. [CrossRef]
29. Kliem, H.; Tadros-Morgane, R. Extrinsic versus intrinsic ferroelectric switching: Experimental investigations using ultra-thin PVDF Langmuir–Blodgett films. *J. Phys. D Appl. Phys.* **2005**, *38*, 1860–1868. [CrossRef]
30. Gu, Z.; Imbrenda, D.; Bennett-Jackson, A.L.; Falmbigl, M.; Podpirka, A.; Parker, T.C.; Shreiber, D.; Ivill, M.P.; Fridkin, V.M.; Spanier, J.E. Mesoscopic Free Path of Nonthermalized Photogenerated Carriers in a Ferroelectric Insulator. *Phys. Rev. Lett.* **2017**, *118*, 096601. [CrossRef]
31. Kretschmer, R.; Binder, K. Surface effects on phase transitions in ferroelectrics and dipolar magnets. *Phys. Rev. B* **1979**, *20*, 1065. [CrossRef]
32. Ducharme, S.; Fridkin, V.M.; Bune, A.V.; Palto, S.P.; Blinov, L.M.; Petukhova, N.N.; Yudin, S.G. Intrinsic ferroelectric coercive field. *Phys. Rev. Lett.* **2000**, *84*, 175. [CrossRef]
33. Stolichnov, I.; Cavalieri, M.; Colla, E.; Schenk, T.; Mittmann, T.; Mikolajick, T.; Schroeder, U.; Ionescu, A.M. Genuinely ferroelectric Sub-1-Volt-Switchable nanodomains in $Hf_xZr_{(1-x)}O_2$ ultrathin capacitors. *ACS Appl. Mater. Interfaces* **2018**, *10*, 30514–30521. [CrossRef] [PubMed]
34. Hoffmann, M.; Fengler, F.P.; Herzig, M.; Mittmann, T.; Max, B.; Schroeder, U.; Negrea, R.; Lucian, P.; Slesazeck, S.; Mikolajick, T. Unveiling the double-well energy landscape in a ferroelectric layer. *Nature* **2019**, *565*, 464–467. [CrossRef] [PubMed]
35. Kim, S.J.; Mohan, J.; Lee, J.; Lee, J.S.; Lucero, A.T.; Young, C.D.; Colombo, L.; Summerfelt, S.R.; San, T.; Kim, J. Effect of film thickness on the ferroelectric and dielectric properties of low-temperature (400 °C) $Hf_{0.5}Zr_{0.5}O_2$ films. *Appl. Phys. Lett.* **2018**, *112*, 172902. [CrossRef]
36. Kim, S.J.; Mohan, J.; Summerfelt, S.R.; Kim, J. Ferroelectric $Hf_{0.5}Zr_{0.5}O_2$ Thin films: A Review of recent advances. *JOM J. Miner. Met. Mater. Soc.* **2018**, *71*, 246–255. [CrossRef]
37. OriginLab. Data Analysis and Graphing Software. Available online: https://www.originlab.com/doc/Origin-Help/Boltzmann-FitFunc (accessed on 14 September 2020).

nanomaterials

Review

Simulation and Computer Study of Structures and Physical Properties of Hydroxyapatite with Various Defects

Vladimir Bystrov [1,*], Ekaterina Paramonova [1], Leon Avakyan [2], José Coutinho [3] and Natalia Bulina [4]

1 Institute of Mathematical Problems of Biology, The Branch of Keldysh Institute of Applied Mathematics, RAS, 142290 Pushchino, Russia; ekatp@yandex.ru
2 Physics Faculty, Southern Federal University, 344090 Rostov-on-Don, Russia; laavakyan@sfedu.ru
3 Department of Physics & I3N, University of Aveiro, Campus Santiago, 3810-193 Aveiro, Portugal; jose.coutinho@ua.pt
4 Institute of Solid State Chemistry and Mechanochemistry, Siberian Branch of RAS, 630128 Novosibirsk, Russia; bulina@solid.nsc.ru
* Correspondence: vsbys@mail.ru or bystrov@impb.ru

Abstract: Simulation and computer studies of the structural and physical properties of hydroxyapatite (HAP) with different defects are presented in this review. HAP is a well-known material that is actively used in various fields of medicine, nanotechnology, and photocatalytic processes. However, all HAP samples have various defects and are still insufficiently studied. First of all, oxygen and OH group vacancies are important defects in HAP, which significantly affect its properties. The properties of HAP are also influenced by various substitutions of atoms in the HAP crystal lattice. The results of calculations by modern density functional theory methods of HAP structures with these different defects, primarily with oxygen and hydroxyl vacancies are analyzed in this review. The results obtained show that during the structural optimization of HAP with various defects, both the parameters of the crystallographic cells of the HAP change and the entire band structure of the HAP changes (changes in the band gap). This affects the electronic, optical, and elastic properties of HAP. The review considers the results of modeling and calculation of HAP containing various defects, the applied calculation methods, and the features of the effect of these defects on the properties of HAP, which is important for many practical applications.

Keywords: hydroxyapatite; modeling; density functional theory; defects; vacancies; substitutions; structural and optical properties; band gap; electronic density of states; nanomaterials

Citation: Bystrov, V.; Paramonova, E.; Avakyan, L.; Coutinho, J.; Bulina, N. Simulation and Computer Study of Structures and Physical Properties of Hydroxyapatite with Various Defects. *Nanomaterials* **2021**, *11*, 2752. https://doi.org/10.3390/nano11102752

Academic Editor: Francisco Torrens

Received: 8 September 2021
Accepted: 13 October 2021
Published: 17 October 2021

1. Introduction

Hydroxyapatite (HAP) is a mineral compound from the apatite group, the basic salt of calcium and phosphoric acid (calcium phosphate) with the formula $Ca_5(OH)(PO_4)_3$, a widely employed multifunctional material in biomedicine, health care, biology, ecology, catalysis, and for environmental remediation [1–10]. First of all, HAP is the main mineral component of mammalian hard tissues (bone and teeth). Along with the organic component (collagen) and living bone cells (osteoclasts, osteoblasts, and osteocytes), HAP crystallizes within the gaps of stacked tropocollagen fibrils, forming and strengthening the bone structure [1–4,9]. Due to innate bioactivity and biocompatibility, HAP is a widely used material in medicine for bone and dental surgery. Here, hydroxyapatite is used as a filler to replace parts of lost bone (in traumatology, orthopedics, hand surgery) and as a coating for implants to promote new bone growth.

Since the mechanical properties of HAP differ from bone material, implants are made from the most suitable material (usually titanium is used), but its surface is covered with HAP [3], which provides better adhesion and reproduction of bone cells on its surface [2,4]. It is important to note here that biological HAP (bio-HAP) found in living organisms differs from an ideal synthetic mineral crystal of HAP in stoichiometric imbalance and various defects [5]. It is the presence of such defects that creates inhomogeneities in the structure of HAP, leading to local changes in the electrical potential and topography of the HAP surface, which ensures successful survivability of bone and growth of bone tissue.

Therefore, it is so important to study and understand the role of defects in HAP: what defects arise here and how they affect the properties of HAP. Then, we will be able to purposefully synthesize HAP with the necessary defects that provide the desired properties.

Besides medical applications, HAP can be used in environmental remediation for heavy metals absorption and removal, especially for bivalent cations such as Pb (II), Cd (II), and Zn (II) [6]. Another feature exhibited by HAP is the capability of degrading hazardous organic and inorganic chemicals, both in air and water, upon exposure to ultraviolet (UV) light, demonstrating the photocatalytic activity [6].

HAP also exhibits a strong affinity for organic compounds and living organisms, which is very useful, for instance, for waste water treatment and for environmental remediation, and even in cancer treatment [4,9]. Again, defects and dopants play a central role here, which requires their investigation. One effective way to study defects in HAP is precisely computer simulation and computational studies.

Recent studies of the structure and various properties of hydroxyapatite (HAP) have convincingly shown that actual samples (natural and synthetic), especially of biological origin (from the bone tissues of mammals, fish, and mollusks), are quite different from the ideal crystallographic HAP models, which are usually used in modeling and calculations [1–6].

It turned out that these actual HAP structures with defects promote better biocompatibility, adhesion, attachment of the bone cells, and their proliferation, since they are closer to the biological HAP, which has its own structural heterogeneities [1–4]. It is important to establish which HAP properties with defects are most essential here.

HAP samples with defects close to biological ones have not only "incorrect" stoichiometry (their composition, as a rule, differs from the stoichiometric Ca/P ratio ~1.67 [9–12]), but also have a number of additional numerous defects (oxygen vacancies and whole OH groups, impurities, interstitials, and substitutions of a number of atoms in the crystallographic lattice) [13–22]. Moreover, the distribution of all these defects in the sample volume is uneven, stochastic, and depends on many conditions of its fabrication: temperature, pressure, heating and cooling rates, the composition of the environment, humidity, etc. [19]. It is clear that any theoretical models and high-precision calculations using a variety of modern methods reflect only part of these complex heterogeneous HAP structures. Nevertheless, by developing these models and calculation methods, and carrying out more and more detailed precise calculations, we can determinate the peculiarities of different studied structural defects and correlate their properties with the set of the experimentally observed data [4–9].

It is known that the basic structure of HAP has such specific features as extended structural channels formed by chains of the OH groups (OH-channels) [5–9,13,16]. This feature leads to the possibility of proton movement along these channels [16,17] and the appearance in HAP of the defects such as vacancies of protons, oxygen, and whole OH groups [2,4,5,7,9,14–20]. Various defects have different effects on the change in the properties of a regular HAP periodic lattice. So, for example, the HAP samples during fabrication and treatment are heated to certain temperatures (about ~500–700 °C or about ~773–973 K [16–20]), and the OH groups leave the samples quite easily. As a result, HAP is dehydrated [7,19,20]. During subsequent cooling, OH groups from the environment are reintroduced into these channels. However, their concentration changes (depending on the humidity of the environment, etc.) and a partial concentration of the OH vacancies will remain in the HAP samples, in some its lattice unit cells, which significantly affects all their properties [2,4,19,20].

Another important and specific defect in HAP are oxygen vacancies [2,4–6,8,9,21,22]. They can arise from both the OH group and the PO_4 group. Moreover, it could be of different types depending on the position of the oxygen atom and the symmetry of the given atomic PO_4 group [8]. Their formation is possible at the higher temperatures (of the order of ~1100–1300 °C or about ~1373–1573 K) [19,20] or also under radiation exposure [23].

Further important defects are interstitials (in particular, the insertion of hydrogen atoms or protons into HAP lattice structure [2,4,5]) and substitution of some atoms in the HAP structure by others atoms in the HAP unit cell lattice (for example, Fe/Ca, Sr/Ca [5,24–29]) and some others (such as Mg/Ca, Se/Ca, Si/P, etc.) [5]. It leads to many important practical applications, because it changes the HAP surface electrical potential [2,4] that initiates higher adhesion and proliferation of various osteo-cells on HAP surface, and this improves the biocompatibility of the implant [2–5,28].

Basically, the main physical and chemical properties of HAP are determined precisely by the presence of various structural defects, such as oxygen and hydroxyl vacancies, interstitials and substitutions of ions, atoms in the structure of HAP [1–4]. Various modern experimental and theoretical methods are used to study them. Among them, computer modeling and calculations, including first principles ab initio and density functional theory (DFT) methods [5–9,30–51] are the most important.

In previous research, using some modern DFT methods (including several types of the hybrid functionals and the theory of many-particle perturbations) and developed software tools (AIMPRO, VASP, and Quantum Espresso (QE) [7,8,38–52]) in the modeling and computer studies of the properties of HAP structures with these various types of the defects, it was found that the forbidden band gap Eg of an ideal (defect-free) stoichiometric HAP lattice turns out to be much larger than the data of measurements of the optical absorption characteristics of the usual HAP samples [2,4,18,19]. These calculations show, that Eg can reach values of the order of ~7.4–7.9 eV [7,35]. This significantly exceeds the experimentally observed values for the HAP samples used in practice (Eg ~3.8–4.5 eV) [2,4,9,20]. These recent computational studies of the HAP structure with defects have convincingly shown that the deviations of the calculated band gap Eg of the ideal stoichiometric HAP crystal from the experimentally observed Eg values are caused precisely by the presence of the various defects in the HAP structure [2,4–9,36,37].

The calculations of oxygen vacancies (from PO_4 and OH groups with different symmetry) and total OH-group vacancies in the HAP structure were carried out recently, showing very interesting results [5–9].

These oxygen and OH vacancies were investigated both in the local charge density approximation (LDA) of the DFT [5,6,38,42,43] and in the DFT generalized gradient approximation (GGA) [48–50], including the use of the modern types of hybrid DFT functionals and the theory of many-particle perturbations [7,8,39,41,44–54]. These studies of the structural and electronic properties of oxygen vacancies presented in the plane wave formalism [46,47,51] were carried out both in the single unit cell model (in the LDA DFT [5,6]) and in the model of large HAP supercell model ($2 \times 2 \times 2 = 8$ unit cells) [7,8,36,37] and (in the GGA DFT [48–50]) by using hybrid DFT functionals.

These calculations and studies have shown that, under equilibrium conditions, oxygen vacancies arise not only in the form of a simple vacant oxygen site (in a neutral charge state) [8], but can also be in the form of extended structures occupying several crystal fragments or chemical units (especially in a double plus charge state) [8,36,37]. In this case, the connection between the presence of defects and changes in the optical properties of HAP and the level of its photoelectronic work function is clearly traced [2,4,5]. In fact, here was shown, that for any different approaches and methods of DFT calculations, it is oxygen vacancies that play their decisive role here, changing the optical absorption width Eg of an ideal stoichiometic HAP crystal from its large characteristic values to a lower level of the order of ~3.6–4.2 eV, determined by these introducing HAP defects (oxygen vacancies), which is as usually recorded on HAP experimental samples.

In general, these studies are still far from complete and require further deep continuation.

In this review, we present some recent data on the study of defects, primarily oxygen vacancies (of the various types) and the OH vacancies, obtained by modern methods of computer simulation and first principle of quantum-mechanical calculations based on DFT.

In addition, in this work, the calculations are compared with the available experimental data, and an analysis of this comparison is given.

2. Computational Details, Main Models, and Methods

2.1. Main Methods and Used Software

The calculations of pristine stoichiometric and defective HAP structures and its properties were carried out from the first principles using DFT by AIMPRO code in Local Density Approximation (LDA) [5,6,38,42,43] and by VASP [7,8,39,44–48] in Generalized Gradient Approximation (GGA) within Kohn-Sham formalism [45–47] according to the Perdew, Burke, Ernzerhof (PBE) approach [49] or one of the following hybrid density functionals: the HSE of Heyd, Scuseria, and Ernzerhof [54]; the Becke three-parameter (B3LYP) functional [49,50]; the PBE0 functional [55]. These are called *semi-local and hybrid density functional calculations* [7].

The LDA method also involves a pseudopotential approach for calculations of each atomic potential and wave function [42,43]. Standard valence configurations were used for this purpose in [5,6]: hydrogen ($1s^1$), phosphorus ($3s^2 3p^3$), and oxygen ($2s^2 2p^4$). For calcium atoms, two different valence configurations were used [5]: (1) with a two-electron ($4s^2$) shell, and with a ten-electron ($3s^2 3p^6 4s^2$) shell. The charge density and potential terms were Fourier-transformed using plane waves, with a cut-off energy of *E*cut = 300 a.u. [5].

For the GGA approach, the projector augmented-wave (PAW) method [51] was used [7,8] to account for [Ca]:$1s^2 2s^2 2p^6 3s^2$, [P]:$1s^2 2s^2 2p^6$, and [O]:$1s^2$ core electrons. Valence electrons were described using plane waves with kinetic energy up to *E*cut = 400 eV [7,8]. For presentation, some structural details and properties calculations of the transformation of DFT files into HyperChem workspace using OpenBabel software [56] as well as the combination with semi-empirical quantum-chemical (QM) PM3 method from HypemChem 8.0 package [40] were used. More computational details are all presented in works [5–8] and are described below. Some recent calculations were made also using Quantum ESPRESSO [41].

2.2. Main Details of HAP Crystal Structure

Hydroxyapatite (HAP) have the general chemical formula $Ca_5(PO_4)_3OH$ [10]. HAP solidifies in the form of an ionic molecular crystal, either with hexagonal ($P6_3/m$) or monoclinic ($P2_1/b$) symmetry, depending on conditions, and has unit cells enclosing two or four formula units, respectively [10–13]. The crystal structure and unit cell parameters of HAp with both $P6_3/m$ and $P2_1/b$ symmetries, and its atomic positions, were determined using X-ray diffraction [11] (Table 1).

Table 1. Unit cell parameters *a,b,c* [Å] (from [11]).

Phase	Group	*a*, Å	*b*, Å	*c*, Å
Hexagonal	$P6_3/m$	9.417	9.417	6.875
Monoclinic	$P2_1/b$	9.480	18.960	6.830

HAP usually crystallizes in a hexagonal crystal system (space group $P6_3/m$) under normal conditions, according to temperature and stoichiometry, but can also exist as a monoclinic structure (space group $P2_1/b$) [1–5,9–13]. As noted by Elliot [10], direct synthesis of monoclinic phase usually requires very strict adherence to the correct stoichiometry, with a Ca:P ratio of 1.67 [4,9–13].

One of the common structural peculiarities of HAP structure is connected with the pseudo-one-dimensional character of the apatite structure. Ca and OH^- ions form a long chain along the main structural *c* axis [9–19], often being named as an 'OH-channel' in HAP. For the $P6_3/m$ phase consisting from two chemical formula units and the HAP general formula is [Ca_{10} $(PO_4)_6(OH)_2$], where the hydroxyl units OH shows stochastic orientation along OH channels.

Thus, according to X-ray data, this actually makes the material mirror-symmetric along the main axis, which runs along the OH channel. Conversely, when all OH blocks show the same alignment along the hexagonal axis, the mirror plane is lost and the space group symmetry drops to $P6_3$. The ordering of the dipoles in the OH-channels, which interact with the chains of the OH-channel, also affects the property of the monoclinic phase. If OH^- ions are oriented in parallel, HAP has non-centrosymmetric ordered structure (space group $P2_1$), which could reveal piezoelectric properties [17,18,36,37].

The structure of HAP, used to study the effect of defects on its properties, is primarily based on its initial pristine stoichiometric structural phase–hexagonal $P6_3$, with a unit cell consisting of 44 atoms and containing structural OH channels with two hydroxyl OH groups in each unit cell [5–9,13–20]. Depending on the orientation of these OH groups, the cells can have different symmetry groups (Table 1): $P6_3/m$—for the hexagonal disordered phase (when the orientation of the OH groups is random) and $P6_3$—for the hexagonal ordered phase (when the orientation of the OH groups is parallel and directed in the same direction, which creates its own internal polarization, similar to ferroelectrics (see below, for example, in [36,37]).

Some earlier calculations were performed using one unit cell model of hexagonal HAP $P6_3$ [5,6] (see Figure 1). The main peculiarity of the next study is the introduction of the supercells model made up of $2 \times 2 \times 2 = 8$ HAP unit cells (space group $P6_3$) for hexagonal HAP phase both for pristine and defective HAP with OH and oxygen vacancies [7,8] (see below in Section 2.4).

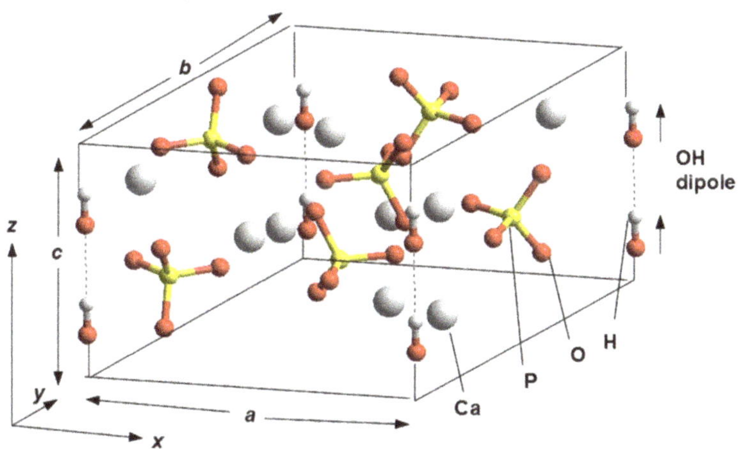

Figure 1. One unit cell model of the Hydroxyapatite (HAP: $Ca_{10}(PO_4)_6(OH)_2$)– in $P6_3$ hexagonal phase. All OH groups are oriented in the same direction. They are positioned at the four corners of the unit cell, but only one pair in one corner belongs to this unit cell, the other three pairs belonging to neighboring unit cells (e.g., one OH per unit cell). It also shows the axes of the Cartesian coordinates *x, y, z* and unit cell parameters ***a*, *b*, *c***. (Adapted with permission from ref. [5]; IOP Publishing, 2015).

2.3. HAP One Unit Cell Model

The stoichiometric HAP unit cell model for pure (defect-free) lattice in $P6_3$ hexagonal phase is presented on Figure 1 [5]. This model was used for the calculations firstly for LDA method [5,6] and then for GGA approach too in several cases [7]. It was used as an initial unit cell model of hexagonal HAP ($P6_3$) lattice and for the cases with several of its defects, such as, OH-vacancy, and various O-vacancies.

This one HAP unit cell consists of 44 atoms and contains two OH groups in each periodical unit cell. It should be noted that only one OH-channel with two OH groups is included in the unit cell, while the remaining three OH-channels (with six OH groups) belong to other periodically repeating unit cells.

For monoclinic HAP nanostructures, the unit cell consists of 88 atoms, because the lattice unit cell is doubled in the monoclinic case along the ***b*** axis. Besides this basic hexagonal unit cell of 44 atoms, the supercells totaling 176 atoms, made with two and four unit cells along the ***a*** and ***c*** directions of the hexagonal lattice were constructed to investigate defects in the work [5].

For the unit cell with 44 atoms, we used a 6 × 6 × 8 grid of **k** points (288 points), following the recipe proposed by Mokhorst and Pack [57], while for the large supercells, the grid used was 2 × 2 × 4. All **k**-point sets were folded according to the symmetry of the problem. All atom positions were allowed to relax along the forces acting upon them, until the total energy converged below 10^{-5} eV. After convergence test calculations, a 2 × 2 × 4 **k**-point mesh was found good enough to sample the Brillouin zones (BZs) of both hexagonal and monoclinic cells.

The HAP unit cell model, containing several defects in the atomic positions of all the vacancies used in this work, is shown in Figure 2. Here, we indicate OH-vacancy and various types of O-vacancies in the HAP unit cell, as deviations from the initial stoichiometric hexagonal HAP structure [6]: (1) an O-vacancy in the OH group of an OH-channel structure (blue circle in Figure 2); (2) a full OH-vacancy from this OH-channel structure (purple circle in Figure 2); (3) various cases of O-vacancies from different positions in PO_4 groups (green circles in Figure 2).

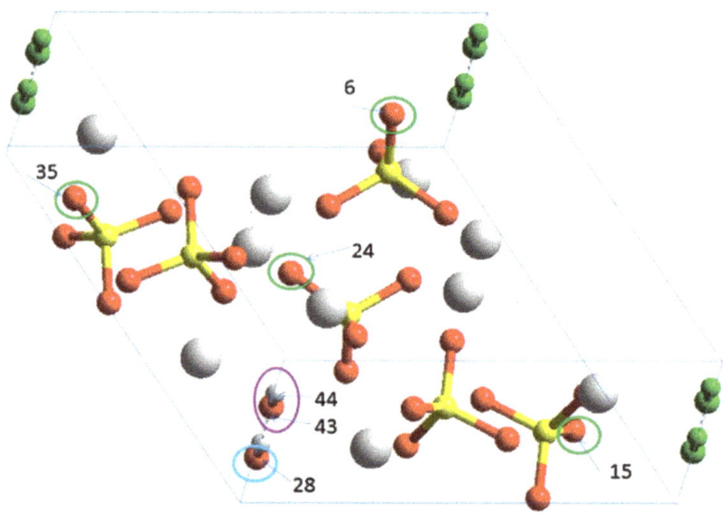

Figure 2. Models and scheme of the atomic positions selected and deleted for modelling presentation of various vacancy-defects in the HAP hexagonal unit cell with initial 44 atoms. Indicated in the colored circles are: (1) in blue—oxygen atom (number 28) from an OH group for creation of an O-vacancy in the OH-channel; (2) in purple—OH group full (with atom numbers 43 and 44) for creation of a complete OH-vacancy; (3) in green—oxygen atoms (with numbers 6, 15, 24, 35) from various differently positioned PO_4 groups corresponding to an O-vacancy in PO_4 groups. The green atoms on the other three corners indicate other OH groups, which belong to neighboring periodical repeated unit cells. (Adapted with permission from ref. [6]; Elsevier, 2016).

For GGA approach, the Brillouin zone (BZ) was sampled using a Γ-centered 2 × 2 × 3 mesh of **k**-points. The Hartree-Fock exact exchange was evaluated at the same **k**-point grid used for the DFT potential and stored on a real-space grid of 128, 128, 96 points along **a**1, **a**2, and **a**3 lattice vectors, corresponding ***a, b, c*** (see Figure 1). The experimental lattice constants are $a1 = a2 = a$ = 9.417 Å and $a3 = c$ = 6.875 Å [11], corresponding to a grid density of about 14 points/Å along all three directions.

In all these calculations—except for the elastic constant calculations (bulk modulus B) and the DOS calculations—full relaxation of the cell, including changes in volume and adjusting the atomic positions and cell shape (total optimization of structure) was allowed. When calculating the elastic constants, only the relaxation of ionic positions within the strained cells was allowed. In several calculations, e.g., when defects as substitutions of atoms (Ca on Sr, Fe, Mg etc.) were explored, the calculations for fixed former atomic positions were also used, and then their relaxation to the new ones. The optimized data obtained for each HAp structure were taken (files with optimized and fixed coordinates for all unit cell atoms in atomic and Cartesian units were used) and transformed further to another, special HyperChem (*.hin) format (using the Babel program [56]). Then, all these atomic coordinate data files were uploaded into HyperChem software tools for further molecular modeling (with visual options for all studied HAp structures), calculations, and exploration of its physical properties using HyperChem package [40].

2.4. HAP Supercell Model

The main feature of our more detailed study is the introduction of the super-cells model made up of $2 \times 2 \times 2 = 8$ polar HAP unit cells (with space group $P6_3$) for hexagonal HAP ordered phase (both for initial pristine stoichiometric and defective HAP with oxygen vacancies and with full OH group vacancy) [7,8].

It should be noted that this phase demonstrates polar ferroelectric state with the total polarization along c-axis, due to OH groups oriented parallel and along the OH-channels. While the disordered phase of hexagonal HAP (with space group $P6_3/m$) forms non-polar paraelectric state, due to compensation of an opposite anti-parallel orientation of OH groups, which are along these OH channels [36,37].

Oxygen vacancies were introduced in hexagonal supercells made up of $2 \times 2 \times 2 = 8$ polar HAP unit cells (space group $P6_3$), comprising a total of 352 atoms (Figure 3). Although the monoclinic phase $P2_1/b$ was found to be more stable [12,33], they only differ by a few tens of meV per unit cell, with their electronic structures being essentially identical [34].

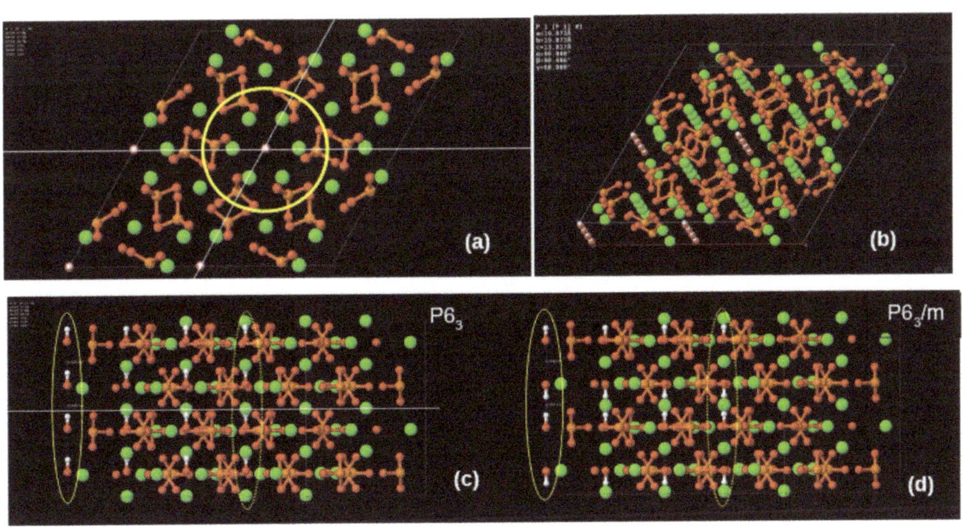

Figure 3. Images of supercell models with $2 \times 2 \times 2 = 8$ unit cell of HAP (visualization using Jmol after DFT calculations: (**a**) top view—along the c axis of the OH channel (a yellow circle highlights the main central part around one OH channel); (**b**) supercell model in isometric projection; (**c**) HAP supercell model in lateral projection for the $P6_3$ phase (OH groups are parallel oriented (ordered) in OH channel along c axis and therefore can create a summarily polar state); (**d**) HAP supercell model in lateral projection for the $P6_3/m$ phase (OH groups are anti-parallel oriented (disordered) in OH channel along c axis and therefore can create a summarily non-polar state).

For the HAP unit cell, we have shown previously that convergence of the electron density and energy is obtained when sampling the band structure with a Γ-centered $2 \times 2 \times 3$ mesh of k-points within the first Brillouin zone. By doubling the size of the cell along all principal directions, reciprocal lattice vectors are contracted by a factor of two so that a $1 \times 1 \times 2$ k-point grid would actually improve on the sampling quality. From convergence tests, we found that the total energy of the 352-atom bulk supercell obtained with a $1 \times 1 \times 1$ grid (Γ-point sampling) differs by less than 0.1 eV from a $1 \times 1 \times 2$-sampled calculation. More importantly, relative energies and ionization energies of defects differ by about 1 meV only. Therefore, all defect calculations employed a Γ-point sampling [7,8].

A full relaxation of such large supercells using plane wave hybrid-DFT is prohibitively expensive. Instead, in works [7,8], defect structures were first found by relaxing all atomic coordinates within PBE, until the maximum force became less than 10 meV/Å. The resulting structures were employed on a second step, where the total energy was obtained within hybrid-DFT by means of a single-point calculation. This procedure was necessary due to the sheer size of the Hamiltonian at hand combined with the use of a plane-wave method. As it was recently shown in [7,52], the relative energies obtained within this methodology are usually affected by error bars of the order of 10 meV or lower.

3. Main Results and Discussions

3.1. Main Structural and Mechanical Properties of HAP

We start by analyzing the structural and mechanical properties of HAP as initial stoichiometric as well as HAP with considering defects: OH-vacancy and various O-vacancy types. Tables 2 and 3 compare the calculated structural data (unit cell lattice parameters and volume) and bulk modulus (for some cases) with the respective experimental data. Table 2 presents the data of pristine stoichiometric HAP (computed by various methods and from different experimental data) and data of HAP with one full OH-vacancy (per one HAP unit cell). Table 3 consists of calculated data for various types of O-vacancies and it is shown below.

Table 2. Data of HAP lattice parameters and bulk modulus for pristine HAP and HAP with OH-vacancy.

Property	Experim. [11]	Experim. [58]	Experim. [59] ([60])	AIMPRO (LDA) [(3)] [5,6]		PBE VASP [(3)] (GGA)		PBE (GGA)/SuperCell [(4)] [7,8]	B3LYP/Super-Cell [(4)] [7,8]	HSE [7]	PBE0 [7]
				OH-OH $P6_3$	OH-HO $P6_3/m$	OH-OH $P6_3$	OH-HO $P6_3/m$				
Initial stoichiometric HAP (hexagonal $P6_3$)											
a, Å	9.417	9.4236	9.4205 (9.4248)	9.4732	9.4624	9.3628	9.3640	9.537	9.5770	9.481	9.477
c, Å	6.875	6.8802	6.8828 (6.8860)	6.9986	7.0182	6.8454	6.8621	6.909	6.8767	6.859	6.851
V, Å [(3)]	527.99	529.13	528.99 (529.71)	543.92	544.20	519.69	521.09	546.07	546.22	533.95	532.88
B,GPa	89 ± 1 [(1)]			81.6 ± 2	82.35 ± 2	-		82 ± 3	86 ± 2	83 ± 3	82.8 ± 0.3
E, a.u. E, eV				−467.0992	−467.0944	−311.82	−311.39				
ΔE,eV				0.132		0.43					
HAP with OH-vacancy											
a, Å		9.4155 [(2)]		9.4883		9.3685	9.4210	9.537	9.5770	-	-
c, Å		6.8835 [(2)]		7.0018		6.8351	6.8800	6.909	6.8767	-	-
V, Å [(3)]		528.48 [(2)]		545.905		519.53	528.83	546.07	546.22	-	-
B,GPa				78 ± 2		−297.78	−297.99				
E, eV						0.21					

[(1)] References [61,62]. [(2)] Data from Bulina N.V. experiment (cooling from 1100 °C in He atm.—comment is in the Section 3.1.1 below). [(3)] OH-OH is the ordered $P6_3$ phase and OH-HO is the disordered $P6_3/m$ phase—comment 1 in the text. [(4)] Two-steps calculation for supercell models—comment 2 in the text.

Table 3. The calculated HAP lattice parameters with various type of the O-vacancy.

Property	Type of O-Vacancy	AIMPRO (LDA) [5,6]	PBE–VASP (GGA)
HAP with O-vacancy from OH (one unit cell HAP $P6_3$ model)			
a, Å	-	9.4539	9.3437
c, Å	-	7.0028	6.8463
V, Å (2)	-	542.03	517.64
HAP with O-vacancy from PO_4 (one unit cell model HAP $P6_3$ model)			
	Atom O in different positions in PO_4 (1)		
a, Å,	V_O1 (O6) V_O2 (O15) V_O3 (O30,O35) V_O4 (O24)	9.4599 9.4630 9.4581 9.47295	9.3570 9.3520 9.3544
aver. *a*, Å		9.4635 ± 0.005	9.3545 ± 0.005
c, Å	V_O1 (O6) V_O2 (O15) V_O3 (O30,O35) V_O4 (O24)	6.9884 6.9890 6.9893 6.97822	6.8139 6.8242 6.8402
aver. *c*, Å		6.9890 ± 0.005	6.8261 ± 0.005
V, Å (2)	V_O1 (O6) V_O2 (O15) V_O3 (O30,O35) V_O4 (O24)	541.60 542.01 541.47 542.31	517.86 518.33 517.45
aver. V, Å (2)		541.85 ± 0.3	517.88 ± 0.1

(1) O-vacancy from PO_4: Atom O—is in various positions of PO_4 (and from different PO4 groups, according with Figure 2). (2) OH-OH is the ordered $P6_3$ phase and OH-HO is the disordered $P6_3/m$ phase—comment 1 in the text.

The results obtained with LDA method (in earlier calculations using AIMPRO [5,6]) and with GGA approximation on the various DFT exchange-correlation functionals: PBE, HSE, B3LYP, and PBE0 (using VASP calculations [7,8]) are presented in Tables 2 and 3. The calculated lattice parameters for initial pristine stoichiometric HAP lattice within the PBE level approach are in good agreement with the various previously known PBE results (see [48–51] and references therein).

Comment 1 for OH groups orientation difference. Recently computed data have shown, first, that HAP can coexist at different stages, both hexagonal and monoclinic [5,36,37], as well as in ordered and disordered hexagonal due to the orientation of the OH groups in one unit cell by aligning the OH units in opposite directions: OH-OH is the ordered $P6_3$ phase and OH-HO is the disordered $P6_3/m$ phase. The energy difference between the two phases is DE = E ($P6_3/m$)—E ($P6_3$) [5,7,8]. This difference, obtained from our various calculations and presented here in Table 2, is: $\Delta E = \sim 0.43$ eV for PBE (GGA) and $\Delta E = \sim 0.132$ eV for LDA (AIMPRO) [5]. In a similar PBE calculation in [34], this $\Delta E \sim 0.4$ eV, and in the most developed model of a large supercell [7,8] is $\Delta E = 0.39$ eV, and the last two energies were obtained from calculations of the completely relaxed B3LYP level [23].

Consequently, during heating (temperature above ~1000 K) and cooling, many OH groups can change their orientation in different unit cells, which ultimately can lead to some stochastic redistribution of them, and the sample will contain a mixture of both types $P6_3$ and $P6_3/m$ hexagonal phase of HAP. It is very difficult to distinguish them experimentally. Therefore, the comparison of the results of our calculations and experimental data must be carried out very carefully, taking into account the possible errors arising also due to this factor of different random orientations of OH groups in experimental samples.

Comment 2 for two-steps calculation for supercell models. For supercell models, structural optimization was made using PBE(GGA) only for initial stoichiometric HAP and after that the calculation in B3LYP was done in the fixed atomic positions in single point (SP) mode. All further calculations of the HAP with defects were performed using PBE(GGA) optimization only for atomic/ionic positions keeping the lattice parameters fixed as it was founded (established) for initial stoichiometric HAP lattice.

The results obtained show the usual ~1% overestimation in relation to the experimental data. This is known to be mostly due to an artificial over-delocalization of the electronic density when the GGA approach is employed. It is shown that the calculations using B3LYP results in improvement of ***c*** lattice parameter, but parameter ***a*** is still overestimated by ~1% [7]. This confirms previous reports that B3LYP generally overestimates the experimental lattice parameters as well [35]. On the other hand, our result differs from the previous B3LYP calculations of HAP, where ***a*** was underestimated by ~1% [35]. More discussion about the reasons of these discrepancy was given in [7].

Table 2 also indicates that the lattice parameters calculated within hybrid DFT are generally closer to the experiments than those obtained using the semi-local functional. The best results are obtained for PBE0 with a deviation of <0.6% in relation to the experiments [7].

It should be noted, that the results obtained earlier in the LDA approximation are the closest to the experimental data on the lattice parameter ***a*** (even slightly better than in the case of a PBE0 with deviation less than 0.6%), while the value of the parameter ***c*** turned out to be more overestimated here (the deviation is almost 1.8%).

The bulk modulus (B) were obtained here usually by fitting the Birch-Murnaghan equation of state [5,61–64]. The calculated bulk modulus show reasonable agreement with the experiments (see data in 2nd column of Table 2, Refs. [61,62]). The errors (shown in the table) were obtained from the standard fitting procedure. B3LYP calculations show deviation from the measurements by about 3% only, while other methods underestimate the experimental value of B by 7–8%. This level of accuracy is in line with typical discrepancies found for many other insulating materials [63]. It is interesting again to note, that LDA approximation shows here the same value of B as on PBE (GGA) level.

Experimental data on stoichiometric natural and synthetic samples should be used and analyzed rather carefully, since there may be impurity ions (carbonate, nitrate) that affect the lattice parameters. Therefore, the composition of stoichiometric hydroxyapatite should be confirmed by other analytical methods (for example, IR data), in addition to diffraction methods, which is usually used to determine the lattice parameters. Also, synthetic stoichiometric HAP should be well annealed at high temperatures in humid air. Table 2 provides experimental data that mostly meet these requirements [11,58–62]. Below in Section 3.3.1, we discuss this data in comparison with our calculations.

The results of more detailed calculations for the presence of the various types of oxygen vacancies in HAP are presented in Table 3. Here, a complete optimization of the HAP structure (cell parameters and ion relaxation) was carried out only by methods LDA (AIMPRO) and PBE (GGA) in the one unit cell model, since such an optimization has not yet been performed for the supercell model.

3.1.1. Defects (OH-Vacancy and Various O-Vacancies) Influence on Structural and Mechanical HAP Properties

Let us now consider the effect of defects (oxygen vacancies and OH group vacancy) in the HAP structure, which changed in their structural and mechanical properties, and which turns out to be important for many practical and technological applications.

(1) ***Influence of OH-vacancy.*** The data calculated by LDA (AIMPRO) show that the relative changes in the lattice parameters of HAP after the appearance of an OH vacancy (in one HAP unit cell) both increase by small values: $\delta \boldsymbol{a}$~0.15% and δc~0.05% (see Table 2).

At the same time, calculations using PBE (GGA) show that the value of the lattice parameter also relatively increases by $\delta \boldsymbol{a}$~0.6%, and the parameter c increases by δc~0.5% (see Table 2). For comparison with experimental data, we use here the results of experiments on heating and cooling HAP samples, which should lead to the escape of some OH groups from the OH HAP channel and the formation of OH vacancies in HAP ([58–60] in Table 2).

In situ diffractometric studies carried out in the group of Dr. N.V. Bulina showed that the cooling of apatite sample in helium (He) atmosphere, heated to a temperature of 1100 °C, leads to a decrease in parameter ***a*** and an increase in parameter ***c***. Table 2 shows these obtained experimental data. In such conditions, OH vacancies formed at high temperatures remain up to room temperature and the cooled sample is now the oxyhydroxyapatite (OHAP) containing vacancies of OH groups. The changes obtained in the lattice parameters in the presence of OH vacancies in comparison with those known for pristine stoichiometric HAP turned out to be as follows: the relative changes were the largest for the parameter ***a***, the decrease was up to $\delta a \sim -0.016\%$, and for the parameter ***c***, an increase by about $\delta c \sim 0.5\%$.

If we compare with the initial experimental data [11], then the changes are similar. Thus, these obtained data are very close to the calculated data in the main trend of changes in the lattice parameters during the formation of an OH vacancy: a slight decrease in the lattice parameter ***a*** (and in the calculations there is an increase) and a noticeable increase in ***c*** value for all cases due to the formation of an OH vacancy in the HAP sample. As for the volumes of the cells, with both methods of calculation, a relative increase in volume was obtained (Table 2). For Bulina's experimental data, the volume slightly decreased. However, if we compare it with the initial volume of [11], then it also increased.

It is important that the calculation results show a similar character of the cell parameter measurement. This means that the chosen model and calculation method are correct and allows us to investigate in more detail the physical reasons for this phenomenon [19].

The calculated values of the modulus B (performed after the LDA method optimized calculation) showed its decrease in this case of the presence of the OH vacancy in HAP sample, which means a decrease in the mechanical strength of HAP samples (Table 2).

It seems that the presence of any vacancy type, including the OH group, should increase the lattice parameters, which is what we get. However, it is likely that in the experiment, out of two vacancies of OH groups, one disappears, and it is occupied by the O^{2-} ion (that is, the hydrogen atom H also disappears here and one oxygen ion remains—this is already forming now the oxyapatite (OAP or OA)). This is possible and leads to the following observed changes in the lattice parameters: HAP (a = 9.4236 Å, c = 6.8802 Å) goes into OHAP (a = 9.4155 Å, c = 6.8835 Å). Moreover, some recent experimental data show that if the OAP is formed here, then the lattice parameters should be as follows: OAP (a = 9.4057 Å, c = 6.8938 Å).

Unfortunately, we have not yet performed calculations of the exact OAP models, containing two OH vacancies per one HAP unit cell. It would be very interesting and useful, and we will definitely do it in the near future. In this case, we suppose to reach decreases of both lattice parameters as experimentally observed by Bulina.

(2) ***Influence of only one O-vacancy from OH group.*** In this case (see in Table 3), the calculated values of the HAP unit cell parameters give some relative decrease in the parameter ***a*** (both for the LDA (AIMPRO) and GGA (PBE/VASP) methods), while the ***c*** parameter also decreases when calculated by the LDA method, but slightly increases when calculated by the GGA (PBE) method. Unfortunately, we have no experimental data for this case.

(3) ***Influence of the different O-vacancy from PO4 group***. In this case (presented in Table 3), the calculated values of the unit cell parameters were considered for different positions of oxygen atoms that create O-vacancies in various place in the HAP unit cell (for example, see Figure 2). For the model of one unit cell, these positions were selected and calculated by the LDA method in the works [5,6], and in Table 3 these data are given. Similarly, the calculation using GGA (PBE) method was performed and presented here too.

These O-vacancies were computationally investigated in more detail by the GGA (PBE) and B3LYP methods in the work [8] in the HAP 2 × 2 × 2 supercell model. We will analyze these data separately in Section 3.3 below, devoted to supercell model (especially devoted to its electronic and optical properties).

For the cases of O-vacancies from the PO4 group indicated in Table 3, the results obtained by both LDA and GGA (PBE) methods in comparison with initial stoichiometric HAP (hexagonal $P6_3$) from Table 2 show that both the unit cell parameters *a* and *c* decrease (as well the unit cell volume), despite some of their differences for the different positions of these selected oxygen atoms, create these O-vacancies. We do not have experimental data for structural analysis in this case, but these cases demonstrate very interesting and important results of their electronic and optical properties, which will be analyzed below.

3.2. Electronic and Optical Properties

3.2.1. Electronic Properties of Pure HAP by Various Methods

To study the electronic and optical properties of HAP, both the initial defect-free and those containing various defects, the densities of states of electrons (DOS) at the different energy levels were calculated. This makes it possible to determine the effect of various defects on the energy band structure of these HAP properties.

As a result of LDA calculations (using AIMPRO software [38]) and GGA calculations (using VASP with functional PBE [39,48]), the distributions of the DOS corresponding to the filling of energy levels in the band structure with electrons and the energies of HAP according to the band theory were obtained [5–9,36,37]. This makes it possible to determine the main energy band structure's parameters—the position of the top of the valence band Ev, the bottom of the conduction band Ec, and the value of the forbidden band gap Eg = Ec − Ev, as well as the position of additional energy levels Ei induced by defects in the internal range of the Eg.

The question now is how defects change this entire band energy structure of the DOS and how close these changes it will be to the experimental values, when creating certain defects in the HAP structure. For these calculations at the first stage [5,6], the defects simulated were created by vacancies into one hexagonal unit cell of HAP consisting of 44 atoms (Figure 1). Figure 2 schematically shows examples of simulated HAP structures with defects such as oxygen vacancies from the OH and PO_4 groups, as well as a vacancy of the entire OH group. For more details and precise calculations with various DFT hybrid functional we use supercell model consisting of 352 atoms (eight unit cells) (Figure 3) [7,8].

Let us first consider HAP without any defect in the initial pristine stoichiometric hexagonal phase $P6_3$ symmetry. Figure 4 (Left figures: a,b,c) shows examples of DOS for perfect stoichiometric defect-free HAP (taking into account both deeper energy levels in the valence band and in the vicinity of the forbidden band) obtained in LDA calculations [5,6]. Similar GGA (PBE) calculations were carried out using VASP software [39]. Figure 4c shows results of DOS from GGA (with PBE functional). Figure 4a–c denotes Eg_0 = Eg because it refers to a defect-free HAP (see details also Section 3.2.3 below).

Note that the calculations using the LDA and GGA (PBE) approximations generally do not fundamentally differ here, the difference is only in the energies—GGA (PBE) gives a larger value of the band gap Eg~5.26 ± 0.05 eV (for perfect stoichiometric HAP) compared to calculations by the LDA method, which give band gap value Eg ~ 4.6 ± 0.05 eV. These data are close to results of other authors [30–35].

Further, Figure 4 (Right figures: (a−e)) shows the total DOS of a HAP unit cell obtained within PBE to be more precise with detailed calculations [7,8], including also local density of states (LDOS) [7]. The shadow plots on subsequent figures [right Figure 4b−e] depict the LDOS projected on several atomic species.

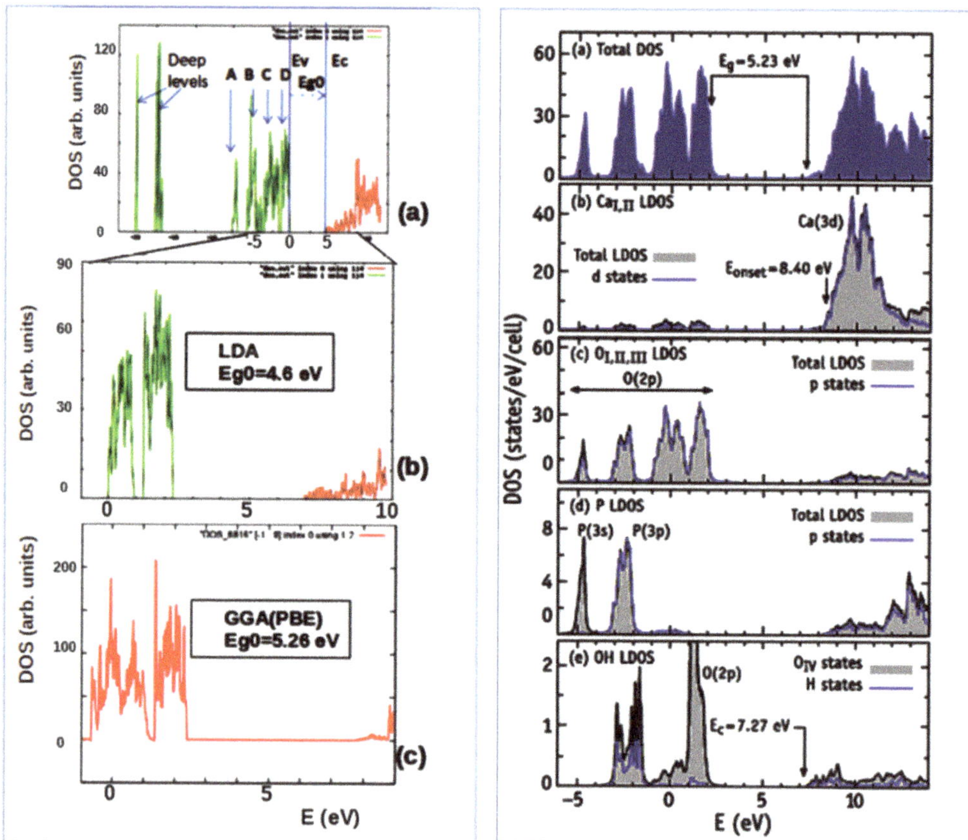

Figure 4. Left figures: Density of electronic states (DOS) for HAP unit cell: (**a**) Initial perfect stoichiometric HAP with 44 atoms in hexagonal $P6_3$ unit cell lattice, including deep levels and A,B,C,D peaks in valence band computed in LDA by AIMPRO; (**b**) the same with main energies around Eg; (**c**) the same computed in GGA (PBE) by VASP; (Data presented here obtained after calculations using AIMPRO (LDA) [38] and using VASP (PBE-GGA) [39]); **Right figures:** Total density of states of bulk HAP (**a**) and local densities of states projected on calcium (**b**), oxygen in PO4 units (**c**), phosphorous (**d**), and oxygen and hydrogen in OH units (**e**). In (**b–d**), the contribution to the LDOS with the dominant angular momentum is represented as a thick line. The results were obtained using the PBE exchange-correlation functional. (Adapted with permission from [7]; AIP Publishing, 2018).

Here, we can see that the calculations using both GGA (PBE) approach do not much differ between one another, the difference is only small in the energies; first GGA (PBE) gives a value of the band gap Eg ~ 5.26 ± 0.05 eV, while second PBE gives a value Eg = 5.23 eV, that is in the accuracy frame and in line with calculations of other authors by this method [9,30–37].

Along with the LDOS of right Figure 4b–d, we plot a thick line representing the dominant angular-momentum component for the corresponding species. In right Figure 4e, we distinguish states projected on O_{IV} and H atoms that form OH molecules. At first glance, Figure 4b,c suggest that the upper end of the valence band is mostly made of O(2p) states, while the conduction band bottom is mostly made of Ca(3d) states. From Figure 4c,d, we find phosphorous 3s-3p states mixing with oxygen 2s-2p states between 5 and 2 eV and they are far below the band gap region. All these results are in line with reports of other authors [9,18,19,30–37].

3.2.2. Electronic Band Structure of Perfect Stoichiometric HAP

Further calculations of HAP perfect stoichiometric structure and properties were developed and performed in [7] using various DFT hybrid functionals. These results obtained demonstrate rise of Eg for perfect stoichiometric HAP using a more developed functional and method. We discuss this point below.

Interestingly, the energy scale of above DOS calculations is directly comparable to the band structure obtained in [7] calculations (see Figure 5a for PBE).

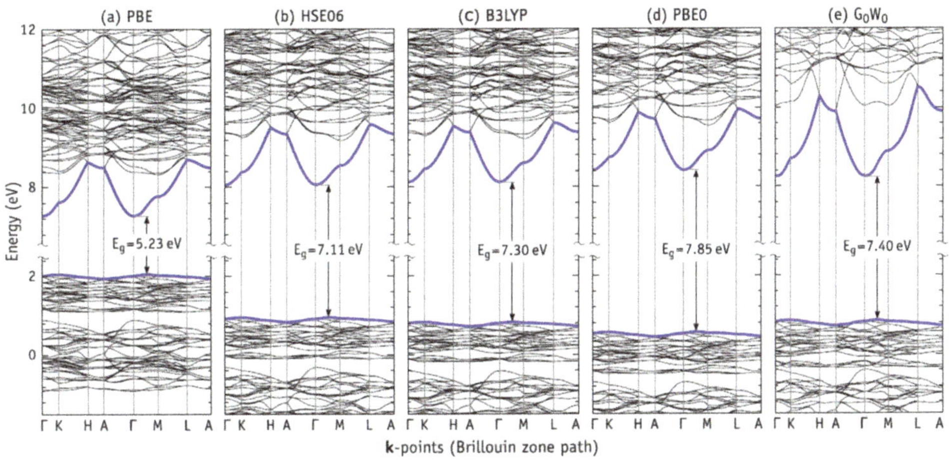

Figure 5. Electronic band structure of bulk perfect stoichiometric hydroxyapatite along a path with breaks at k-points of high symmetry. The calculations were carried out using DFT using functionals: (**a**) PBE—in the generalized gradient approximation (GGA), and hybrid functionals: (**b**) HSE, (**c**) B3LYP, and (**d**) PBE0, as well as (**e**) using the method multi-particle perturbations G0W0 using the wave functions of the PBE functional. The maximum valence and minimum conduction bands are shown in bold lines. Indirect transitions determining the band gap Eg are also shown for each case. (Adapted with permission from [7]; AIP Publishing, 2018).

Figure 5a–d compare the electronic band structure obtained using different exchange-correlation functionals (PBE, HSE, B3LYP, and PBE0) with the analogous *G0W0* quasi-particle calculation shown in Figure 5e [7]. The band energies along the several high-symmetry directions were obtained by interpolation of the first-principles data using Wannier90 [65].

The shape of the PBE band structure in Figure 3a is indistinguishable from that reported by Slepko and Demkov [34] displaying a low-dispersive valence band top and a high dispersive conduction band bottom (thick bands). Dispersion of the conduction band minimum states is considerably more pronounced along directions parallel to the *c*-axis (Γ–*A*, *K*–*H*, and *M*–*L*), indicating a stronger carrier delocalization and mobility along the main axis. This property could be explored for tuning HAP electrical conductivity through n-type doping or for photo-current measurements. On the other hand, p-type doping is not expected to be beneficial. The valence band top states show very little dispersion, and their heavy holes imply a relatively lower mobility.

Also, in agreement with [34] we find HAP to be an indirect-gap material with *Eg* = 5.23 eV at the PBE level. The conduction band minimum is located at **k** = Γ, while the valence band top energy was found somewhere along Γ–*K* or Γ–*M*. The valence band maximum along Γ–*M* is only 0.1 meV higher than the one along ~Γ–*K*. We note that this picture was the same regardless of the functional used, including when using the G_0W_0 method.

The band structure obtained within the HSE06 level is shown in Figure 5b. The increase in the band gap width by more than 30% with respect to the PBE result is self-evident. Using HSE06, we obtain *Eg* = 7.11 eV. Often, the band energies are offset in order to lock the valence band top at the origin of the energy scale. We did not follow this procedure and that allowed us to disclose how the gap change depends on the shift of both valence band and conduction band states.

Figure 5b–d show that admixing a fraction of Fock exchange with the semi-local exchange energy has a profound effect on both valence and conduction band states. Consequently, the use of hybrid functionals has implications not only to the accuracy of calculated defect-related or inter-band transitions (e.g., observed in luminescence or UV-VIS absorption), but also to transitions involving core or vacuum states (e.g., observed in electron photoemission or core electron energy loss spectroscopy) [7].

The gap of the B3LYP band structure depicted in Figure 5c is 0.6 eV narrower than that reported by Corno et al. [35] using the same functional. Again, the difference is likely to be due to the unsuitability of the atomic-like basis employed in [35], which resulted in the under-screening of the band structure. The band gap width and the band gap edge energies obtained at the B3LYP-level are closer to the *G0W0* results than any other functional.

Thus, here it should be noted and concluded that with the advent of each of the next developed types of DFT functional, the total calculated electronic band structure of pure HAP is change: the energies of valence band top is down and the bottom of conductive band is up. As result, the width of the forbidden band gap Eg = Ec − Ev increases (see Figure 5a–e).

This trend of the calculated band gaps agrees with that obtained by Garza and Scuseria [66] for an eclectic mix of semiconductors and insulators. Accordingly, HSE06 and B3LYP showed a closer correlation with the experiments (the former giving slightly smaller gaps overall), whereas inclusion of larger fractions of exact exchange, like in PBE0, led to an overestimation of Eg. This ordering also suggests that the gap width obtained within G_0W_0 (Eg = 7.4 eV) should be close to the real figure.

Another interesting point should be noted here, regarding the origin of the bottom of the conduction band, in particular, the highly dispersive bands shown in Figure 5, the situation here is somewhat controversial. In [34], an analysis of LDOS at the bottom of the conduction showed that the lowest energy bands originate from Ca(4s) states.

The authors in [7] argue that this view finds several difficulties. First, in [7], the authors find the onset of the Ca-LDOS (E_{onset} = 8.40 eV) located ~1 eV above the conduction band minimum energy. Within that energy range, all that can be related to Ca are the flat Ca(3d) bands above 8.4 eV. Second, right Figure 4e shows that states just above Ec have a considerable localization on OH molecules.

Interestingly, in this case the highly dispersive bands that form the bottom of the conduction band of HAP are anti-bonding states from an infinite . . . OH–OH– . . . hydrogen bridge sequence (along OH-channel through all HAP crystal structure), much like a 1D-ice phase. This statement finds support in the band structure of hexagonal ice (see, e.g., Figures 4 and 6 in [67,68]). From comparison of these band structure images with Figure 5a, it becomes immediately evident that the dispersion shape of the lowest conduction band of HAP is analogous to that of the hexagonal ice.

In this relation, it is interesting to note that nano-confined water [66] inside hydrophilic cavity (or inner channel) inside so-called peptide nanotube (such as, e.g., the diphenylalanine nanotube) has structure close to the one-line hexagonal ice structure [69,70]. So, we should note here that it is a common property in tubular or columnar molecular-like periodical crystal structure, which could be observed in these related structures.

3.2.3. Electronic Band Structure of Defective HAP (with OH-Vacancy and O-Vacancies)

In Figure 6, examples of DOS in the results of GGA (PBE) calculations are given for defects in HAP structure, such as O vacancies from the OH group, O vacancies from the PO_4 group, and a full vacancy of the OH group, based on HAP unit cell model with 44 atoms.

The data of calculated energies of the electronic band structure's basic states are presented in Table 4.

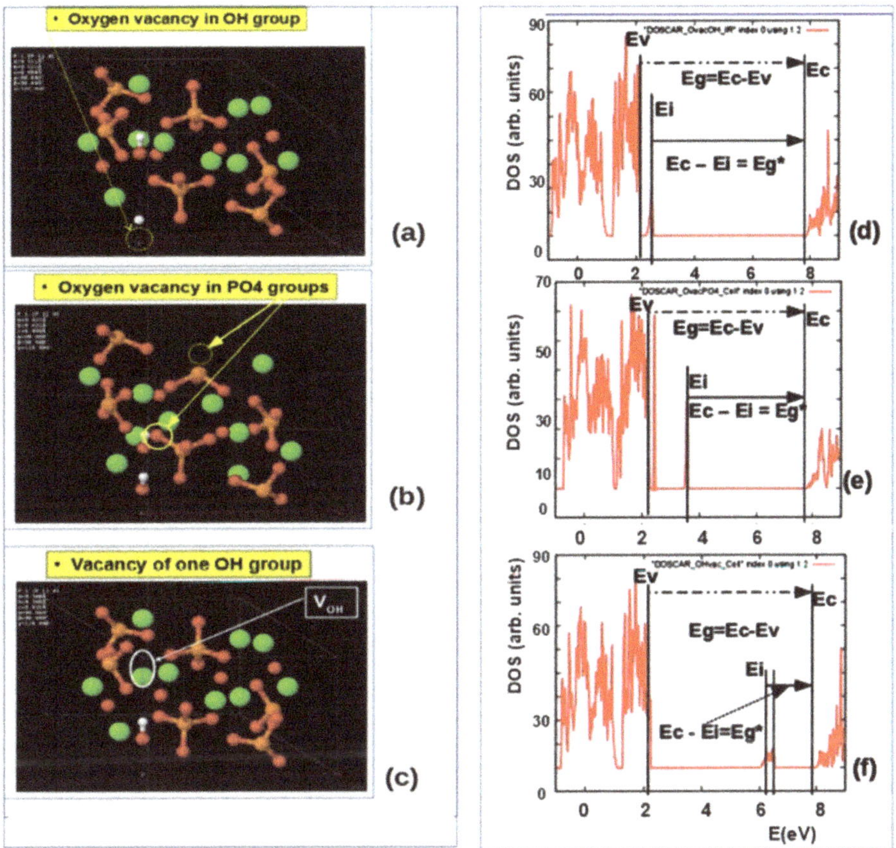

Figure 6. Density of electronic states (DOS) for HAP unit cell with defects: (**a**) HAP unit cell model with O vacancies from the OH group; (**b**) HAP unit cell model with O vacancies from the PO_4 groups (this may be with O atom from different positions); (**c**) full vacancy of the OH group from HAP unit cell model; (**d**) DOS for case of the O vacancy in OH group; (**e**) DOS for case of the O vacancy in PO_4 group; (**f**) DOS for case of the full OH vacancy. (Data presented here obtained by calculations firstly using AIMPRO (LDA) [5,6,38] and then VASP (GGA) [39], similar to our works [4–8]). (Adapted with permission from [5]; IOP Publishing, 2015).

These data are close to results of other authors [19,30–34]. The results obtained also lead to a small shift in all level energies for defects in the calculations of the GGA (PBE) methods relative to LDA (Table 3). Both methods predict changes in the band gap Eg and the electron work function $\Delta\phi$, which can be measured experimentally [2,4,9,53]. To compare this change in the band gap Eg in the presence of a defect in the HAP and to distinguish it from defect-free HAP, we here introduced the designation for the case of a defect-free HAP as Eg0, and then the relative change in the band width will be $\Delta Eg = Eg - Eg0$, where Eg is the band gap width in the case of a HAP with a defect. The electron work function ϕhas approximately the same changes $\Delta\phi \approx \Delta Eg$ (it is important for comparison of such changes with experimental data of the photoelectron emissions) [4,5,9]. All these computed changes are summarized in Table 4 below.

Table 4. Calculated data for HAP defects by two various approaches (aver. error = ±0.05, ±0.1eV) [4–6].

Defect Type	LDA				GGA (PBE)			
	Eg = Ec − Ev, eV	ΔEg = Eg − Eg0 ~ Δϕ, eV	Ei − Ev, eV	Eg* = Ec − Ei, eV	Eg = Ec − Ev, eV	ΔEg = Eg−Eg0~ ~ ΔϕeV	Ei − Ev, eV	Eg* = Ec − Ei, eV
HAP in $P6_3/m$, Eg0	4.6	-	-	-	5.26	-	-	-
$O_{(OH)}$ vac	5.15	+0.55	0.1 (1 occ.)	5.05	5.72	+0.46	0.27	5.45
OH vac	5.49	+0.89	3.11–3.82 peaks: 3.40 3.53 3.66 ($\frac{1}{2}$ occ.)	2.38–1.67 peaks: 2.09 1.96 1.83	5.75	+0.49	3.66–4.28 peaks: 3.96 4.11 4.17	1.97–1.35 peaks: 1.78 1.63 1.57
Ovac(PO4) [(1)] V_O1 (O6) V_O2 (O15) V_O3 (O30,O35) V_O4 (O24)	4.734 4.768 4.735 4.5614		1.346 1.300 1.347 0.9557	3.388 3.468 3.388 3.6057	5.416 5.246 5.326		1.045 1.212 1.142	4.115 4.034 4.184
aver. $O_{(PO4)}$ -vac	4.70 ± 0.2	+0.15	1.14 ± 0.3	3.52 ± 0.3	5.34 ± 0.2	+0.08	1.13 ± 0.2	4.11 ± 0.2

[(1)] $OvacPO_4$: atom O in various positions of PO_4 and from the different PO_4 group, according with Figure 2.

In addition, we will denote the energy of photoexcitation of an electron from the local level Ei to the conduction band Ec, which arises in the presence of a defect, as Eg* = Ec − Ei, and this will correspond to the new effective width of the band gap Eg* for optical excitation of an electron and/or photoabsorption of light, which are usually observed in the experiment. These designations are also used in Figure 6 and others figures and tables below.

Similarly, here we introduce the excitation energy of an electron from the valence band Ev to this level Ei, as Ev − Ei, which is presented in Table 4.

It is important to note that the levels of optical absorption for OH vacancies in HAP, established here (Table 3), lie in the region of ~1.6–2.3 eV and they are practically not observed in the experiment, even at a significant concentration of OH vacancies. The calculated density of electronic states (DOS) from the HAP data shows that the intensity of the DOS peaks for the energy levels Ei caused by OH vacancies (located approximately in the middle of the band gap Eg) is very low, compared with the intensity of the main DOS peaks (the top of the valence band Ev and the bottom of the conduction band Ec, which determine the band gap Eg = Ec − Ev) [5]. Therefore, this transition is difficult to detect experimentally by photoelectron excitation. This turns out to be possible only by the photo-luminescence method, upon excitation of a significant number of electrons and from deep valence band levels into the conduction band by the method of synchrotron irradiation [53]. However, in this case, namely the changed value of the optical absorption can be experimentally recorded, which rise up to the value Eg* ~5.3–5.5 eV, due to rise of HAP band gap caused by OH vacancies (Table 3).

At present, work on the experimental detection and observation of these spectral characteristics of optical absorption in this energy range is underway, created by such defects, as the vacancies of OH groups in HAP, are being actively carried out and are in progress.

Figure 7, Figure 8, and Figure 9 below show examples of the band structure and the partial charge density for electrons with their eigenvalues in the range specified around the energy level Ei of corresponding defect structure for main defects in HAP, obtained after optimized VASP calculations in one unit cell model.

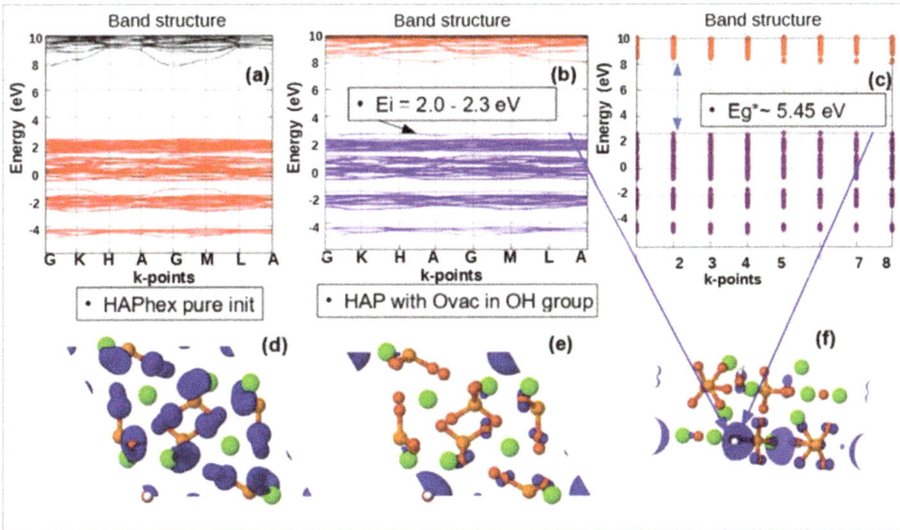

Figure 7. Band structure and partial charge density distribution for O-vacancy from OH group in hexagonal HAP (for one unit cell model): (**a**,**d**) band structure and partial charge density (top view) of initial pure HAP; (**b**,**e**) band structure and partial charge density (top view) for HAP with O-vacancy in OH group (local energy level Ei ~2.0–2.3 eV is close to the top of valence band ΔE = Ei − Ev ~0.1–0.3 eV); (**c**) band structure distribution along k-points with local energy level leaded to optical Eg*~5.45 eV; (**f**) partial charge density for electrons with their eigenvalues around the energy level Ei of the O-vacancy in HAP (side view of the unit cell).

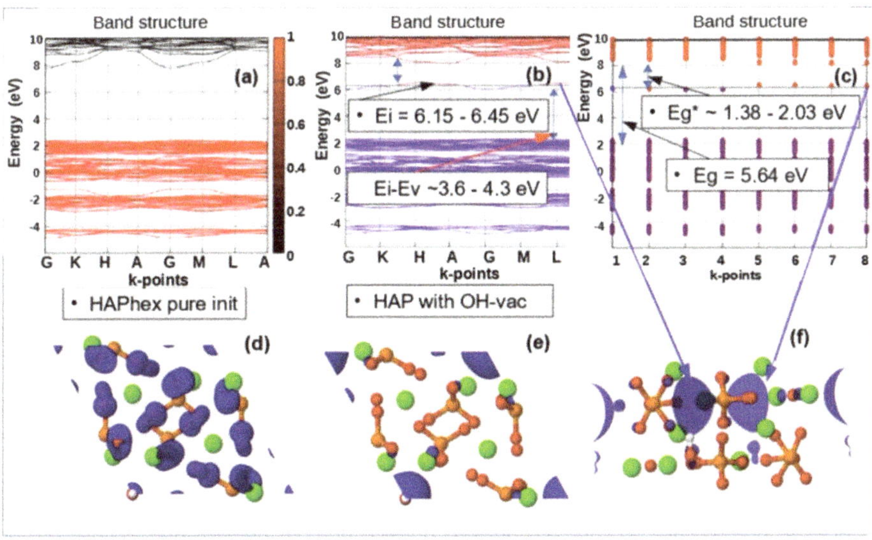

Figure 8. Band structure and partial charge density distribution for OH-vacancy in hexagonal HAP (for one unit cell model): (**a**,**d**) band structure and partial charge density (top view) of initial pure HAP; (**b**,**e**) band structure and partial charge density (top view) for HAP with OH-vacancy (local energy level Ei ~6.15–6.45 eV is the middle of the forbidden band and with ΔE = Ei-Ev~3.6–4.3 eV); (**c**) band structure distribution along k-points with local energy level leaded to optical Eg*~1.38–2.03 eV in comparison with total Eg ~5.64 eV; (**f**) partial charge density for electrons with their eigenvalues around the energy level Ei of the OH-vacancy in HAP (side view of the unit cell).

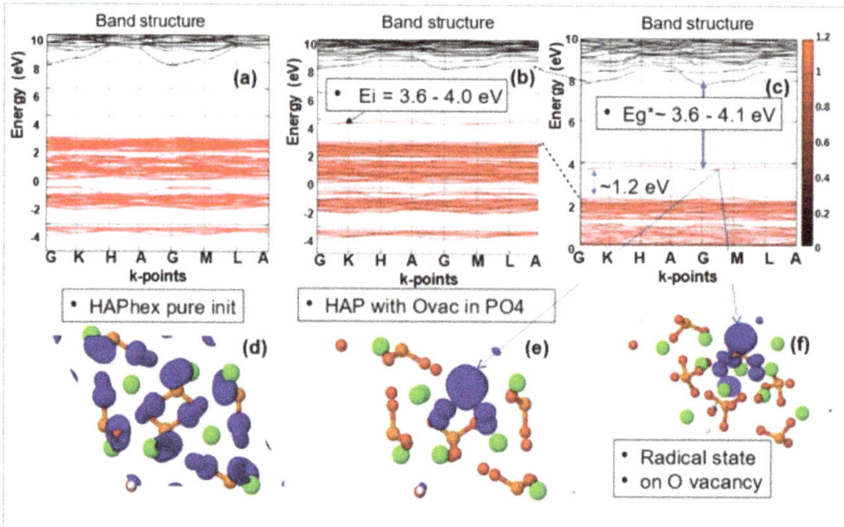

Figure 9. Band structure and partial charge density distribution for O-vacancy from PO_4 group in hexagonal HAP (for one unit cell model): (**a**,**d**) band structure and partial charge density (top view) of initial pure HAP; (**b**,**e**) band structure and partial charge density (top view) for HAP with O-vacancy from PO4 (local energy level Ei ~3.6–4.0 eV is the middle of the Eg band and with ΔE = Ei − Ev ~1.1–1.3 eV); (**c**) band structure distribution along k-points with local energy level leaded to optical Eg*~3.6–4.1 eV in comparison with total Eg ~5.34 eV; (**f**) partial charge density for electrons with their eigenvalues around the energy level Ei of the O-vacancy of PO_4 group in HAP (side view of the unit cell).

These results clearly show that it is precisely the oxygen vacancies of the PO_4 group that provide the optical band with the energy Eg* = ~3.6–4.2 eV, which is usually observed in many experimentally prepared samples [4–6,19–22,36,37,53]. All these samples in the manufacturing process undergo heat treatment at temperatures T = 700–1200 ° C (that is ~973–1473 K) and this leads to the formation of not only OH vacancies, but also a multitude of O vacancies [19–22] (especially from the PO_4 group). This leads to the observed optical absorption and optical band width Eg* ~3.6–4.2 eV, while the band gap of the defect-free perfect stoichiometric HAP turns out to be wider and equal to Eg~5.4 eV and more (in the GGA (PBE) calculations).

Note that the existing various experimental data [6,9,20] demonstrate optical absorption in HAP samples just in this energy range. In addition, these calculation data correlate with other, more accurate calculations using more advanced methods and models, considered below in the Section 3.3 (Section 3.3.1, Section 3.3.2). This means that even simpler models and methods correctly reflect the essence of changes in the physical properties of HAP in the presence of defects in its structure.

Further development of the calculations was the transition to calculations from one unit cell to a super-cell model consisting from of 2 × 2 × 2 = 8 unit cells, (space group $P6_3$, comprising a total of 352 atoms), as well as the application of the hybrid functional PBE (for 1st step optimization) in combination with B3LYP (for2nd step calculation) using DFT method [7,8]. This made a possibility to classify different types of oxygen vacancies of the PO_4 group more correctly and accurately, which increases the calculation accuracy and to *highlight not only neutral defects, but new more complex types of defects—extended charged complex oxygen vacancies.*

These new structural extended (or bridging structures) defects can play an important role in changing the properties of electrically charged HAP.

3.3. Electronic and Optical Properties of HAP with Defects in Supercell Model

3.3.1. Oxygen Originated Complex HAP Defects in Supercell Model

Performing these series of calculations, it was found that different types of defects arise, if the symmetry of the atomic group associated with different spatial arrangement of various atoms in the PO_4 group. Besides, usual oxygen vacancy V_O, here arises complex extended defects, which depend on their charge state Q = 0, +1, +2. (Figure 10) [8,36,37]. The O vacancy of OH group influences the formation of new complex defects because of charge Q variation.

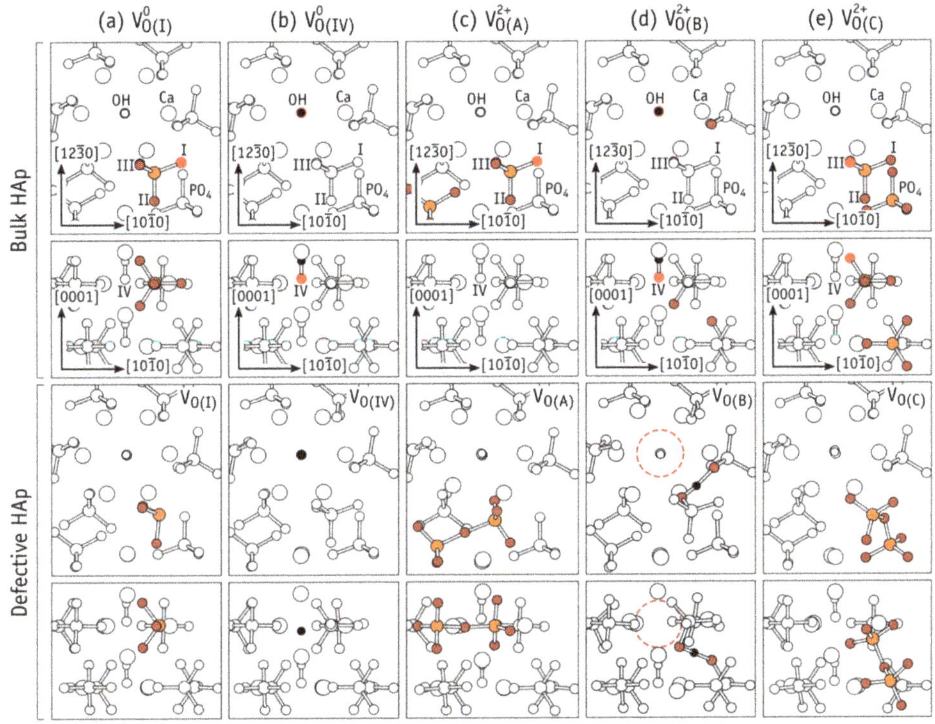

Figure 10. Diagrams showing "Bulk" (upper half) and "Defective" (lower half) HAP. For each row pair, upper and lower rows display the same structures viewed along the [0001] and [1230] directions of the hexagonal lattice, respectively. Formation of structures I, IV, A, B, and C of V_O defects is explained in columns (**a**–**e**), respectively. Only atoms belonging to the core of the defect are colored (P, O, and H atoms are shown in orange, red, and black, respectively). Vacancies were created by removing the bright red O atom shown in the "Bulk" figures. Upon atomic relaxation, the resulting structures are those in the corresponding "Defective" figures. Reprinted with permission from [8]. (Adapted with permission from [8]. Copyright (2019) American Chemical Society).

Crystalline HAP has oxygen atoms on four symmetries in equivalent sites, which are referred to as oxygen types I—IV. Types I, II, and III in the phosphate units, while type IV oxygen atoms are located in the OH^- anions. These are denoted as O(I), . . . , O(IV) and are shown in the upper half of Figure 10, where portions of bulk HAP are depicted. We note that PO_4 groups have two nearly symmetric O(III) atoms which are superimposed in the upper view of bulk HAP in Figure 10.

Among the many vacancy structures investigated, those shown in Figure 10a—e are the most relevant as they showed lower energy. Metastable structures with more than 2 eV above the ground state (for each particular charge state) will not be discussed. Figure 10a describes the formation of a pyramidal PO_3 structure, where removal of the O(I) atom in

"Bulk HAP" highlighted using a bright red color, leads to $V_{O(I)}$ as shown in the lower part of the figure, where a "Defective HAP" region is shown.

Analogous structures for $V_{O(II)}$ and $V_{O(III)}$ were obtained as well. In the neutral charge state, the resulting $[PO_3^{3-}]_{PO4}^{0}$ structures display a fully occupied sp^3 orbital on the P atom, resembling the phosphine molecule. This is shown in Figure 11a, where an isosurface of the electron density corresponding to the highest occupied state is represented in blue for the specific case of $V_{O(III)}^{0}$.

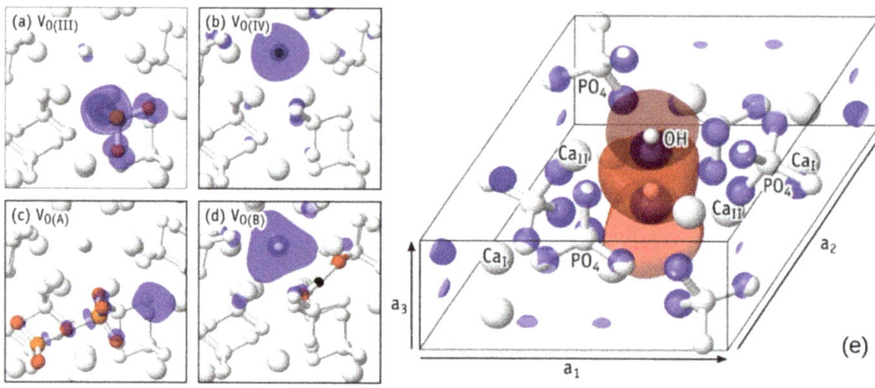

Figure 11. Electron density isosurfaces from the highest occupied Kohn−Sham level of neutral V_O defects in HAP. The density of $V_{O(III)}$ in (**a**) is representative of $V_{O(I)}$ and $V_{O(II)}$ as well. The density of $V_{O(IV)}$ in (**b**) shows the case of a missing O(IV) atom, leaving an isolated H atom in the OH channel. The density of extended charged defects $V_{O(A)}$ and $V_{O(B)}$ are shown in (**c**,**d**) (description is in text). Isosurfaces are drawn at constant electron density n = 0.001 $Å^{-3}$. Reprinted with permission from [8]. Copyright (2019) American Chemical Society. (**e**) Lowest unoccupied Kohn-Sham state (bottom of the conduction band) of a HAP at **k** = G. Blue and red isosurfaces represent y(**r**) = +0.02 and y(**r**) = 0.02 phases of the wave function, respectively. All atoms are shown in white. Reprinted with permission from [7]. (Adapted with permission from [7]. CCC (2018) AIP Publishing).

Figure 11b shows the case of a missing O(IV) atom, leaving an isolated H atom in the OH channel. This defect can be rationalized as the removal of neutral O from OH^-, leaving a hydride anion, and maintaining charge neutrality of the whole system. In the neutral charge state, we have a $[H^-]_{OH}^{0}$ structure, where after structural relaxation, the hydride species becomes located close to the site of the missing O(IV) atom (compare lower diagrams of bulk and defective HAP in Figure 10b).

The density corresponding to the highest occupied state of $V_{O(IV)}^{0}$ is represented in Figure 11b, which clearly shows the formation of a hydride anion in the OH channel. Figure 11e shows the OH-vacancy defect [7]. In $V_{O(I)}-V_{O(IV)}$ defects, all atoms (except the missing oxygen) remain essentially close to their original crystalline coordinates, hence the use of subscripted O(I)−O(IV) labels to identify their structure. However, additional structures, hereafter referred to as ***extended structures***, were also found for the oxygen vacancy in HAP [8]. One type of such extended structures can be described as a pair of neighboring oxygen vacancies connected by an O-interstitial, $2V_O$ + O. Another 2nd type is best described as a complex made of an OH-vacancy next to an H-interstitial, V_{OH} + H. Two defects of type $2V_O$ + O are singled out and labeled $V_{O(A)}$ and $V_{O(C)}$. They are shown in Figure 10c,e, respectively. One V_{OH} + H defect is shown in Figure 10d and is referred to as $V_{O(B)}$. These new defect structures are very important for practices too.

A dashed circle is used in the figure to highlight the missing OH unit. The highest occupied state of the extended structures is shown in Figure 11c,d. They either overlap the void regions of the HAP crystal or the vacant volume of the OH-channel, thus suggesting that they are donors with antibonding character or strong resonance with conduction band states [7].

For relaxations in charge state +2 that started from structures I and III, the final structures were, respectively, A and C. Here, the P atom of the PO_3 unit in $V_{O(I)}$ (or $V_{O(III)}$) moved across the plane defined by the three O atoms to connect to the O atom from the nearest PO_4 moiety. Such a severe relaxation can be explained by electron transfer from a neighboring $PO_4{}^{3-}$ anion to the empty $P(sp^3)$ orbital of $[PO_3{}^-]_{PO4}{}^{2+}$ in $V_{O(I)}{}^{2+}$ or $V_{O(III)}{}^{2+}$ (see Figure 11a) and the subsequent formation a new P−O bond. The result is an extended $[PO_3{}^{2-}-O-PO_3{}^{2-}]_{2(PO4)}{}^{2+}$ structure shown in Figure 10c,e.

When initiating the relaxation in charge state +2 from structures II and IV, the resulting configuration was in both cases $V_{O(B)}{}^{2+}$. In this charge state, the defect comprises an interstitial H+ next to a positively charged OH vacancy, that is, $[PO_4{}^{3-}-H^+-PO_4{}^{3-}]_{2PO4}{}^+ + V_{OH}{}^+$. The proton is located on a high electron density site between two oxygen anions. The net positive charge of the OH vacancy follows from depletion of two electrons from the channel-state represented by the isosurface of Figure 11d. When starting from structure II, Coulomb attraction and subsequent reaction between neighboring OH^- and $[PO_3{}^-]_{PO4}{}^{2+}$ leads to the formation of the $V_{O(B)}{}^{2+}$ extended structure. Alternatively, when starting from structure IV, a proton in the initial $[H^+]_{OH}{}^{2+}$ configuration is strongly attracted by O-anions in neighboring $PO_4{}^{3-}$ moieties, also ending up in $V_{O(B)}{}^{2+}$ as depicted in Figure 10d.

The study of transitions between the energy levels of such more complex defects (including at different levels of the charge state of the HAP crystal, and different types of its conductivity: electron n-type, or hole p-type) showed that the transition of electrons can also include a phononless transition with energy ~3.6–3.9 eV [8,36,37]. This mechanism can also explain the onset of absorption at 3.4–4.0 eV in the experimental observation of photocatalysis in HAP under constant UV illumination [6,8].

The results obtained convincingly show the primary role of defects such as oxygen vacancies and vacancy of OH group in the formation of actually observed values in the range approximately of ~3.6–4.3 eV for oxygen vacancies Ovac {VO(I), VO(II), VO(III)} from PO_4 groups, and at ~5.06 eV for oxygen vacancy VO(IV) from OH group (see below in Section 3.3.3) for the effective optical absorption and excitation bands Eg* in various samples of HAP materials [5–9]. At the same time, for example, the main experimentally observed values of the optical absorption band according to data measured UV spectra for the HAP samples of marine origin, and the commercial ones, respectively, (presented in Tables 2 and 4 in work [6]) lie in the close range approximately of ~3.4–4.2 eV (which could be arisen by oxygen vacancy from PO_4 groups) and ~5.05 eV (which could be arisen by oxygen vacancy from OH group). These data directly correspond to the data calculated in our computational studies presented and discussed in this article.

Nevertheless, the issues related to obtaining more accurate, fair, and correct numerical results, which are very important for many practical applications, as well as the issues of statistics (and dynamics) of accumulation and annealing of defects in various unit cells in the entire volume of the crystal, remain rather complex problems and require further research.

In addition, it is obvious that all these points require additional experimental observations and research. This work must be carried out in a complex.

These works are now in progress in the computational direction, and new results will soon be obtained in this direction of calculating defects in HAP, including, in addition to the types of vacancies here considered, also other defects of the type of insertions and substitutions of some atoms in HAP unit cell lattice.

3.3.2. Kohn−Sham Energy Levels of Neutral Oxygen Vacancy V_O Defects

Inspection of the Kohn−Sham eigenvalues at $\mathbf{k} = \Gamma$ confirmed that neutral vacancies V_O defects are all donors [8,37]. The calculations performed in these cases, taking into account the even more accurate B3LYP calculation scheme, showed similar shifts in the energy levels created by defects such as oxygen vacancies of the PO_4 group, and their main contribution, which determines the change in optical properties and the change in the work function, remains at a level of ~ 1 eV or less (Table 5), despite the differences in the calculations of the band gap Eg for initially perfect stoichiometric defect-free HAP.

Table 5. Data for O vacancy from OH and PO_4 in different position and symmetry (errors ±0.05 eV).

Defect Type	PBE				B3LYP			
	Eg = Ec − Ev, eV	ΔEg = Eg − Eg0 ~Δφ, eV	Ei − Ev, eV	Ec − Ei = Eg*, eV	Eg = Ec − Ev, eV	ΔEg = Eg − Eg0 ~Δφ, eV	Ei − Ev, eV	Ec − Ei =Eg*, eV
HAP in $P6_3/m$, Eg0	5.23	-	-	-	7.3	-	-	-
$A0=A^0_I$ (V_O(I))	5.0674	−0.1626	1.1496 ~1.15	3.9178	7.0497	−0.2503	1.4291 ~1.43	5.6206
$A0=A^0_{II}$ (V_O(II))	5.2004	−0.0296	1.3167 ~1.32	3.8837	7.2311	−0.0689	1.6512 ~1.65	5.5799
$A0=A^0_{III}$ (V_O(III))	5.1393	−0.0907	1.3811 ~1.38	3.7582	7.1333	−0.1667	1.685 ~1.68	5.4488
$D0=D^0_I$ (V_O(IV))	5.3004	+0.0704	0.4189 ~0.42	4.8815	7.3842	+0.0842	0.7347 ~0.73	6.6495

As can be seen, regardless of the method of calculation, oxygen vacancies of the OH and PO_4 groups (in the absence of charge in these defects, i.e., at Q = 0) form a group of energy levels (Figure 12) located close to the top of the valence band and they are donor electrons [8,48] (electron acceptors, that is, levels with a negative charge were not found).

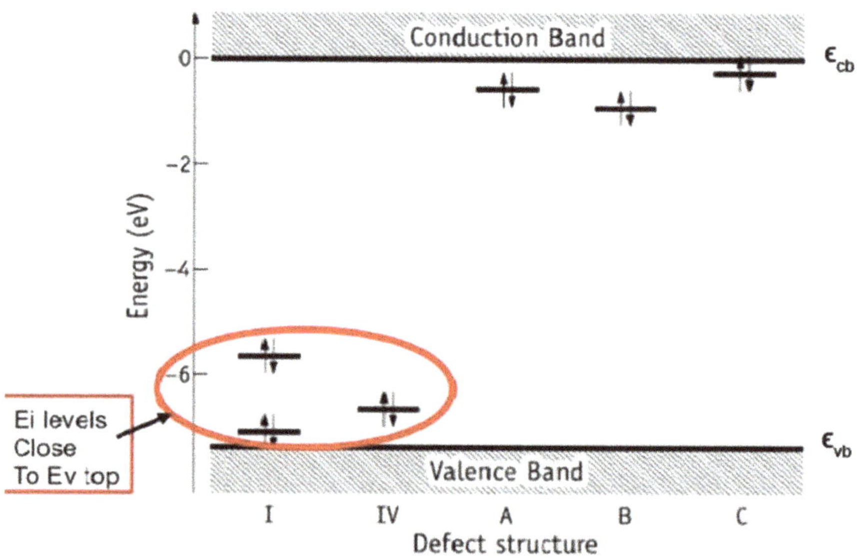

Figure 12. Kohn−Sham energy levels of neutral V_O defects in a HAP supercell at the **k** = Γ point. The defect structure I is also representative of structures II and III (see text). The latter have gap states that deviate from those of $V_{O(I)}$ by less than 0.2 eV [33]. Adapted with permission from [8]. (Copyright (2019) American Chemical Society).

In this case, the shift of the levels of oxygen vacancies in the PO_4 group is ~1.15–1.65 eV upward from Ev, and for a vacancy from the OH group, ~0.4–0.7 eV, and this also corresponds to a change in the electron work function during the formation of such defects. As everyone can see, the positions of these energy levels are not very different (especially in the case of PBE calculations), although in the case of B3LYP these levels are slightly higher than Ev in the direct case. In any case, these deviations are within 1 eV.

It is also important to note that such levels close at the top of the valence band have recently been observed in experiments on photoelectron emission spectroscopy, and the work function of a photoelectron from HAP was measured for various external influences [2,4,5,53]. In addition, in [53] it was noted that such energy levels (and photoelectrons emitted from these energy levels) can arise under a number of actions on HAP samples (heating and annealing, gamma irradiation, microwave effects, and combined hydrogenation with microwave radiation).

In these cases, a sufficiently large number of oxygen vacancies (also in the OH group) can be induced having the lowest energies levels measured from the top of the valence band (see Table 4 and Figure 12).

3.3.3. OH-Vacancy in HAP Supercell Model and Some General Remarks

The vacancy of one OH group in the HAP supercell model was considered in a similar way. In this case, calculation with the GGA approximation in the hybrid functional PBE and B3LYP was performed using VASP [39] and Quantum ESPRESSO [41].

Figure 13 shows the results obtained in this case. After PBE optimization and B3LYP calculation using VASP energy band structure for HAP with OH-vacancy (Figure 13b), the following data are presented: Ev = 0.726 eV, Ec = 8.066 eV and Eg = Ec − Ev = 8.066 − 0.726 = 7.34 eV; the defective energy level due to OH-vacancy Ei = 5.3586 eV (occupied, spin up) and resulted optical gap from this energy level is equal Eg* = Ec − Ei = 8.066 − 5.3586 = 2.7074 eV. This data is in line with previous data obtained using another similar method in one unit cell model.

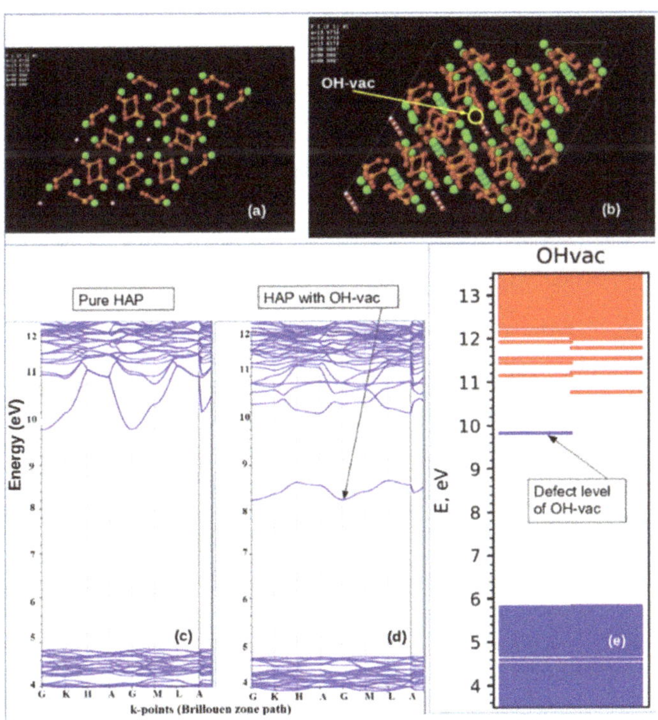

Figure 13. Model of OH-vacancy in HAP supercell: (**a**) top view on HAP supercell; (**b**) OH-vacancy in HAP supercell (iso projection); (**c**) band structure of pure HAP; (**d**) band structure of HAP with OH-vacancy computed using VASP; (**e**) band structure of HAP with defect level from OH-vacancy computed using Quantum ESPRESSO.

In addition, it is important to note that the energy of formation of defects in the HAP crystal [7,8] was considered and investigated also in the formalism with a chemical potential based on work [52], which describes the procedure for calculating the energies of formation of defects in a similar study. This formalism also takes into account the formation of defects with an excess of electrons Δn_e (with respect to the neutral state) when they are capable of capturing/emitting electrons from/to an electronic reservoir with a chemical potential μ_ε. The results obtained in this approach allow more correct calculations taking into account the thermodynamics of the formation of a defect with a screened charge as ***a quasi-particle*** in a crystal [7,8]. As a result, we obtain corrections to the energies, including the energies of optical properties and transitions, which, in our opinion, turn out to be even closer to those observed experimentally. These generalized data are shown here in Table 6.

Table 6. HAP electronic properties with various vacancies Ovac {$V_{O(I)}$, $V_{O(II)}$, $V_{O(III)}$ from PO_4, $V_{O(IV)}$ from OH} as well as V_{OH} vacancy of OH group—in supercell model; for one unit cell model vacancies Ovac notation are the same as in Table 3: V_O1 (O6), V_O2 (O15), V_O3 (O30,O35), V_O4 (O24) (all energy in eV).

Unit Cell	AIMPRO (LDA), Eg*	VASP-PBE (GGA), Eg*	Super-Cell	PBE opt (Eg* = Ec − Ei) (GGA-Supercell)		B3LYP opt (Eg* = Ec − Ei) (GGA-Supercell)		Spectr.
Ovac from Figure 2			Ovac from Figure 10	Kohn–Sham	Defect as quasiparticle in crystal	Kohn–Sham	Defect as quasiparticle in crystal	
V_O1	3.3880 (Eg = 4.734)	4.115 (Eg = 5.416)	$V_{O(I)}$	3.9178 (Eg = 5.067)	3.7052	5.6206 (Eg = 7.05)	4.3375	UVA–UVB
V_O2	3.6057 (Eg = 4.562)	4.034 (Eg = 5.246)	$V_{O(II)}$	3.8837 (Eg = 5.200)	3.6575	5.5799 (Eg = 7.23)	4.3486	
V_O3	3.4677 (Eg = 4.768)	4.184 (Eg = 5.326)	$V_{O(III)}$	3.7582 (Eg = 5.139)	3.5166	5.4488 (Eg = 7.13)	4.1291	
V_O4	3.6057 (Eg = 4.562)	-	-	-	-	-	-	
Aver. (O1_4)	3.52 ± 0.3 (Eg = 4.70)	4.11 ± 0.3 (Eg = 5.34)	Aver. (I-III)	3.8532 ± 0.2 (Eg = 5.14)	3.6264 ± 0.2	5.5498 ± 0.2 (Eg = 7.14)	4.2717 ± 0.2	UVA
V_O of OH	5.05 (Eg = 5.15)	5.45 (Eg = 5.72)	$V_{O(IV)}$	4.8815	5.0630	6.6495	5.8563	UVC
V_{OH}	2.38-1.67 2.09 1.96 1.83 (Eg = 5.49)	1.97- 1.35 1.78 1.63 1.57 (Eg = 5.75)	V_{OH}	1.73	1.7372	2.9174 and 2.7074	1.7491 and 2.200	Green-Red
Eg0	4.6	5.26		5.4	5.237	7.34	6.849	

The results obtained convincingly show the primary role of defects such as oxygen vacancies and vacancy of OH group in the formation of actually observed values of the effective optical absorption and excitation bands Eg* in various samples of HAP materials. These works are in progress and new results will soon be obtained in this direction of calculating defects in HAP, including, in addition to the types of vacancies here considered, also other defects of the type of insertions and substitutions of some atoms in HAP unit cell lattice.

4. Conclusions

Computer studies of defects in hydroxyapatite, carried out by various methods, show their significant influence both on the structural and mechanical, as well as on the electronic and optical properties of hydroxyapatite with such defects. This is very important for practical applications. In this review, we consider in more detail defects such as OH group vacancies and oxygen vacancies from different PO4 groups and OH groups as well. Calculations performed by different methods showed that these vacancies significantly change, first of all, the electronic and optical properties of HAP. In this case, the vacancy of the entire OH group leads to an entire absorption band in the range ~1.4–2.4 eV (with several DOS peaks). It is close to the red color of optical spectrum. However, the intensity of these DOS is small compared to the DOS of the top of the valence band. As a result, it is practically impossible to register this range of optical absorption (it is at the level of measurement errors). At the same time, an OH vacancy leads to a noticeable increase in the band gap of the order of ΔEg ~0.5–0.9 eV. This leads to a change in the optical absorption and also a change in the work function of the electron exit, which is recorded in the experiment. For example, it can be the level Eg ~5.5–5.75 eV, which belongs to the ultraviolet spectrum and is close to the observed optical absorption of HAP samples with OH vacancies (created in them experimentally upon heating and subsequent cooling).

Oxygen vacancies have very different optical properties. Oxygen vacancies arising from various oxygen atoms of the PO_4 group and the OH group turn out to be of different types and depend on the symmetry of the corresponding group. As it was shown in [7] that besides neutral O-vacancy here may exist charged oxygen vacancies also form not only point defects, but more complex defects—extended or bridging (in the case of a charge Q = +2, Figure 10). The latter reconstructing to point-like defects (at Q = +1 and = 0). This rearrangement from extended to point defects occurs with bond breaking and causes also optical absorption effects. This transition leads to spontaneous rupture of bridging P-O-P or O-H-O bonds at extended defects and, most likely, explains the onset of absorption at 3.4–4.0 eV for observing photocatalysis under constant ultraviolet illumination [6]. It is important that these types of structure and the stability of defects strongly depend on the charge states.

It was found that the oxygen vacancies essentially occur in such two distinct forms, either as a simple vacant oxygen site (referred to as structures I-IV, Figure 10a), or as an oxygen atom replacing two neighboring oxygen vacancies (bridge or extended structures named as "A–C", Figure 10c–e). The former type of vacancies is deep donors, while the latter are shallow donors with rather low ionization energies. No acceptor states (stable negatively charged defects) were found. Vacancy structures I-IV are more stable in the neutral charge state, while bridge structures A-C are preferred in the double plus charge state. This means that the oxygen vacancy adopts rather different configurations on samples where the Fermi energy is in the upper or the lower half of the band gap. As regards to the neutral O-vacancy, corresponding inspection of the Kohn-Sham eigenvalues at k = Γ confirmed that neutral oxygen vacancy defects are all donors involving the luminescence and absorption of ~3.6–4.2 eV.

Further development and more accurate calculation of these electronic properties and optical photoexciting and photocatalytic processes can be made also by addition-correct calculations of the electron–electron correlation of the excited electron states taking into account the Frank Condon (FC) relaxation. In [36,37], these contributions are calculated for the case of the V_O(IV) vacancy in HAP $P6_3$ supercell model and their results showed that the FC shift obtained has reasonable values. These investigations should be continued for all other oxygen vacancy types.

Despite the need to continue further comprehensive study of HAP defects, however, based on the results already obtained so far, we must conclude that for any case, irrespective of these FC relaxation processes, the formation of the various types defects in HAP through different oxygen vacancies gives rise to an opportunity of the photo-excitation processes in the close ultraviolet (UV-C) and in the visible light region. These photo-excitation effects obviously produce the photo-catalytic activity of HAP, as well as provide the changes in the photoelectron work function and surface electric potential, which is important for implant covered by HAP. One only needs to introduce a sufficient amount of these oxygen vacancies by some external actions, such as heating/annealing, gamma-irradiation, and combined hydrogenation with microwave irradiations [71]. The electron photoexcitation from additional energy levels (with Eg* = ~3.4–4.2 eV), which arose due to oxygen vacancies inside the forbidden band gap, can provide this electronic and optical properties of the treated HAP samples.

Author Contributions: V.B.: conceptualization and wrote the manuscript; E.P. and L.A.: performed the calculations and made analysis of main computed data; J.C.: supervision main methodology for DFT calculations; N.B.: participation in the processing of experimental data and comparison analysis with calculated data. All authors have read and agreed to the published version of the manuscript.

Funding: This research was funded by the Russian Science Foundation (RSF), grant No. 21-12-00251.

Institutional Review Board Statement: Not applicable.

Informed Consent Statement: Not applicable.

Data Availability Statement: The data presented in this study are available on request from the corresponding author.

Acknowledgments: The authors are grateful to the Faculty of Physics of the University of Aveiro (UA), Portugal, for the opportunity to use a Linux cluster for computational research on this topic of the HAP research within the framework of the non-commercial scientific agreement between IMPB RAS (the branch of KIAM RAS) and UA for the period 2015–2021 of our collaboration. In addition, the authors express their gratitude for the opportunity to partially use in this study the equipment of the general research facilities of high-performance computing (HPC) resources at Lomonosov Moscow State University.

Conflicts of Interest: The authors declare no conflict of interest.

References

1. Ratner, B.D.; Hoffman, A.S.; Schoen, F.J.; Lemons, J.E. *Biomaterials Science*, 3rd ed.; Academic Press: Oxford, UK, 2013.
2. Bystrov, V.; Bystrova, A.; Dekhtyar, Y.; Khlusov, I.A.; Pichugin, V.; Prosolov, K.; Sharkeev, Y. Electrical functionalization and fabrication of nanostructured hydroxyapatite coatings. In *Bioceramics and Biocomposites: From Research to Clinical Practice*; Jiulian, A., Ed.; John Wiley & Sons, Inc.: Hoboken, NJ, USA, 2019; pp. 149–190.
3. Leon, B.; Janson, J.A. *Thin Calcium Phosphate Coatings for Medical Implants*; Springer: Berlin, Germany, 2009.
4. Baltacis, K.; Bystrov, V.; Bystrova, A.; Dekhtyar, Y.; Freivalds, T.; Raines, J.; Rozenberga, K.; Sorokins, H.; Zeidaks, M. Physical fundamentals of biomaterials surface electrical functionalization. *Materials* **2020**, *13*, 4575. [CrossRef]
5. Bystrov, V.S.; Coutinho, J.; Bystrova, A.V.; Dekhtyar, Y.D.; Pullar, R.C.; Poronin, A.; Palcevskis, E.; Dindune, A.; Alkan, B.; Durucan, C. Computational study of the hydroxyapatite structures, properties and defects. *J. Phys. D Appl. Phys.* **2015**, *48*, 195302. [CrossRef]
6. Bystrov, V.S.; Piccirillo, C.; Tobaldi, D.M.; Castro, P.M.L.; Coutinho, J.; Kopyl, S.; Pullar, R.C. Oxygen vacancies, the optical band gap (Eg) and photocatalysis of hydroxyapatite: Comparing modelling with measured data. *Appl. Catal. B Environ.* **2016**, *196*, 100–107. [CrossRef]
7. Avakyan, L.A.; Paramonova, E.V.; Coutinho, J.; Öberg, S.; Bystrov, V.S.; Bugaev, L.A. Optoelectronics and defect levels in hydroxyapatite by first-principles. *J. Chem. Phys.* **2018**, *148*, 154706. [CrossRef] [PubMed]
8. Bystrov, V.S.; Avakyan, L.A.; Paramonova, E.V.; Coutinho, J. Sub-Band Gap Absorption Mechanisms Involving Oxygen Vacancies in Hydroxyapatite. *J. Chem. Phys. C.* **2019**, *123*, 4856–4865. [CrossRef]
9. Bystrov, V.S. Computational Studies of the Hydroxyapatite Nanostructures, Peculiarities and Properties. *Math. Biol. Bioinform.* **2017**, *12*, 14–54. [CrossRef]
10. Elliott, J. *Structure and Chemistry of the Apatites and Other Calcium Orthophosphates, Studies in Inorganic Chemistry*; Elsevier Science: Amsterdam, The Netherlands, 2013; p. 404.
11. Hughes, J.M.; Cameron, M.; Crowley, K.D. Structural variations in natural F, OH, and Cl apatites. *Am. Mineral.* **1989**, *74*, 870–876.
12. Ma, G.; Liu, X.Y. Hydroxyapatite: Hexagonal or monoclinic? *Cryst. Growth Des.* **2009**, *9*, 2991–2994. [CrossRef]
13. Kay, M.I.; Young, R.A.; Posner, A.S. Crystal structure of hydroxyapatite. *Nature* **1964**, *204*, 1050. [CrossRef]
14. Hitmi, N.; LaCabanne, C.; Young, R.A. OH-dipole reorientability in hydroxyapatites: Effect of tunnel size. *J. Phys. Chem. Solids* **1986**, *47*, 533–546. [CrossRef]
15. Hitmi, N.; LaCabanne, C.; Young, R.A. Oh–reorientability in hydroxyapatites: Effect of F– and Cl–. *J. Phys. Chem. Solids* **1988**, *49*, 541–550. [CrossRef]
16. Nakamura, S.; Takeda, H.; Yamashita, K. Proton transport polarization and depolarization of hydroxyapatite ceramics. *J. Appl. Phys.* **2001**, *89*, 5386. [CrossRef]
17. Horiuchi, N.; Nakamura, M.; Nagai, A.; Katayama, K.; Yamashita, K. Proton conduction related electrical dipole and space charge polarization in hydroxyapatite. *J. Appl. Phys.* **2012**, *112*, 074901. [CrossRef]
18. Tofail, S.A.M.; Haverty, D.; Stanton, K.T.; McMonagle, J.B. Structural order and dielectric behaviour of hydroxyapatite. *Ferroelectrics* **2005**, *319*, 117–123. [CrossRef]
19. Matsunaga, K.; Kuwabara, A. First-principles study of vacancy formation in hydroxyapatite. *Phys. Rev. B Condens. Matter Mater. Phys.* **2007**, *75*, 014102. [CrossRef]

20. Aronov, D.; Chaikina, M.; Haddad, J.; Karlov, A.; Mezinskis, G.; Oster, L.; Pavlovska, I.; Rosenman, G. Electronic states spectroscopy of Hydroxyapatite ceramics. *J. Mater. Sci. Mater. Med.* **2007**, *18*, 865–870. [CrossRef]
21. Nishikawa, H. A high active type of hydroxyapatite for photocatalytic decomposition of dimethyl sulfide under UV irradiation. *J. Mol. Catal. A Chem.* **2004**, *207*, 149–153. [CrossRef]
22. Nishikawa, H. Photo-induced catalytic activity of hydroxyapatite based on photo-excitation. *Phosphorus Res. Bull.* **2007**, *21*, 97–102. [CrossRef]
23. Silin, A.R.; Trukhin, A.I. *Point Defects and Elementary Excitations in Crystalline and Glassy SiO_2*; Zinatne: Riga, Latvia, 1985. (In Russian)
24. Paramonova, E.V.; Avakyan, L.A.; Bystrov, V.S.; Coutinho, J. Hybrid density functional study of iron impurities in hydroxyapatite. In *Book of Abstracts, Proceedings of the 4th International Conference on Nanomaterials Science and Mechanical Engineering (ICNMSME-2021), University of Aveiro, Aveiro, Portugal, 6–9 July 2021*; UA Editora, Universidade de Aveiro: Aveiro, Portugal, 2021; p. 115.
25. Paramonova, E.V.; Avakyan, L.A.; Bystrov, V.S.; Coutinho, J. Magnetic iron substitutions in hydroxyapatite: Density functional study. In *Book of Abstracts, Proceedings of the 2nd International Conference on Nanomaterials Science and Mechanical Engineering, University of Aveiro, Aveiro, Portugal, 9–12 July 2019*; UA Editora, Universidade de Aveiro: Aveiro, Portugal, 2019; p. 67.
26. Bulina, N.V.; Chaikina, M.V.; Prosanov, I.Y. Mechanochemical Synthesis of Sr-Substituted Hydroxyapatite. *Inorg. Mater.* **2018**, *54*, 820–825. [CrossRef]
27. Šupova, M. Substituted hydroxyapatites for biomedical applications: A review. *Ceram. Int.* **2015**, *41*, 9203–9231. [CrossRef]
28. Capuccini, C.; Torricelli, P.; Boanini, E.; Gazzano, M.; Giardino, R.; Bigi, A. Interaction of Sr-doped hydroxyapatite nanocrystals with osteoclast and osteoblast-like cells. *J. Biomed. Mater. Res. Part A* **2009**, *89*, 594–600. [CrossRef] [PubMed]
29. Bystrov, V.S.; Paramonova, E.V.; Bystrova, A.V.; Avakyan, L.A.; Coutinho, J.; Makarova, S.V.; Bulina, N.V. Structural and physical properties of Sr-substituted hydroxyapatite: Modeling and experiments. In *Book of Abstracts, Proceedings of the 4th International Conference on Nanomaterials Science and Mechanical Engineering (ICNMSME-2021), University of Aveiro, Aveiro, Portugal, 6–9 July 2021*; UA Editora: Aveiro, Portugal, 2021; pp. 82–83.
30. Calderin, l.; Stott, M.J.; Rubio, A. Electronic and crystallographic structure of apatites. *Phys. Rev. B Condens. Matter Mater. Phys.* **2003**, *67*, 134106. [CrossRef]
31. Rulis, P.; Ouyang, L.; Ching, W.Y. Electronic structure and bonding in calcium apatite crystals: Hydroxyapatite, fluorapatite, chlorapatite, and bromapatite. *Phys. Rev. B Condens. Matter Mater. Phys.* **2004**, *70*, 155104. [CrossRef]
32. Rulis, P.; Yao, H.; Ouyang, L.; Ching, W.Y. Electronic structure, bonding, charge distribution, and x-ray absorption spectra of the (001) surfaces of fluorapatite and hydroxyapatite from first principles. *Phys. Rev. B Condens. Matter Mater. Phys.* **2007**, *76*, 245410. [CrossRef]
33. Haverty, D.; Tofail, S.A.M.; Stanton, K.T.; McMonagle, J.B. Structure and stability of hydroxyapatite: Density functional calculation and Rietveld analysis. *Phys. Rev. B Condens. Matter Mater. Phys.* **2005**, *71*, 094103. [CrossRef]
34. Slepko, A.; Demkov, A.A. First-principles study of the biomineral hydroxyapatite. *Phys. Rev. B Condens. Matter Mater. Phys.* **2011**, *84*, 134108. [CrossRef]
35. Corno, M.; Busco, C.; Civalleri, B.; Ugliengo, P. Periodic ab initio study of structural and vibrational features of hexagonal hydroxyapatite $Ca_{10}(PO_4)_6(OH)_2$. *Phys. Chem. Chem. Phys.* **2006**, *8*, 2464. [CrossRef]
36. Bystrov, V.S.; Coutinho, J.; Avakyan, L.A.; Bystrova, A.V.; Paramonova, E.V. Piezoelectric, ferroelectric, optoelectronic phenomena in hydroxyapatite by first-principles and with various defects. *Nanomater. Sci. Eng.* **2019**, *1*, 10–21. [CrossRef]
37. Bystrov, V.S.; Coutinho, J.; Avakyan, L.A.; Bystrova, A.V.; Paramonova, E.V. Piezoelectric, ferroelectric, and optoelectronic phenomena in hydroxyapatite with various defect levels. *Ferroelectrics* **2020**, *550*, 45–55. [CrossRef]
38. AIMPRO. 2010. Available online: http://aimpro.ncl.ac.uk/ (accessed on 28 February 2016).
39. VASP (Vienna Ab initio Simulation Package). Available online: https://www.vasp.at/ (accessed on 31 July 2019).
40. HyperChem. *Tools for Molecular Modeling (Release 8)*; Hypercube, Inc.: Gainnesville, FL, USA, 2011.
41. Quantum Espresso. Available online: https://www.quantum-espresso.org/ (accessed on 7 September 2021).
42. Britney, P.R.; Jones, R. LDA Calculations Using a Basis of Gaussian Orbitals. *Phys. Status Solidi B Basic Res.* **2000**, *217*, 131–171.
43. Briddon, P.R.; Rayson, M.J. Accurate Kohn–Sham DFT with the Speed of Tight Binding: Current Techniques and Future Directions in Materials Modeling. *Phys. Status Solidi B.* **2011**, *248*, 1309–1318. [CrossRef]
44. Martin, R.M.; Reining, L.; Ceperley, D.M. *Interacting Electrons: Theory and Computational Approaches*; Cambridge University Press: Cambridge, UK, 2016.
45. Kresse, G.; Hafner, J. *Ab initio* molecular-dynamics simulation of the liquid-metal-amorphous-semiconductor transition in germanium. *Phys. Rev. B Condens. Matter Mater. Phys.* **1994**, *49*, 14251–14269. [CrossRef]
46. Kresse, G.; Furthmüller, J. Efficient iterative schemes for *ab initio* total-energy calculations using a plane-wave basis set. *Phys. Rev. B Condens. Matter Mater. Phys.* **1996**, *54*, 11169–11186. [CrossRef]
47. Kresse, G.; Joubert, D. From ultrasoft pseudopotentials to the projector augmented-wave method. *Phys. Rev. B Condens. Matter Mater. Phys.* **1999**, *59*, 1758–1775. [CrossRef]
48. Perdew, J.P.; Burke, K.; Ernzerhof, M. Generalized Gradient Approximation Made Simple. *Phys. Rev. Lett.* **1996**, *77*, 3865–3868. [CrossRef] [PubMed]
49. Lee, C.; Yang, W.; Parr, R.G. Development of the Colle-Salvetti correlation-energy formula into a functional of the electron density. *Phys. Rev. B Condens. Matter Mater. Phys.* **1988**, *37*, 785–789. [CrossRef]

50. Becke, A.D. A new mixing of Hartree-Fock and local density functional theories. *J. Chem. Phys.* **1993**, *98*, 1372–1377. [CrossRef]
51. Blöchl, P.E. Projector augmented-wave method. *Phys. Rev. B Condens. Matter Mater. Phys.* **1994**, *50*, 17953–17979. [CrossRef] [PubMed]
52. Qian, G.-X.; Martin, R.M.; Chadi, D.J. First-principles study of the atomic reconstructions and energies of Ga- and As-stabilized GaAs(100) surfaces. *Phys. Rev. B Condens. Matter Mater. Phys.* **1988**, *38*, 7649–7663. [CrossRef]
53. Bystrova, A.V.; Dekhtyar, Y.; Popov, A.I.; Coutinho, J.; Bystrov, V.S. Modified Hydroxyapatite Structure and Properties: Modeling and Synchrotron Data Analysis of Modified Hydroxyapatite Structure. *Ferroelectrics* **2015**, *475*, 135–147. [CrossRef]
54. Heyd, J.; Scuseria, G.E.; Ernzerhof, M. Hybrid functionals based on a screened Coulomb potential. *J. Chem. Phys.* **2003**, *118*, 8207. [CrossRef]
55. Adamo, C.; Barone, V. Toward reliable density functional methods without adjustable parameters: The PBE0 model. *J. Chem. Phys.* **1999**, *110*, 6158. [CrossRef]
56. Open Babel. Available online: https://openbabel.org/docs/dev/Installation/install.html (accessed on 30 August 2021).
57. Monkhorst, H.J.; Pack, L.D. Special points for Brillouin-zone integrations. *Phys. Rev.* **1976**, *13*, 5188. [CrossRef]
58. Bulina, N.V.; Makarova, S.V.; Prosanov, I.Y.; Vinokurova, O.B.; Lyakhov, N.Z. Structure and thermal stability of fluorhydroxyapatite and fluorapatite obtained by mechanochemical method. *J. Solid State Chem.* **2020**, *282*, 121076. [CrossRef]
59. Pogosova, M.A.; Provotorov, D.I.; Eliseev, A.A.; Jansen, M.; Kazin, P.E. Synthesis and characterization of the Bi-for-Ca substituted copper-based apatite pigments. *Dyes Pigment.* **2015**, *113*, 96–101. [CrossRef]
60. Gomes, S.; Renaudin, G.; Mesbah, A.; Jallot, E.; Bonhomme, C.; Babonneau, F.; Nedelec, J.M. Thorough analysis of silicon substitution in biphasic calcium phosphate bioceramics: A multi-technique study. *Acta Biomater.* **2010**, *6*, 3264–3274. [CrossRef] [PubMed]
61. Katz, J.L.; Ukraincik, K. On the anisotropic elastic properties of hydroxyapatite. *J. Biomech.* **1971**, *4*, 221–227. [CrossRef]
62. Gilmore, R.S.; Katz, J.L. Elastic properties of apatites. *J. Mater. Sci.* **1982**, *17*, 1131–1141. [CrossRef]
63. Paier, J.; Marsman, M.; Kresse, G. Why does the B3LYP hybrid functional fail for metals? *J. Chem. Phys.* **2007**, *127*, 024103. [CrossRef] [PubMed]
64. Birch, F. Finite Elastic Strain of Cubic Crystals. *Phys. Rev.* **1947**, *71*, 809. [CrossRef]
65. Mostofi, A.A.; Yates, J.R.; Pizzi, G.; Lee, Y.-S.; Souza, I.; Vanderbilt, D.; Marzari, N. An updated version of wannier90: A tool for obtaining maximally-localised Wannier functions. *Comput. Phys. Commun.* **2014**, *185*, 2309. [CrossRef]
66. Garza, A.J.; Scuseria, G.E. Predicting Band Gaps with Hybrid Density Functionals. *J. Phys. Chem. Lett.* **2016**, *7*, 4165. [CrossRef] [PubMed]
67. Prendergast, D.; Grossman, J.C.; Galli, G. The electronic structure of liquid water within density-functional theory. *J. Chem. Phys.* **2005**, *123*, 014501. [CrossRef] [PubMed]
68. Engel, E.A.; Monserrat, B.; Needs, R.J. Vibrational renormalisation of the electronic band gap in hexagonal and cubic ice. *J. Chem. Phys.* **2015**, *143*, 244708. [CrossRef]
69. Salehli, F.; Aydin, A.O.; Chovan, D.; Kopyl, S.; Bystrov, V.; Thompson, D.; Tofail, S.A.M.; Kholkin, A. Nanoconfined water governs polarization-related properties of self-assembled peptide nanotubes. *Nano Sel.* **2021**, *2*, 817–829. [CrossRef]
70. Bystrov, V.; Coutinho, J.; Zelenovskiy, P.; Nuraeva, A.; Kopyl, S.; Zhulyabina, O.; Tverdislov, V. Structures and properties of the self-assembling diphenylalanine peptide nanotubes containing water molecules: Modeling and data analysis. *Nanomaterials* **2020**, *10*, 1999. [CrossRef] [PubMed]
71. Bystrov, V.S.; Paramonova, E.V.; Dekhtyar, Y.D.; Bystrova, A.V.; Pullar, R.C.; Kopyl, S.; Tobaldi, D.M.; Piccirillo, C.; Avakyan, L.A.; Coutinho, J. Surface modified hydroxyapatites with various functionalized nanostructures: Computational studies of the vacancies in HAp. *Ferroelectrics* **2017**, *509*, 105. [CrossRef]

nanomaterials

MDPI

Article

Off-Resonant Absorption Enhancement in Single Nanowires via Graded Dual-Shell Design

Wenfu Liu [1,2,*], Xiaolei Guo [1], Shule Xing [1], Haizi Yao [1], Yinling Wang [1], Liuyang Bai [1], Qi Wang [1], Liang Zhang [1], Dachuan Wu [2], Yuxiao Zhang [2], Xiao Wang [2] and Yasha Yi [2,3,*]

1 School of Mechanical and Energy Engineering, Huanghuai University, Zhumadian 463000, Henan, China; guoxiaolei@huanghuai.edu.cn (X.G.); xingshule@huanghuai.edu.cn (S.X.); yaohaizi@huanghuai.edu.cn (H.Y.); wangyinling@huanghuai.edu.cn (Y.W.); bailiuyang@huanghuai.edu.cn (L.B.); wangqi@huanghuai.edu.cn (Q.W.); zhangliang@huanghuai.edu.cn (L.Z.)

2 Integrated Nano Optoelectronics Laboratory, University of Michigan, Dearborn, MI 48128, USA; dachuanw@umich.edu (D.W.); yuxiaoz@umich.edu (Y.Z.); xwa@umich.edu (X.W.)

3 Energy Institute, University of Michigan, Ann Arbor, MI 48109, USA

* Correspondence: liuwenfu@huanghuai.edu.cn (W.L.); yashayi@umich.edu (Y.Y.)

Received: 8 August 2020; Accepted: 1 September 2020; Published: 2 September 2020

Abstract: Single nanowires (NWs) are of great importance for optoelectronic applications, especially solar cells serving as powering nanoscale devices. However, weak off-resonant absorption can limit its light-harvesting capability. Here, we propose a single NW coated with the graded-index dual shells (DSNW). We demonstrate that, with appropriate thickness and refractive index of the inner shell, the DSNW exhibits significantly enhanced light trapping compared with the bare NW (BNW) and the NW only coated with the outer shell (OSNW) and the inner shell (ISNW), which can be attributed to the optimal off-resonant absorption mode profiles due to the improved coupling between the reemitted light of the transition modes of the leak mode resonances of the Si core and the nanofocusing light from the dual shells with the graded refractive index. We found that the light absorption can be engineered via tuning the thickness and the refractive index of the inner shell, the photocurrent density is significantly enhanced by 134% (56%, 12%) in comparison with that of the BNW (OSNW, ISNW). This work advances our understanding of how to improve off-resonant absorption by applying graded dual-shell design and provides a new choice for designing high-efficiency single NW photovoltaic devices.

Keywords: single nanowires; silicon; dual shells; off-resonance; absorption; photocurrent

1. Introduction

Single nanowire (NW) solar cells have attracted more and more research interests as nanoelectronic power sources in recent years due to their unique characteristics, such as enhanced light-harvesting capability, efficient carrier collection, ultra-compact volume, large surface area and convenience of integrating with optoelectronic nanosystems [1–8]. Besides its potential applications as power sources for nanoelectronic devices, single NW solar cell is also helpful to understand the mechanism of the self-assembled NW-based solar cells [7–16]. It is well known that light-harvesting ability is one of the most critical factors for photovoltaic applications, which determines the photoelectric conversion efficiency of a solar cell. Surprisingly, there is a strong interaction of the incident light with a single NW due to the leaky mode resonances (LMRs), which leads to a much higher absorption cross-section than its physical geometry [17–19]. However, the light absorption of single NWs is still far from the expectation due to the sharp resonant peak and narrow width, which can only achieve superior light absorption at the peak position.

Therefore, various strategies have been implemented to improve the light absorption in the whole solar spectrum range. It was shown that the light absorption could be readily tuned by controlling the size, geometry and orientation of the NWs [20–28]. Moreover, our previous studies [29–33] showed that the light absorption could be further improved by introducing a non-absorbing dielectric shell as the antireflection coating, which was experimentally and numerically demonstrated in the recent studies [34–38]. Comparing to the bare NWs (BNWs), semiconductor core-dielectric shell NWs (CSNWs) not only provides the possibility to tune the position of the resonant peak but also enhance the light-harvesting capability at the off-resonant wavelengths by adjusting the thickness and the refractive index of the shell, which is attributed to the high nanofocusing effect [30,38]. However, a further increase of the shell thickness has little contribution to the enhancement of the light absorption of the CSNW structure because the incident light will be mainly concentrated in the dielectric shell. Recently, some new strategies have been employed to improve the light-trapping capability of the CSNW, including the off-axial core-shell design [39] and partially capped design [40]. At the same time, the graded-index concept has been employed to enhance light absorption in the two-dimensional (2D) structures [41–43]. However, to the best of our knowledge, the dual shells with the graded refractive index have not been applied to improve the light absorption of single NWs.

In this work, we report a single dual-shell coated NW (DSNW), in which the dual shells have a graded refractive index. We demonstrate that graded dual-shell design can lead to the giant enhancement of off-resonant absorption. The detailed analysis of the absorption mode and photogeneration rate profiles shows that this enhancement results mainly from the optimal off-resonant absorption mode profiles under an improved coupling between the reemitted light of the transition modes of the LMRs of the Si core and the nanofocusing light from the dual shells with the graded refractive index. Simulation results indicate that the photocurrent density is significantly enhanced by 134%, 56% and 12% in comparison with that of the BNW and that of the nanowire only coated with the outer shell (OSNW) and the inner shell (ISNW), respectively.

2. Model and Methods

2.1. Model

The cross-sectional schematic diagram of the DSNW is shown in the insets of Figure 1. It should be noted here that the OSNW and the ISNW are also shown for comparison. The geometrical parameters of the DSNW are characterized by the radius r (=100 nm) of the Si core, the thickness t_1 of the outer shell and the thickness t_2 of the inner shell of the DSNW which varies from $t_2 = 0$ nm (i.e., OSNW) to $t_2 = 180$ nm (i.e., ISNW), the total shell thickness $t = t_1 + t_2 = 180$ nm and the total radius $R = r + t = 280$ nm. Note here that the radius of the Si core is chosen to be 100 nm as a representative nanoscale size and the total shell thickness is chosen to be 180 nm as the DSNW shows the optimal absorption in this study and the OSNW (or ISNW) approaches the optimal absorption at this thickness in our previous study [31–33]. The incident light indicated by colorful arrows in the insets of Figure 1 is assumed to be illuminated perpendicularly to the axial from the top, the wavelength range is from 300 to 1100 nm with a step size of 5 nm considering solar radiation and the bandgap of Si. The wavelength-dependent complex refractive index of Si fitted with the experimental data [44] and that of the inner shell, the outer shell and the surrounding medium (air) are set to be 2.5, 1.5 and 1.0, respectively. Note here that for simplicity, we have neglected the wavelength dependence of the refractive indices to neatly determine the impact of their size on the absorption spectra [30,31,45], as discussed later; however, as long as the wavelength dependence is negligible, our results could apply to any other dielectric with similar refractive index, such as SiO_2, Si_3N_4 and so forth.

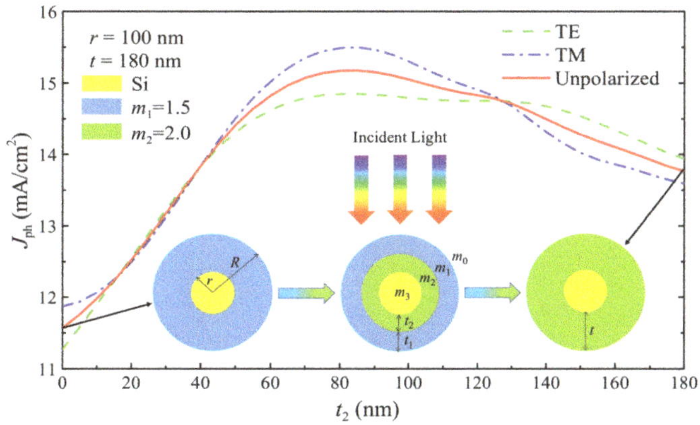

Figure 1. J_{ph} versus t_2 of the dual-shell coated nanowire (DSNW) for TE (Transverse electric field to the nanowire axis), TM (Transverse magnetic field to the nanowire axis) and unpolarized light illumination. The insets are the cross-sectional views of the NW only coated with the outer shell (OSNW) (left), DSNW (middle) and the NW only coated with the outer shell (ISNW) (right), where J_{ph} is the photocurrent density; t_1, t_2, $t = t_1 + t_2 = 180$ nm, $r = 100$ nm and $R = r + t = 280$ nm are the outer shell thickness, the inner shell thickness, the total shell thickness, the core radius and the total radius; m_3, m_2, m_1 and m_0 are the complex refractive indices of the Si core, the inner shell, the outer shell and air, respectively.

2.2. Methods

The light absorption performance of the DSNW was performed by solving the corresponding Maxwell's equations based on 2D finite difference time domain (FDTD) method [46–48] by assuming that the length of the NW is far greater than the radius, which can be referred to the work of Kim and co-workers for details [26–28,36]. In this simulation, the perfectly matched layers (PML) boundary conditions are used to avoid any non-physical reflection with the boundaries, the total-field scattered-field (TFSF) method was applied to ensure that a single NW interacts with an infinite plane wave. Also, the minimum cell size of the FDTD mesh is set to 1 nm to guarantee the accuracy of the simulation results.

2.2.1. The Absorption Efficiency and Mode Profile

To qualify the light absorption performance of the DSNW, we define the absorption efficiency Q_{abs} of the Si core as [38,49–51]:

$$Q_{abs} = C_{abs}/C_{geo}, \tag{1}$$

where C_{geo} is the geometric cross-section (i.e., the projected area of the Si core) and C_{abs} is the absorption cross-section calculated by [38,49–51]

$$C_{abs} = \frac{\int_V P_{abs} dV}{I_0} = k_0 \varepsilon''_r \int_V |E_r|^2 \mathrm{d}V, \tag{2}$$

where k_0 is the wave vector in air, ε''_r is the imaginary part of the relative permittivity, E_r is the normalized electric field intensity, V is the volume of the Si core, I_0 is the solar incident light intensity and P_{abs} describes the wavelength-dependent absorption mode profile calculated from Poynting theorem, which can be expressed as [38,49–51]

$$P_{abs} = \frac{1}{2}\omega\varepsilon'' |E|^2 \tag{3}$$

$$I_0 = \frac{1}{2} c\varepsilon_0 |E_0|^2 \tag{4}$$

$$\varepsilon''_r = \varepsilon'' / \varepsilon_0 = 2nk, E_r = E/E_0, \tag{5}$$

where ω, c and E_0 is the angular frequency, the speed of light and the electric field intensity of the solar incident light; ε_0 and ε'' are the permittivities in air and the imaginary part of the permittivity of Si; n and k are the real and imaginary part of the complex refractive index of Si (i.e., $m = n + ik$, $m^2 = \varepsilon_r = \varepsilon'_r + i\varepsilon''_r$); E is the electric field intensity in the Si core, respectively.

2.2.2. The Photogeneration Rate

With the assumption that each photon absorbed inside the Si core has a contribution to the photocurrent, the spatially dependent photogeneration rate G is readily calculated by [52,53]

$$G = \int_{300}^{1100} \frac{P_{abs}}{\hbar\omega} d\lambda = \int_{300}^{1100} \frac{\varepsilon'' |E|^2}{2\hbar} \mathrm{d}\lambda, \tag{6}$$

where $\hbar$ is the reduced Planck's constant and λ is the wavelength of the incident light.

2.2.3. The Photocurrent Density

To evaluate the light-harvesting capability as single NW solar cells, we can calculate the photocurrent density J_{ph} by:

$$J_{\mathrm{ph}} = \frac{q}{C_{geo}} \int_V G dV = q \int_{300}^{1100} \Gamma(\lambda) Q_{abs}(\lambda) d\lambda, \tag{7}$$

where q is the element charge, Γ is the AM1.5G standard solar photon flux density spectrum. It should be noted here that 100% collection efficiency is assumed, which has been widely employed to evaluate the ultimate photocurrent [18,53].

2.2.4. The Photocurrent Enhancement Factor (PEF)

To evaluate the photocurrent enhancement of the DSNW, we calculate the photocurrent enhancement factor (PEF) using the relation:

$$\mathrm{PEF} = \left(J_{\mathrm{ph,DSNW}} - J_{\mathrm{ph,rNW}}\right) / J_{\mathrm{ph,rNW}}, \tag{8}$$

where $J_{\mathrm{ph,DSNW}}$ and $J_{\mathrm{ph,rNW}}$ are the photocurrent density for the DSNW and the reference NWs (BNW, OSNW and ISNW), respectively.

Finally, it is important to stress that the unpolarized illumination (e.g., sunlight) is regarded as the average of transverse electric (TE, electric field normal to the NW axis) and transverse magnetic (TM, magnetic field normal to the NW axis) illumination [22,39,40].

$$Q_{\mathrm{abs}} = \left(Q_{\mathrm{abs}}^{\mathrm{TE}} + Q_{\mathrm{abs}}^{\mathrm{TM}}\right)/2 \tag{9}$$

$$J_{\mathrm{ph}} = \left(J_{\mathrm{ph}}^{\mathrm{TE}} + J_{\mathrm{ph}}^{\mathrm{TM}}\right)/2. \tag{10}$$

3. Results and Discussion

3.1. The Absorption Mechanism

To understand the absorption mechanism responsible for the improved light-harvesting performance of the DSNW, we investigate the photocurrent density (J_{ph}), the absorption efficiency (Q_{abs}), the absorption mode profile (P_{abs}) and the photogeneration rate (G), respectively. Note here that r = 100 nm, t = 180 nm, $R = r + t$ = 180 nm, $t_2 = 0 \rightarrow 180$ nm, where $t = 0$, $t_2 = 0$ and t_2 = 180 nm denote the cases of the BNW, OSNW and ISNW, respectively; m_3 is the complex refractive index of the Si core, $m_2 = 2.0$, $m_1 = 1.5$ and $m_0 = 1.0$, as shown in the insets of Figure 1.

3.1.1. The Photocurrent Density (J_{ph})

To evaluate the light-harvesting performance of the DSNW, we first study the effect of the inner shell thickness on the photocurrent density (J_{ph}) obtained by Equation (7). In Figure 1, we show t_2-dependent J_{ph} under normally-incident TE, TM and unpolarized light illumination, respectively. It is observed that J_{ph} increases rapidly first when initially increasing t_2, reaches a peak at t_2 = 85 nm and then decreases when continuing to increase t_2. More importantly, J_{ph} of the DSNW is always bigger than that of the OSNW as long as the inner shell is adopted and higher than that of the ISNW in a broad inner thickness range of $t_2 > 40$ nm. For a direct comparison, we list the J_{ph} values of the considered BNW, OSNW, ISNW and DSNW configurations under TE, TM and unpolarized illumination, as shown in Table 1. The maximum J_{ph} values for TE and TM light are 14.85 and 15.50 mA/cm^2, respectively. Under unpolarized light illumination (e.g., sunlight), the maximum J_{ph} reaches 15.18 mA/cm^2, which is 96.6%, 31.2% and 10.2% higher than that of the BNW (7.72 mA/cm^2), OSNW (11.57 mA/cm^2) and ISNW (13.77 mA/cm^2), respectively. It is found that this photocurrent enhancement is mainly ascribed to the improvement of J_{ph} under any polarized situations (especially TM light), indicating the potential of the DSNW in improving the light absorption of single NWs.

Table 1. Photocurrent densities (in mA/cm^2) of the four typical configurations (in nm) of the bare NW (BNW), OSNW, ISNW and DSNW under TE (Transverse electric field to the nanowire axis), TM (Transverse magnetic field to the nanowire axis) and unpolarized light illumination, where r, t_2, and t are the core radius, the inner shell thickness and the total shell thickness, respectively.

Configuration	Parameter	J_{ph}^{TE}	J_{ph}^{TM}	J_{ph}
BNW	$r = 100, t_2 = 0, t = 0$	7.24	8.20	7.72
OSNW	$r = 100, t_2 = 0, t = 180$	11.27	11.87	11.57
ISNW	$r = 100, t_2 = 180, t = 180$	13.94	13.59	13.77
DSNW	$r = 100, t_2 = 85, t = 180$	14.85	15.50	15.18

3.1.2. The Absorption Efficiency (Q_{abs})

To understand the physical mechanism of the improved photocurrent, we then examine the absorption spectra of the DSNW. In Figure 2a,b, we present 2D maps of λ-dependent Q_{abs} as a function of t_2 under TE and TM light illumination, which is given by Equation (1). It is clear that the dual-shell design can lead to absorption enhancement in the whole spectrum range compared to the OSNW and almost the entire spectrum range (except several narrow peaks) compared to the ISNW under both TE and TM light illumination (especially at the off-resonant wavelengths), as discussed later. Moreover, Q_{abs} can be divided into three regions using two vertical dashed lines by employing the characteristic wavelengths λ_{c1} (~430 nm) and λ_{c2} (~525 nm) for both TE and TM lights, as labeled in Figure 2a,b. Firstly, in the wavelength range of $\lambda < \lambda_{c1}$, Q_{abs} periodically changes with increasing t_2. Note here that Q_{abs} can be divided into two regions using a horizontal dashed line by employing the characteristic inner shell thickness t_{2c} (~90 nm) for both TE and TM lights, as labeled in Figure 2a,b. Q_{abs} reaches the maximum absorption near t_2 = 50 nm and t_2 − 130 nm for the first ($t_2 < t_{2c}$) and second ($t_2 > t_{2c}$) period, respectively. In other words, the excellent absorption can be obtained in the inner shell thickness range of $40 < t_2 < 60$ nm and $110 < t_2 < 140$ nm for two periods, respectively. Secondly, in the wavelength range of $\lambda_{c1} < \lambda < \lambda_{c2}$, Q_{abs} reaches the maximum absorption near t_2 = 50 nm, that is, the superior absorption can be obtained in the inner shell thickness range of $60 < t_2 < 120$ nm. Finally, in the wavelength range of $\lambda > \lambda_{c2}$, Q_{abs} appears to be comparable due to the trade-off between the suppression at the resonant wavelengths and the enhancement at the off-resonant wavelengths, resulting in little contribution to the photocurrent enhancement. Therefore, the photocurrent enhancement of the DSNW with $t_2 < 60$ nm and $t_2 > 120$ nm is attributed to the improved absorption in the wavelength range of $\lambda < \lambda_{c1}$, while that of the DSNW with $60 < t_2 < 120$ nm is mainly attributed to the improved absorption in the wavelength range of $\lambda_{c1} < \lambda < \lambda_{c2}$, which is due to the fact that there is a much higher solar radiation in the wavelength range of $\lambda_{c1} < \lambda < \lambda_{c2}$ than $\lambda < \lambda_{c1}$, leading to a more significant contribution to the photocurrent according to Equation (7).

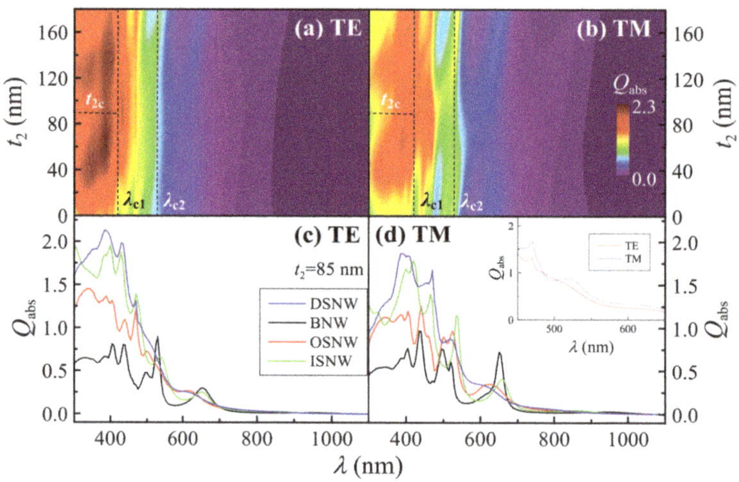

Figure 2. Q_{abs} versus λ and t_2 of the DSNW for (**a**) TE and (**b**) TM light illumination; Q_{abs} versus λ of the DSNW ($t_2 = 85$ nm) under (**c**) TE and (**d**) TM light illumination, together with the BNW ($t = 0$), OSNW ($t_2 = 0$) and ISNW ($t_2 = 180$ nm) as references. The inset in (**d**): Q_{abs} versus λ (450–650 nm) for TE and TM light, where Q_{abs} and λ are the absorption efficiency and the wavelength of the incident light, respectively. Note that λ_{c1} (~430 nm) and λ_{c2} (~525 nm) are the characteristic wavelengths denoted by two vertical dashed lines and t_{2c} (~90 nm) is the characteristic inner shell thickness in the wavelength range of $\lambda < \lambda_{c1}$ denoted by a horizontal dashed line.

To quantitatively characterize the absorption enhancement of the DSNW, we also examine the absorption spectra corresponding to the optimal J_{ph} in Figure 1. In Figure 2c,d, we show λ-dependent Q_{abs} of the DSNW with $t_2 = 85$ nm for TE and TM light, where the results of the BNW, OSNW and ISNW are also included for comparison. It is shown that Q_{abs} of the DSNW is much higher than that of the BNW and OSNW in the wavelength range of $\lambda < \lambda_{c2}$ and that of the ISNW in the wavelength range of $\lambda < \lambda_{c2}$ (except for several narrow peaks, for example, $\lambda = 470$ for TE light) for both TE and TM lights, resulting in a significant photocurrent enhancement. In contrast, although Q_{abs} of the DSNW is weaker at the resonant wavelengths, higher at the off-resonant wavelengths than that of all the other three NW structures in the wavelength range of $\lambda > \lambda_{c2}$, leading to a similar contribution to the photocurrent, as discussed above. It is worth noting that Q_{abs} can be substantially enhanced at the off-resonant wavelengths over the whole wavelength range for both TE and TM lights, especially for TM light (e.g., near $\lambda = 470$ nm), which results in the more prominent photocurrent enhancement for TM than TE light. It should also be noted that the match between the absorption efficiency and the solar spectrum becomes another essential factor in evaluating the photocurrent according to Equation (7). For instance, although Q_{abs} of the DSNW for TE light is much higher than that for TM light in the wavelength range of $\lambda < \lambda_{c1}$, solar radiation is much lower, which leads to a less photocurrent enhancement, while Q_{abs} for TM light is much higher than that for TE light in the wavelength range of $450 < \lambda < 650$ nm (except the narrow wavelength range of $490 < \lambda < 505$ nm), as shown in the inset of Figure 2d and solar radiation is much higher at the same time, which results in a more significant contribution to the photocurrent.

3.1.3. The Absorption Mode Profile (P_{abs})

The absorption behavior presented above can be well described by the absorption mode profiles (P_{abs}) calculated by Equation (3) [22,24,26,50,52]. In Figure 3, we examine the normalized absorption mode profiles inside the Si core corresponding to the wavelengths in Figure 2c,d under TE and TM light illumination (these profiles from left to right columns are related to the evolution of the structure from BNW to OSNW and then to ISNW and finally to DSNW). Figure 3a,c show the off-resonant absorption mode profiles for TE and TM light, while Figure 3b,d show the corresponding resonant absorption mode profiles. Note here that the resonant (or off-resonant) absorption for all the four NW configurations may occur at different wavelengths due to the difference of the thickness and the refractive index of the dielectric shells, that is, t_1 (or t_2) or m_1 (or m_2)-driven shift [30,31], as shown in Figure 2a,b. It is observed that the absorption enhancement is attributed to the excitation of the LMRs, likewise in BNW [17,18], which can capture light by multiple total internal reflections at the Si core/inner shell interface when the wavelength of the incident light matches one of the LMRs supported by the Si core. The LMRs can be noted as TE_{ml} or TM_{ml}, where m and l are the azimuthal mode number and the radial order of the resonances, respectively. Figure 3b,d show that the resonant absorption mode profiles of the DSNW are different from that of the BNW, similar to that of the INSW due to the fact that the LMRs of the Si core occur at the Si core/inner shell interface. Specifically, the modes of the BNW, OSNW, ISNW and DSNW are TE_{12}, TE_{31}, TE_{31} and TE_{31} at λ = 495, 470, 470 and 470 nm for TE light and TM_{12}, TM_{12}, TM_{41} and TM_{41} at λ = 495, 500, 465 and 470 nm for TM light, respectively. The absorption of the DSNW is indeed enhanced compared to the BNW and OSNW for both TE and TM lights and slightly suppressed for TE light and enhanced for TM light compared to the ISNW. Figure 3a,c show that the off-resonant absorption mode profiles of the DSNW exhibit a transition mode referred to the LMRs, such transition modes are very close to the corresponding LMRs, which is attributed to the fact that the presence of the graded dual shells makes more light couple into the Si core, leading to a more significant absorption enhancement compared to all the other three NWs.

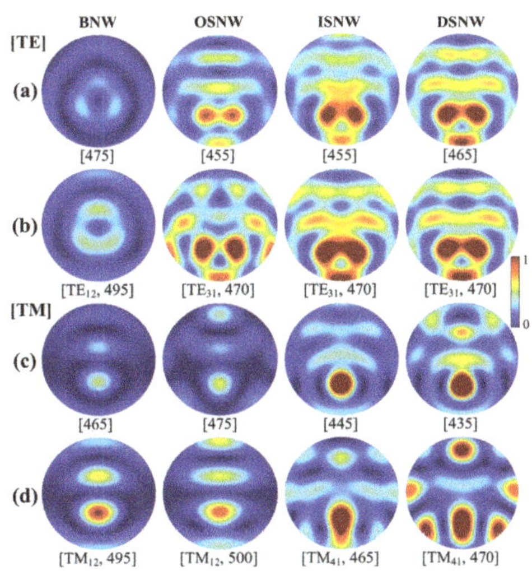

Figure 3. The representative normalized absorption mode profiles inside the Si core corresponding to the wavelengths in Figure 2c,d [from left to right columns are associated with the BNW, OSNW, ISNW and DSNW, respectively]: (**a**,**b**) for TE and (**c**,**d**) for TM light illumination; (**a**,**c**) for off-resonant and (**b**,**d**) for resonant wavelengths. Note that the numbers in square brackets are the wavelengths and the modes of the the leaky mode resonances (LMRs) at the resonant wavelengths are also labeled in the left of square brackets in Figure 3b,d.

The absorption behavior presented above can also be well understood by employing the interference effect of light, which occurs at multiple interfaces due to the difference of the refractive index between Si core/inner shell, inner shell/outer shell and outer shell/air. For weaker LMRs in the wavelength of $\lambda < \lambda_{c2}$, as shown in Figure 2, the resonant absorption is greatly enhanced due to the constructive interference with the reemitted light of the weaker LMRs of the Si core. However, for stronger LMRs in the wavelength of $\lambda > \lambda_{c2}$, the resonant absorption is suppressed due to the destructive interference with the reemitted light of the stronger LMRs of the Si core. In contrast, the off-resonant absorption over the whole spectrum is greatly enhanced due to the constructive interference with the reemitted light of the weaker transition modes of the Si core. In a word, the off-resonant (or weaker resonant) absorption is dramatically enhanced owing to an improved coupling between the reemitted light of weaker transition modes (or weaker LMRs) of the Si core and the nanofocusing light from the graded dual shells at the core/inner shell interface [38,39].

3.1.4. The Photogeneration Rate Profile (G)

To further confirm the physical origin discussed above, we show the photogeneration rate (G) obtained Equation (6). In Figure 4, we present the normalized photogeneration rate profiles for TE and TM polarized light, respectively. It is shown in Figure 4a that the absorption of the DSNW for TE light is much stronger than that of the BNW and OSNW and that of almost all the regions of the ISNW [evidently enhanced in the regions labeled by circles (see Figure 4a) and slightly decreased in the region labeled by the square (see Figure 4a)]. It is shown in Figure 4b that the absorption of the DSNW for TM light is also much stronger than that of the BNW and OSNW and that of almost all the regions of the ISNW [evidently enhanced in the regions by labeled by triangles (see Figure 4b)]. These results further reveal that this enhancement arises mainly from the off-resonant absorption enhancement due to the improved coupling between the reemitted light of the weaker transition modes of the LMRs of the Si core and the nanofocusing light from the graded dual shells. More importantly, the photogeneration rate profiles of the DSNW for both TE and TM lights have similar patterns with that of all the other three NWs, again indicating that the absorption enhancement is mainly attributed to the LMRs, likewise in the BNWs.

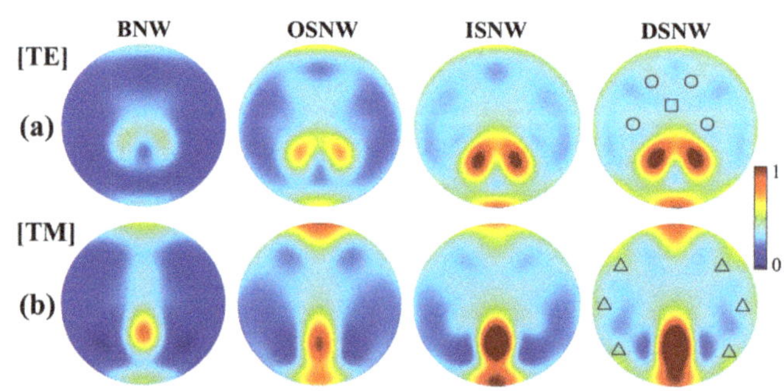

Figure 4. The normalized generation rate profiles of the DSNW for (**a**) TE and (**b**) TM light, together with the BNW, OSNW and ISNW for comparison. Note that the regions labeled by circles and triangles denote that the absorption of the DSNW is significantly enhanced for TE and TM light and the region labeled by the square denotes that the absorption of the DSNW is slightly decreased for TE light, respectively.

3.2. The Optimization of the Light-Harvesting Performance

To evaluate and optimize the light-trapping performance of the DSNW for photovoltaic applications, we now investigate the effect of both the shell thickness and the refractive index on the photocurrent density calculated using Equation (7). Note that except for m_2, all the other structural details of the DSNW are consistent with that shown in the insets of Figure 1. In Figure 4a, we show 2D J_{ph} as a function of t_2 and m_2 of the DSNW and the optimal t_2 as a function of m_2. Figure 4a shows J_{ph} sharply increases with increasing t_2 at a fixed m_2, reaches its maximum and then decreases when continuing to increase t_2. More importantly, J_{ph} of the DSNW is always much larger than that of the OSNW at any t_2 values and higher than that of the ISNW in a broad inner shell thickness range of $t_2 > 40$ nm for $m_2 < 3.5$ and $t_2 > 60$ nm for $3.5 < m_2 < 4.0$. It is observed that the maximum values of J_{ph} can be obtained in the inner shell thickness range of $90 < t_2 < 110$ nm for $3.0 < m_2 < 3.5$. In Figure 5b, we show m_2-dependent J_{ph} of the DSNW (corresponding to the optimal t_2 in Figure 5a), together with that of the BNW, OSNW and ISNW for comparison. Also, in Figure 5c, we show the photocurrent enhancement factors (PEFs) defined by Equation (8). It is readily observed that J_{ph} of the DSNW is much larger than all the other three NWs. In particular, the maximum J_{ph} reaches 18.10 mA/cm^2 at $t_2 = 100$ nm for $m_2 = 3.25$, which is 134.4%, 56.4% and 12.4% much larger than that of the BNW (7.72 mA/cm^2), OSNW (11.57 mA/cm^2) and ISNW (16.10 mA/cm^2), respectively.

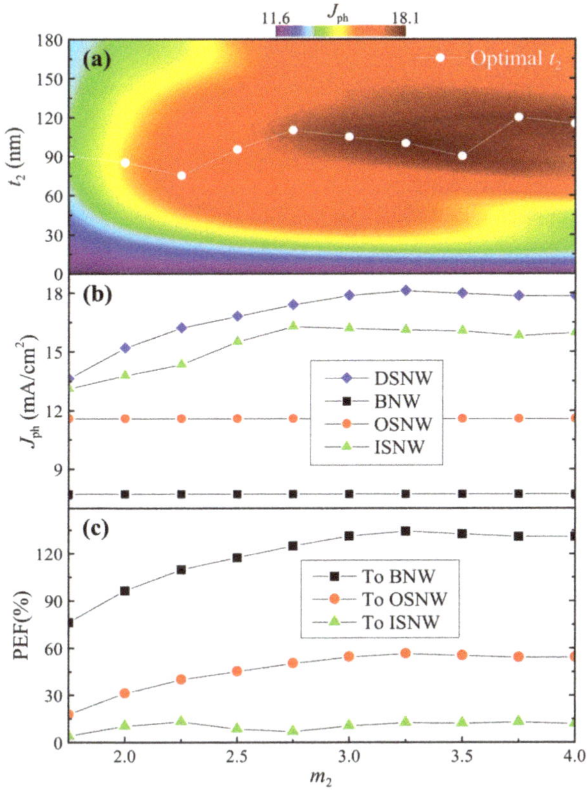

Figure 5. (**a**) J_{ph} versus t_2 and m_2 of the DSNW. The white dashed line represents the position of the maximum J_{ph} at various m_2 values. (**b**) J_{ph} versus m_2 of the DSNW, corresponding to the optimal t_2. Also, J_{ph} versus m_2 of the BNW, OSNW and ISNW are included for comparison. (**c**) photocurrent enhancement factor (PEF) versus m_2 of the DSNW compared to the BNW, OSNW and ISNW, respectively. Note that the PEF is the photocurrent enhancement factor compared to the reference NWs.

4. Conclusions

In summary, we proposed a single NW by coating dual dielectric shells. The influence of the thickness and the refractive index of the inner shell of the DSNW on the light absorption for photovoltaic applications are numerically investigated. It is found that the size and material of the inner shell can lead to significantly improved off-resonant absorption. The examination of the spatial profiles of the absorption mode and photogeneration rate reveals that the enhancement effect is the result of the constructive interference under the improved coupling between the reemitted light of the transition modes of the LMRs of the Si core and the nanofocusing light from the graded dual shells. The simulation results show that the photocurrent density can be enhanced by 134.4%, 56.4% and 12.4% in comparison with that of the BNW, OSNW and ISNW, respectively. Therefore, such a dual shell coated structure can be applied to a variety of semiconductors to improve the off-resonant absorption and provides an effective way to achieve high-efficiency single NW solar cells.

Author Contributions: Conceptualization, W.L. and Y.Y.; methodology, W.L., X.G. and S.X.; software, W.L., H.Y. and D.W; investigation, W.L., D.W., H.Y., Y.W. and L.B.; writing—original draft preparation, W.L.; writing—review and editing, W.L., X.G., S.X., Q.W., L.Z., Y.Z., X.W. and Y.Y. All authors have read and agreed to the published version of the manuscript.

Funding: This research was funded by the National Natural Science Foundation of China (Grant No. 61404057, 51801068 and 11875284), Key disciplines Fund for Optical Engineering in Henan Province (Grant No. 2018119) and Key Youth Teacher of Universities in Henan Province (Grant No. 2016GGJS-145).

Acknowledgments: We would also like to thank the support from the China Scholarship Council (Grant No. 201902720047) and the Construction of Master's Program on Energy and Power Engineering at Huanghuai University.

Conflicts of Interest: The authors declare no conflict of interest.

References

1. Tian, B.; Zheng, X.; Kempa, T.J.; Fang, Y.; Yu, N.; Yu, G.; Huang, J.; Lieber, C.M. Coaxial silicon nanowires as solar cells and nanoelectronic power sources. *Nature* **2007**, *449*, 885–889. [CrossRef] [PubMed]
2. Nehra, M.; Dilbaghi, N.; Marrazza, G.; Kaushik, A.; Abolhassani, R.; Mishra, Y.K.; Kim, K.H.; Kumar, S. 1D semiconductor nanowires for energy conversion, harvesting and storage applications. *Nano Energy* **2020**, *76*, 104991. [CrossRef]
3. Li, Z.; Tan, H.H.; Jagadish, C.; Fu, L. III–V Semiconductor Single Nanowire Solar Cells: A Review. *Adv. Mater. Technol.* **2018**, *3*, 1800005. [CrossRef]
4. Kempa, T.J.; Day, R.W.; Kim, S.-K.; Park, H.-G.; Lieber, C.M. Semiconductor nanowires: A platform for exploring limits and concepts for nano-enabled solar cells. *Energy Environ. Sci.* **2013**, *6*, 719–733. [CrossRef]
5. Christesen, J.D.; Zhang, X.; Pinion, C.W.; Celano, T.A.; Flynn, C.J.; Cahoon, J.F. Design principles for photovoltaic devices based on Si nanowires with axial or radial p–n junctions. *Nano Lett.* **2012**, *12*, 6024–6029. [CrossRef] [PubMed]
6. Zhan, Y.; Li, X.; Li, Y. Numerical Simulation of Light-Trapping and Photoelectric Conversion in Single Nanowire Silicon Solar Cells. *IEEE J. Sel. Top. Quantum Electron.* **2013**, *19*, 1–8. [CrossRef]
7. Tian, B.; Kempa, T.J.; Lieber, C.M. Single nanowire photovoltaics. *Chem. Soc. Rev.* **2009**, *38*, 16–24. [CrossRef] [PubMed]
8. Tang, J.; Huo, Z.; Brittman, S.; Gao, H.; Yang, P. Solution-processed core-shell nanowires for efficient photovoltaic cells. *Nat. Nanotechnol.* **2011**, *6*, 568–572. [CrossRef]
9. Holm, J.V.; Jørgensen, H.I.; Krogstrup, P.; Nygård, J.; Liu, H.; Aagesen, M. Surface-passivated GaAsP single-nanowire solar cells exceeding 10% efficiency grown on silicon. *Nat. Commun.* **2013**, *4*, 1498. [CrossRef]
10. Yuan, X.; Chen, X.; Yan, X.; Wei, W.; Zhang, Y.; Zhang, X. Absorption-Enhanced Ultra-Thin Solar Cells Based on Horizontally Aligned p–i–n Nanowire Arrays. *Nanomaterials* **2020**, *10*, 1111. [CrossRef]
11. Chen, W.; Roca i Cabarrocas, P. Rational design of nanowire solar cells: From single nanowire to nanowire arrays. *Nanotechnology* **2019**, *30*, 194002. [CrossRef] [PubMed]
12. Otnes, G.; Barrigón, E.; Sundvall, C.; Svensson, K.E.; Heurlin, M.; Siefer, G.; Samuelson, L.; Åberg, I.; Borgström, M.T. Understanding InP Nanowire Array Solar Cell Performance by Nanoprobe-Enabled Single Nanowire Measurements. *Nano Lett.* **2018**, *18*, 3038–3046. [CrossRef] [PubMed]
13. Song, K.-D.; Kempa, T.J.; Park, H.-G.; Kim, S.-K. Laterally assembled nanowires for ultrathin broadband solar absorbers. *Opt. Express* **2014**, *22*, A992–A1000. [CrossRef] [PubMed]

14. Zhang, X.; Pinion, C.W.; Christesen, J.D.; Flynn, C.J.; Celano, T.A.; Cahoon, J.F. Horizontal Silicon Nanowires With Radial p-n Junctions: A Platform for Unconventional Solar Cells. *J. Phys. Chem. Lett.* **2013**, *4*, 2002–2009. [CrossRef] [PubMed]
15. Floris, F.; Fornasari, L.; Bellani, V.; Marini, A.; Banfi, F.; Marabelli, F.; Beltram, F.; Ercolani, D.; Battiato, S.; Sorba, L.; et al. Strong Modulations of Optical Reflectance in Tapered Core–Shell Nanowires. *Materials* **2019**, *12*, 3572. [CrossRef]
16. Floris, F.; Fornasari, L.; Marini, A.; Bellani, V.; Banfi, F.; Roddaro, S.; Ercolani, D.; Rocci, M.; Beltram, F.; Cecchini, M. Self-Assembled InAs Nanowires as Optical Reflectors. *Nanomaterials* **2017**, *7*, 400. [CrossRef]
17. Cao, L.; White, J.S.; Park, J.-S.; Schuller, J.A.; Clemens, B.M.; Brongersma, M.L. Engineering light absorption in semiconductor nanowire devices. *Nat. Mater.* **2009**, *8*, 643–647. [CrossRef]
18. Cao, L.; Fan, P.; Vasudev, A.P.; White, J.S.; Yu, Z.; Cai, W.; Schuller, J.A.; Fan, S.; Brongersma, M.L. Semiconductor Nanowire Optical Antenna Solar Absorbers. *Nano Lett.* **2010**, *10*, 439–445. [CrossRef]
19. Brönstrup, G.; Jahr, N.; Leiterer, C.; Csáki, A.; Fritzsche, W.; Christiansen, S. Optical properties of individual silicon nanowires for photonic devices. *ACS Nano* **2010**, *4*, 7113–7122. [CrossRef]
20. Kim, S.; Cahoon, J.F. Geometric Nanophotonics: Light Management in Single Nanowires through Morphology. *Acc. Chem. Res.* **2019**, *52*, 3511–3520. [CrossRef]
21. Choi, J.S.; Kim, K.-H.; No, Y.-S. Spatially localized wavelength-selective absorption in morphology-modulated semiconductor nanowires. *Opt. Express* **2017**, *25*, 22750–22759. [CrossRef] [PubMed]
22. Zhang, C.; Yang, Z.; Wu, K.; Li, X. Design of asymmetric nanovoid resonator for silicon-based single-nanowire solar absorbers. *Nano Energy* **2016**, *27*, 611–618. [CrossRef]
23. Yang, Z.; Cao, G.; Shang, A.; Lei, D.Y.; Zhang, C.; Gao, P.; Ye, J.; Li, X. Enhanced Photoelectrical Response of Hydrogenated Amorphous Silicon Single-Nanowire Solar Cells by Front-Opening Crescent Design. *Nanoscale Res. Lett.* **2016**, *11*, 233. [CrossRef] [PubMed]
24. Yang, Z.; Li, X.; Lei, D.Y.; Shang, A.; Wu, S. Omnidirectional absorption enhancement of symmetry-broken crescent-deformed single-nanowire photovoltaic cells. *Nano Energy* **2015**, *13*, 9–17. [CrossRef]
25. Luo, S.; Yu, W.B.; He, Y.; Ouyang, G. Size-dependent optical absorption modulation of Si/Ge and Ge/Si core/shell nanowires with different cross-sectional geometries. *Nanotechnology* **2015**, *26*, 085702. [CrossRef] [PubMed]
26. Kim, S.-K.; Song, K.-D.; Kempa, T.J.; Day, R.W.; Lieber, C.M.; Park, H.-G. Design of Nanowire Optical Cavities as Efficient Photon Absorbers. *ACS Nano* **2014**, *8*, 3707–3714. [CrossRef]
27. Kim, S.-K.; Day, R.W.; Cahoon, J.F.; Kempa, T.J.; Song, K.-D.; Park, H.-G.; Lieber, C.M. Tuning Light Absorption in Core/Shell Silicon Nanowire Photovoltaic Devices through Morphological Design. *Nano Lett.* **2012**, *12*, 4971–4976. [CrossRef]
28. Kempa, T.J.; Cahoon, J.F.; Kim, S.-K.; Day, R.W.; Bell, D.C.; Park, H.-G.; Lieber, C.M. Coaxial multishell nanowires with high-quality electronic interfaces and tunable optical cavities for ultrathin photovoltaics. *Proc. Natl. Acad. Sci. USA* **2012**, *109*, 1407–1412. [CrossRef]
29. Liu, W.; Oh, J.I.; Shen, W.Z. Light absorption mechanism in single c-Si (core)/a-Si (shell) coaxial nanowires. *Nanotechnology* **2011**, *22*, 125705. [CrossRef]
30. Liu, W.; Oh, J.I.; Shen, W.Z. Light Trapping in Single Coaxial Nanowires for Photovoltaic Applications. *IEEE Electron. Device Lett.* **2011**, *32*, 45–47. [CrossRef]
31. Liu, W. Tunable light absorption of Si nanowires by coating non-absorbing dielectric shells for photovoltaic applications. *Optoelectron. Adv. Mater.* **2011**, *5*, 916–919.
32. Liu, W.; Hao, H. Enhanced Absorption of Single Silicon Nanowire with Si_3N_4 Shell for Photovoltaic Applications. *Adv. Mat. Res.* **2015**, *1090*, 173–177. [CrossRef]
33. Liu, W.; Sun, F.Z. Light Absorption Enhancement in Single Si (core)/SiO_2 (shell) Coaxial Nanowires for Photovoltaic Applications. *Adv. Mat. Res.* **2012**, *391–392*, 264–268. [CrossRef]
34. Zhong, Z.; Li, Z.; Gao, Q.; Li, Z.; Peng, K.; Li, L.; Mokkapati, S.; Vora, K.; Wu, J.; Zhang, G.; et al. Efficiency enhancement of axial junction InP single nanowire solar cells by dielectric coating. *Nano Energy* **2016**, *28*, 106–114. [CrossRef]
35. Solanki, A.; Gentile, P.; Boutami, S.; Calvo, V.; Pauc, N. Dielectric Coating-Induced Absorption Enhancement in Si Nanowire Junctions. *Adv. Opt. Mater.* **2015**, *3*, 120–128. [CrossRef]

36. Kim, S.-K.; Zhang, X.; Hill, D.J.; Song, K.-D.; Park, J.-S.; Park, H.-G.; Cahoon, J.F. Doubling Absorption in Nanowire Solar Cells with Dielectric Shell Optical Antennas. *Nano Lett.* **2015**, *15*, 753–758. [CrossRef] [PubMed]
37. Yu, Y.; Ferry, V.E.; Alivisatos, A.P.; Cao, L. Dielectric Core–Shell Optical Antennas for Strong Solar Absorption Enhancement. *Nano Lett.* **2012**, *12*, 3674–3681. [CrossRef] [PubMed]
38. Li, X.; Zhan, Y.; Wang, C. Broadband enhancement of coaxial heterogeneous gallium arsenide single-nanowire solar cells. *Prog. Photovolt. Res. Appl.* **2015**, *23*, 628–636. [CrossRef]
39. Zhang, C.; Yang, Z.; Shang, A.; Wu, S.; Zhan, Y.; Li, X. Improved optical absorption of silicon single-nanowire solar cells by off-axial core/shell design. *Nano Energy* **2015**, *17*, 233–240. [CrossRef]
40. Zhou, J.; Zhang, Z.; Wu, Y.; Xia, Z.; Qin, X. Significantly enhanced coupling to half-space irradiation using a partially capped nanowire for solar cells. *Nano Energy* **2018**, *45*, 61–67. [CrossRef]
41. Saylan, S.; Milakovich, T.; Hadi, S.A.; Nayfeh, A.; Fitzgerald, E.A.; Dahlem, M.S. Multilayer antireflection coating design for $GaAs_{0.69}P_{0.31}$/Si dual-junction solar cells. *Sol. Energy* **2015**, *122*, 76–86. [CrossRef]
42. Tsai, M.-T.; Yang, Z.-P.; Jing, T.-S.; Hsieh, H.-H.; Yao, Y.-C.; Lin, T.-Y.; Chen, Y.-F.; Lee, Y.-J. Achieving graded refractive index by use of ZnO nanorods/TiO_2 layer to enhance omnidirectional photovoltaic performances of InGaP/GaAs/Ge triple-junction solar cells. *Sol. Energy Mater. Sol. Cells* **2015**, *136*, 17–24. [CrossRef]
43. Yeh, L.K.; Lai, K.Y.; Lin, G.J.; Fu, P.H.; Chang, H.C.; Lin, C.A.; He Jr, H. Giant efficiency enhancement of GaAs solar cells with graded antireflection layers based on syringelike ZnO nanorod arrays. *Adv. Energy Mater.* **2011**, *1*, 506–510. [CrossRef]
44. Palik, E.D. *Handbook of Optical Constants of Solids*; Academic Press: London, UK, 1985; Volume 1, pp. 547–552.
45. Khudiyev, T.; Bayindir, M. Superenhancers: Novel opportunities for nanowire optoelectronics. *Sci. Rep.* **2014**, *4*, 7505. [CrossRef] [PubMed]
46. Kane, Y. Numerical Solution of Initial Boundary Value Problems Involving Maxwell's Equations in Isotropic Media. *IEEE Trans. Antennas Propag.* **1966**, *14*, 302–307. [CrossRef]
47. Taflove, A.; Hagness, S.C. *Computational Electrodynamics: The Finite-Difference Time-Domain Method*, 3rd ed.; Artech House: Norwood, MA, USA, 2005.
48. Ee, H.S.; Song, K.D.; Kim, S.K.; Park, H.G. Finite-Difference Time-Domain Algorithm for Quantifying Light Absorption in Silicon Nanowires. *Isr. J. Chem.* **2012**, *52*, 1027–1036. [CrossRef]
49. Bohren, C.F.; Huffman, D.R. *Absorption and Scattering of Light by Small Particles*; John Wiley & Sons, Inc.: New York, NY, USA, 1998.
50. Zhou, L.; Yu, X.; Zhu, J. Metal-Core/Semiconductor-Shell Nanocones for Broadband Solar Absorption Enhancement. *Nano Lett.* **2014**, *14*, 1093–1098. [CrossRef]
51. Mann, S.A.; Garnett, E.C. Extreme Light Absorption in Thin Semiconductor Films Wrapped around Metal Nanowires. *Nano Lett.* **2013**, *13*, 3173–3178. [CrossRef]
52. Ferry, V.E.; Polman, A.; Atwater, H.A. Modeling light trapping in nanostructured solar cells. *ACS Nano* **2011**, *5*, 10055–10064. [CrossRef]
53. Munday, J.N.; Atwater, H.A. Large integrated absorption enhancement in plasmonic solar cells by combining metallic gratings and antireflection coatings. *Nano Lett.* **2010**, *11*, 2195–2201. [CrossRef]

nanomaterials

MDPI

Article

Tuning the Magnetic Moment of Small Late 3d-Transition-Metal Oxide Clusters by Selectively Mixing the Transition-Metal Constituents

R. H. Aguilera-del-Toro [1], M. B. Torres [2,*], F. Aguilera-Granja [3] and A. Vega [1]

1 Departamento de Física Teórica, Atómica y Óptica, Universidad de Valladolid, 47011 Valladolid, Spain; rodrigohumberto.aguilera@uva.es (R.H.A.-d.-T.); avega@fta.uva.es (A.V.)

2 Departamento de Matemáticas y Computación, Universidad de Burgos, 09006 Burgos, Spain

3 Instituto de Física, Universidad Autónoma de San Luis Potosí, San Luis Potosí 78290, Mexico; faustino@ifisica.uaslp.mx

* Correspondence: begonia@ubu.es

Received: 10 August 2020; Accepted: 4 September 2020; Published: 11 September 2020

Abstract: Transition-metal oxide nanoparticles are relevant for many applications in different areas where their superparamagnetic behavior and low blocking temperature are required. However, they have low magnetic moments, which does not favor their being turned into active actuators. Here, we report a systematical study, within the framework of the density functional theory, of the possibility of promoting a high-spin state in small late-transition-metal oxide nanoparticles through alloying. We investigated all possible nanoalloys $A_{n-x}B_xO_m$ (A, B = Fe, Co, Ni; $n = 2, 3, 4$; $0 \leq x \leq n$) with different oxidation rates, m, up to saturation. We found that the higher the concentration of Fe, the higher the absolute stability of the oxidized nanoalloy, while the higher the Ni content, the less prone to oxidation. We demonstrate that combining the stronger tendency of Co and Ni toward parallel couplings with the larger spin polarization of Fe is particularly beneficial for certain nanoalloys in order to achieve a high total magnetic moment, and its robustness against oxidation. In particular, at high oxidation rates we found that certain FeCo oxidized nanoalloys outperform both their pure counterparts, and that alloying even promotes the reentrance of magnetism in certain cases at a critical oxygen rate, close to saturation, at which the pure oxidized counterparts exhibit quenched magnetic moments.

Keywords: magnetism; transition-metal oxide clusters; DFT calculations; structure; electronic properties

PACS: 75.75+a; 36.40Cg; 75.30.Pd; 75.50.-y

1. Introduction

For a long time, transition-metal oxide (TMO) nanoparticles (NP) have been the matter of intense research due to their relevance in a large variety of technological applications, such as those in medicine [1–9], new generation batteries [10–12], bactericidal agents [13–15], and catalytic processes [16–21]. TMO-NP are easy to obtain and cheap, and are being considered now as good candidates for the replacement of critical materials that are either harmful or scarce and expensive. Among their electronic properties, magnetism is one of the most relevant one. TMO-NP exhibit superparamagnetic behavior and low blocking temperature, even for sizes of tens of nanometers, due to their low magnetic anisotropy energy. Those characteristics make them useful, for instance, in magnetic resonance imaging and as magnetic markers, for which it is fundamental to avoid agglomeration within the environment where they have to act [22].

There is one thing in common for the efficiency of any magnetic NP, independently of the application in which it will be used: it should be as strongly magnetic as possible for it to be turned into active actuators. Unfortunately, this is precisely the weak point of TMO-NP in general, since TM–O interactions induce antiparallel (AP) magnetic couplings which render them to be in a low-spin state, and consequently to have a small total magnetic moment. To overcome this issue, attempts have been made to increase the total moment of TMO-NP by their doping with another element [23,24]. In this context, Szczerba et al. have experimentally investigated the doping of iron oxide nanoparticles with Zn with the goal of avoiding the spin missalignment, thereby enhancing the total moment [24]. The idea behind it is that Zn substitutes Fe atoms at tetrahedral sites, which are were the Fe atoms antiparallelly couple with the rest in the original iron oxide NP. Although this doping works by eliminating the harmful Fe atoms, the substituted Zn atoms do not contribute themselves to the total moment of the resulting doped NP, since they are nonmagnetic. However, if one could transform the AP couplings into P ones by means of a substitutional magnetic dopant instead of a nonmagnetic one, the total moment could be even larger. This is the main idea behind the present work.

TMO-NP of late transition-metal elements (Fe, Co and Ni) have been extensively investigated both experimentally and theoretically [25–35]. Photoelectron spectroscopy (PES) [36,37] allows one to determine the most abundant stoichiometries and cluster products resulting after fragmentation of a parent cluster. Ion mobility mass spectrometry (IMMS) [29,38] allows one to indirectly determine structural transition patterns as a function of cluster size. In the context of those type of experiments, DFT calculations have provided compelling evidence of which are the putative ground states (GS) of TMO-NP of different sizes. Our group has contributed to such knowledge by proposing GS of Co- and Ni-oxide cationic clusters [39,40] and Fe-oxide cationic clusters [41] after checking that those GS reproduced the most favorable fragmentation channels, the most abundant individuals in the mass spectra and the structural patterns consistent with IMMS. We have also investigated the neutral and anionic counterparts of those TMO-NP [39,40]. Once the GS configurations were benchmarked against the experimental results, we analyzed in detail which configurations are optimal from the magnetic point of view for their potential use as nanomagnets. An interesting and surprising result was that, although AP couplings characterize in general the TMO, leading to a low-spin states, we identified several TMO-NP in high-spin states with large total magnetic moments. There are even some TMO clusters with higher total moment than their non-oxidized transition-metal counterparts as a consequence of the robustness of the P magnetic couplings together with a non negligible contribution of the spin-polarization of the oxygen atoms to the total moment. We also verified that while increasing the d-band filling in the transition metal (TM) series (that is, while going from Fe to Ni), the paralell magnetic couplings become more robust against oxidation, although, in general, the AP couplings prevail nevertheless, particularly at high oxidation rates.

The above results mean that, in principle, there exists the possibility of enhancing the total magnetic moment of a TMO-NP by doping it substitutionally with a different TM element, preferably one on its right in the transition-metal series. The mechanism behind such improvement of the magnetic properties of the TMO-NP would be the promotion of P magnetic couplings, with a clear advantage over doping with nonmagnetic elements such as Zn: the total moment would be contributed to a large extent by the dopant. Moreover, there perhaps exists a particular TM mixing that leads to high-spin states for most oxidation rates which would imply the robustness of the total magnetic moment against the unavoidable environmental conditions. Exploring those aspects is appealing and this was the main objective of the present work, since no studies in this respect had been carried out to our knowledge, an exception being our study of oxidized FeCo dimers [42]. The results obtained in that study provide an illustrative example of what one could expect: the iron dimer oxide Fe_2O_m is known to be in a low-spin state due to AP magnetic couplings [42]. In $FeCoO_m$ clusters, the total moment oscillates as a function of m between high-spin states (characterized by P magnetic couplings) and low-spin states (characterized by AP couplings). Oxide clusters in the high-spin state retain the same total moment as the unoxidized FeCo dimer which largely exceeds the value of the Fe_2O_m counterpart. In the context of extended magnetic alloys, Fe–Co ones are certainly interesting for applications in soft magnets. At 30–40% of Co content the Fe–Co alloy is the material with the highest saturation magnetization, exceeding that of pure Fe bcc [43]. We note that cooperative effects in other binary transition-metal systems have been already demonstrated.

In view of the information described above, we found it appealing to explore, in a systematic way, the possibility of inducing ferromagnetic-like order to promote a high-spin state in TMO-NP (TM = Fe, Co, Ni) by their doping with a different TM element of the list. For that purpose, we extended our previous work on oxidized FeCo dimers to all possible pure and bimetallic dimers (the smallest size), trimers and tetramers, allowing us to have a variety of stoichiometries and geometrical arrangements in the range of small sizes. The goal was to identify optimal mixed TM oxides from the magnetic point of view, and to extract qualitative trends, if possible, that might help to choose the best dopant for larger magnetic TMO-NP. Taking into account the database of geometries of the putative ground states and low-energy isomers of Co- and Ni-oxide clusters obtained in our previous works, and those recently obtained for Fe-oxide clusters, we determined the putative ground states of the mixed TM oxide clusters. Before carrying out the present study, a tasting was done for Cr clusters doped with Fe, Co and Ni. Cr is the TM element with the highest spin magnetic moment and the best candidate for a highly magnetic NP if ferromagnetic-like coupling can be promoted. Unfortunately, the AP couplings were robust against doping with a late 3d TM element. Since this was the main conclusion of this initial test, we did not find it necessary to publish the results obtained, but the reader can obtain more details upon request from the authors.

In the next section we describe our theoretical approach and computational method. The results are presented in Section 3. The main conclusions of the work are summarized in the last section.

2. Materials and Methods

We performed fully self-consistent DFT calculations using the SIESTA code [44], which solves the spin-polarized Kohn–Sham equations within the pseudopotential approach. For the exchange and correlation potential we used the Perdew–Burke–Ernzerhof form of the generalized gradient approximation (GGA) [45]. Our DFT setup, and in particular the chosen functional to describe exchange and correlation effects (the PBE functional), is similar to that employed in our previous studies of the pure metal oxide clusters [39–42]. In the one of [42], we conducted benchmark calculations on the diatomic molecules FeCo, FeO, CoO and O_2, obtaining good agreement with experimental results. In those of [39,40] (devoted to Co and Ni oxide clusters, respectively), we obtained a rather good agreement with experiments regarding the fragmentation channels, for which good descriptions of ground states and relative energies are required. We employed norm-conserving scalar relativistic pseudopotentials [46] in their fully nonlocal form [47], generated from the atomic valence configurations $3d^7 4s^1$, $3d^8 4s^1$ and $3d^9 4s^1$ for Fe, Co and Ni, respectively (with core radii 2.00, 2.32 and 2.44 a.u. for s, p and d orbitals), and $2s^2 2p^4$ for O (with core radii 1.14 a.u.). Non-linear partial core corrections [48], which are known to be important for TM pseudopotentials, were included for Fe, Co and Ni at the core radius of 0.7 Å. Valence states were described using double-zeta basis sets for all atomic species. The energy cutoff used to define the real-space grid for numerical calculations involving the electron density was 250 Ry. The Fermi distribution function that enters in the calculation of the density matrix was smoothed with an electronic temperature of 15 meV. We used an energy criterium of 10^{-4} eV for converging the electronic density. The individual clusters were placed in a cubic supercell of $20 \times 20 \times 20$ Å^3, a size large enough as to make the interaction between the cluster and its replicas in neighboring cells negligible, allowing us to consider only the Γ point ($k = 0$) when integrating over the Brillouin zone. The putative global minimum structures resulted from an unconstrained conjugate-gradient structural relaxation using the DFT forces; the input structures were relaxed until interatomic forces were smaller than 0.001 eV/Å.

We note that, although the size of the investigated clusters was small, the problem of locating the global minimum structures was still a formidable task due to the many variables involved which required use to consider many input configurations. First, except for the dimers, one has to consider the different homotops of the heteroatomic clusters, resulting from the different inequivalent positions of the atoms of the two transition-metal elements within the structural skeleton. Second, oxygen can be attached in top, bridge and hollow positions and between TM atoms, and also through atomic or molecular adsorption (particularly in the range of high oxygen concentration). Third, different total spin states have to be checked to find the spin magnetic moment of the putative ground state. A given spin state can result from different magnetic couplings within the system, P or AP in which case the spin homotops have to be also considered. On top of that, all variables are interconnected since they depend on each other. The protocol used for the structural search was as follows. First, we selected several low energy isomers of Fe, Co and Ni oxide clusters that were investigated in our previous works [39–41]. Details of the structural search for these pure TM oxide clusters can be found in those papers, and it is important to state here that an indirect support for them is the fact that consistency with fragmentation and ion mobility experiments was achieved. For all those configurations we mixed the different TM atoms to build inputs for the homotops of each oxide nanoalloy. This creates, in some cases, inequivalent positions for oxygen that in the pure TM oxide were equivalent by symmetry and therefore were not checked. For instance, in Fe_2O, the oxygen atom can be attached only in two positions, in the bridge or on top of Fe, but in FeCoO oxygen can be also attached on top of Co. After adding to the database these new configurations arising from the inequivalent O positions, the next step was to perform, for each one, different DFT calculations for different spin isomers. In order to save computing time, good practice is to ensure selfconsistency in various steps. This allows one to not push to the end of accuracy those cases in which a weak selfconsistency is enough to: (i) determine that a given input has transformed into another input of the database; (ii) a given configuration energetically destabilizes with respect to others already checked.

3. Results and Discussion

We will discuss first the structural properties and stability of the binary TM oxides $A_{n-x}B_xO_m$ (A, B = Fe, Co, Ni; n = 2, 3, 4; $0 \leq x \leq n$) with different oxidation rates, m. These properties are connected with their electronic structures and magnetic properties that will be discussed in a second subsection. There are several factors that contribute in determining the structural properties and stability in these binary TM oxide clusters, in particular the differences in binding energy of Fe, Co and Ni and the relative strengths of the metal–metal and metal–oxygen bonds. The Pauling scale electronegativities [49] of O, Fe, Co and Ni are 3.4, 1.83, 1.88 and 1.91, respectively, so we expected a non negligible electronic charge transfer from the metal atoms towards O atoms in these systems. This effect will strengthen metal–oxygen bonds due to the partial ionic contribution while weakening the metal–metal bonds. Concerning the magnetic properties, it is known that clusters of Fe, Co and Ni have P magnetic moments and larger per atom spin-polarization than their respective bulk counterparts (due to electron localizaton). However, due to the expected loss of electronic charge in the metal atoms, and the weakening of the metal–metal interaction, weakening of the tendency to P magnetic couplings was also expected. Moreover, the contribution of the spin-polarization of the oxygen atoms to the total moment may not be negligible. The interplay between all those factors depends on the composition and size, such that it is very difficult to anticipate which of the oxidized nanoalloys will exhibit the best figures of merit from the point of view of the magnetic properties without carrying out explicit calculations for each particular system.

3.1. Structural Properties and Stability

The putative global minimum structures of $A_{n-x}B_xO_m$ (A, B = Fe, Co, Ni; $0 \leq x \leq n$), with m up to oxygen saturation, are shown in Figures 1–3, 4–6 and 7–9 for n = 2, 3 and 4, respectively. In Figure 10 we plot the average binding energy per atom of the corresponding binary TM oxide clusters as a function of the oxidation rate. This quantity is calculated as:

$$E_b(A_{n-x}B_xO_m) = \frac{-E(A_{n-x}B_xO_m) + (n-x)E(A) + xE(B) + mE(O)}{n+m} \tag{1}$$

where $E(S)$ is the total energy of system S, and the values are also given below the structures shown in Figures 1–9.

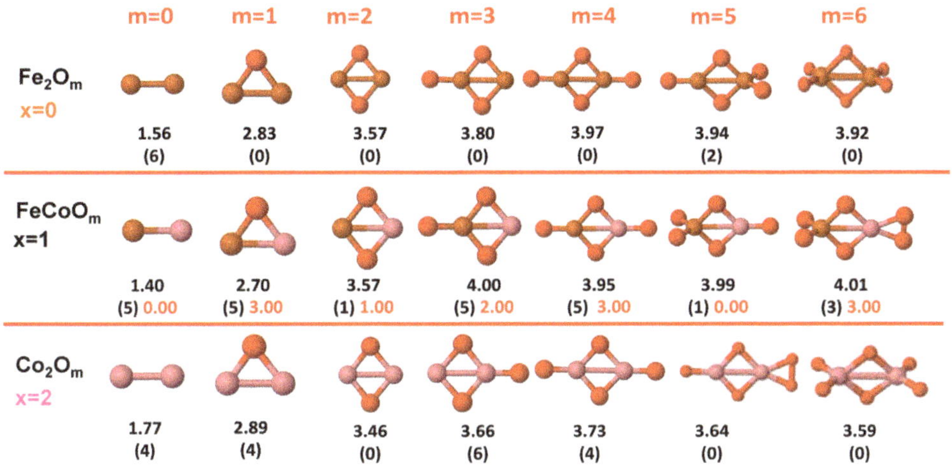

Figure 1. Putative global minimum structures of $Fe_{2-x}Co_xO_m$, x = 0–2, m = 1–6. Numbers below structures are the binding energy per atom (in eV), numbers in parentheses are the magnetic moment (in μ_B) and the third number (when present) is the excess magnetic moment (in μ_B) defined in Equation (2). Fe atoms in brown color, Co atoms in pink and oxygen atoms in red.

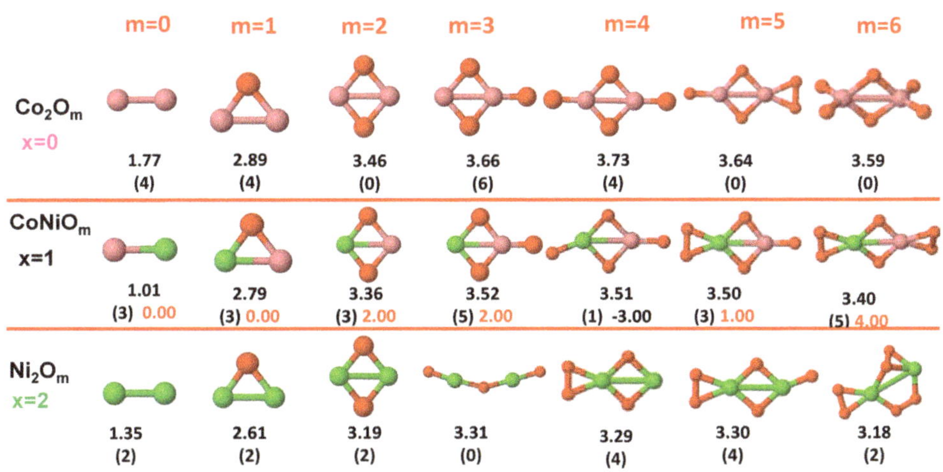

Figure 2. Putative global minimum structures of $Co_{2-x}Ni_xO_m$, x = 0–2, m = 1–6. Numbers below structures are the binding energy per atom (in eV), numbers in parentheses are the magnetic moment (in μ_B) and the third number (when present) is the excess magnetic moment (in μ_B) defined in Equation (2). Co atoms in pink color, Ni atoms in green and oxygen atoms in red.

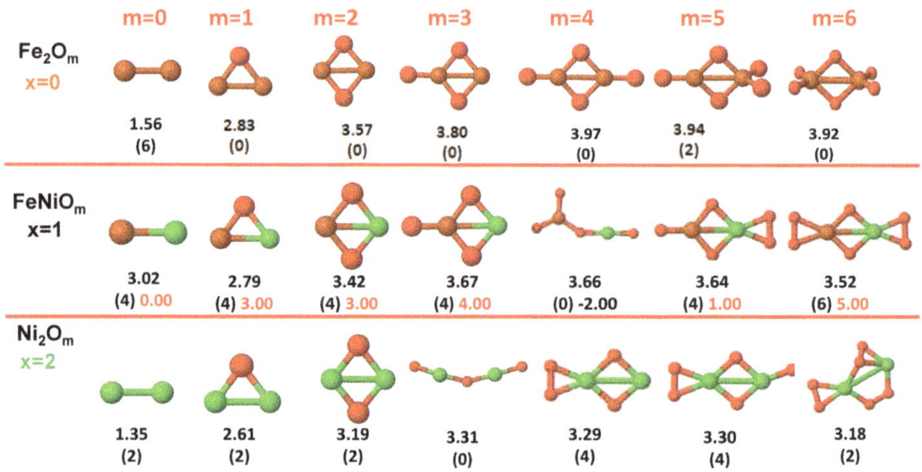

Figure 3. Putative global minimum structures of $Fe_{2-x}Ni_xO_m$, x = 0–2, m = 1–6. Numbers below structures are the binding energy per atom (in eV), numbers in parentheses are the magnetic moment (in μ_B) and the third number (when present) is the excess magnetic moment (in μ_B) defined in Equation (2). Fe atoms in brown color, Ni atoms in green and oxygen atoms in red.

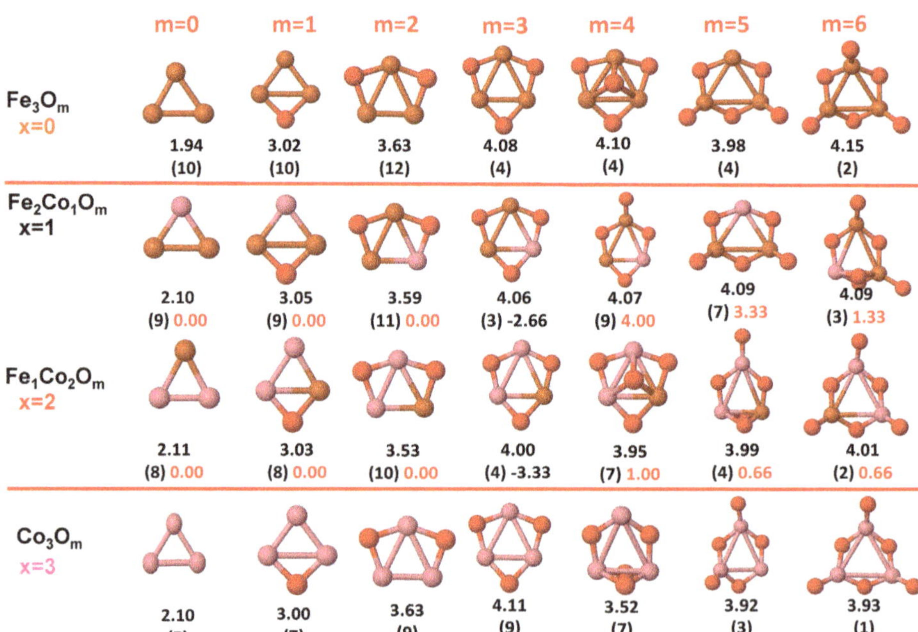

Figure 4. Putative global minimum structures of $Fe_{3-x}Co_xO_m$, x = 0–2, m = 1–6. Numbers below structures are the binding energy per atom (in eV), numbers in parentheses are the magnetic moment (in μ_B) and the third number (when present) is the excess magnetic moment (in μ_B) defined in Equation (2). Fe atoms in brown color, Co atoms in pink and oxygen atoms in red.

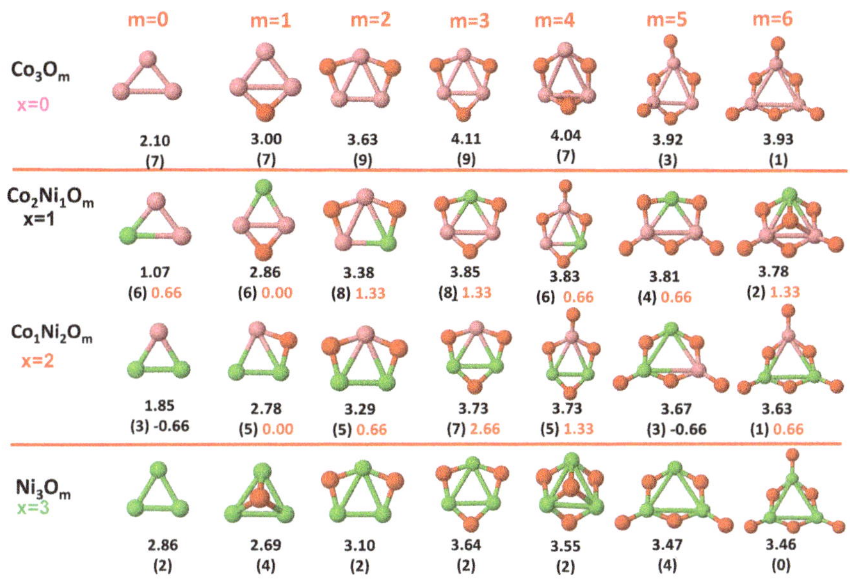

Figure 5. Putative global minimum structures of $Co_{3-x}Ni_xO_m$, x = 0–2, m = 1–6. Numbers below structures are the binding energy per atom (in eV), numbers in parentheses are the magnetic moment (in μ_B) and the third number (when present) is the excess magnetic moment (in μ_B) defined in Equation (2). Co atoms in pink color, Ni atoms in green and oxygen atoms in red.

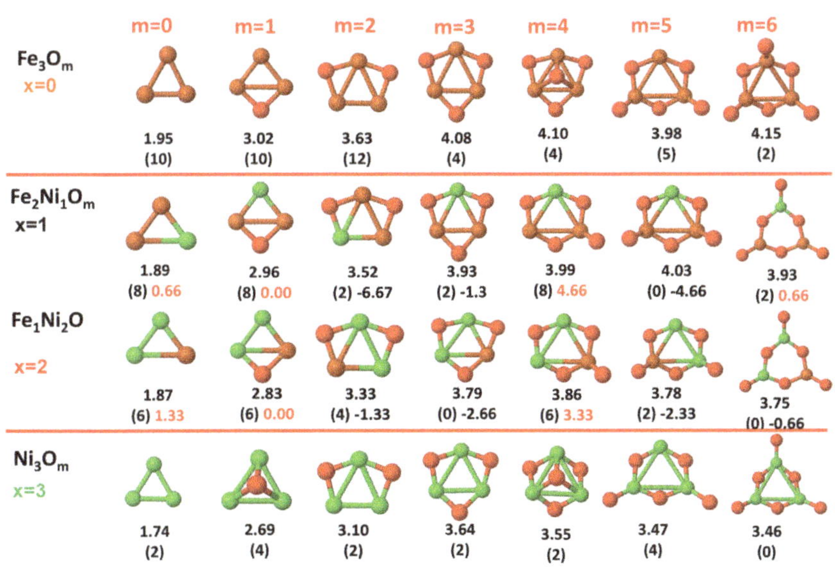

Figure 6. Putative global minimum structures of $Fe_{3-x}Ni_xO_m$, x = 0–2, m = 1–6. Numbers below structures are the binding energy per atom (in eV), numbers in parentheses are the magnetic moment (in μ_B) and the third number (when present) is the excess magnetic moment (in μ_B) defined in Equation (2). Fe atoms in brown color, Ni atoms in green and oxygen atoms in red.

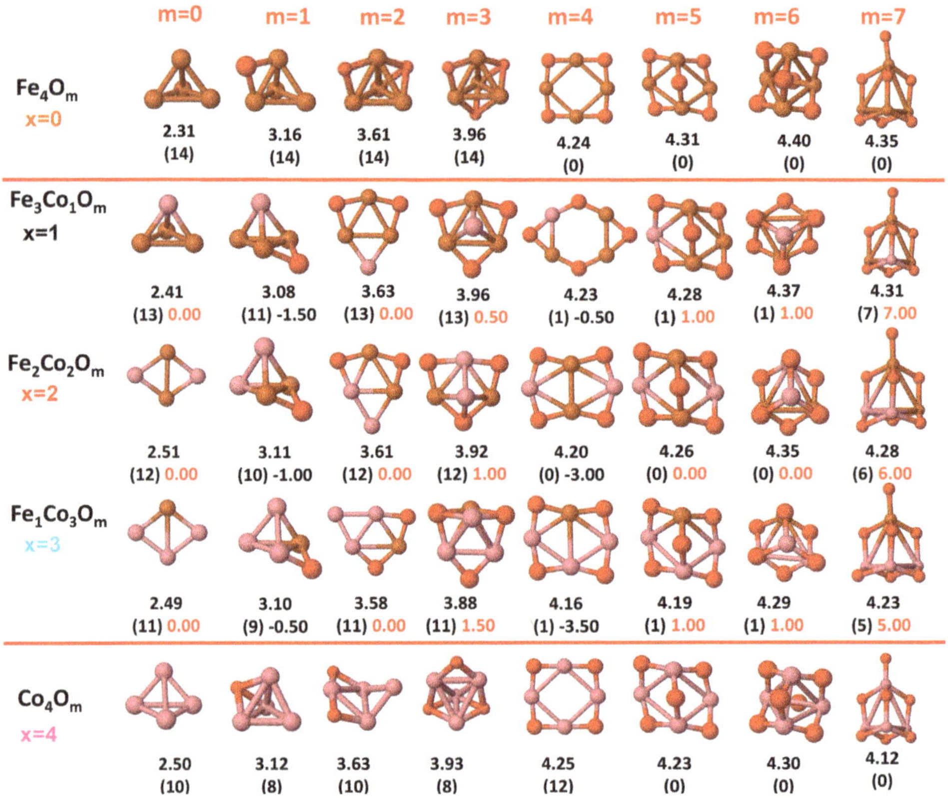

Figure 7. Putative global minimum structures of $Fe_{4-x}Co_xO_m$, x = 0–2, m = 1–6. Numbers below structures are the binding energy per atom (in eV), numbers in parentheses are the magnetic moment (in μ_B) and the third number (when present) is the excess magnetic moment (in μ_B) defined in Equation (2). Fe atoms in brown color, Co atoms in pink and oxygen atoms in red.

Some general trends can be inferred regarding the geometrical structures. Most pure and oxidized TM dimers and trimers, and their related mixtures, have a lineal and triangular subcluster of TM atoms, respectively. The exceptions to this trend are Ni_2O_3 with a bent linear structure with exclusively metal–oxygen bonds (similar to that found by Aguilera del Toro et al. [50] in In_2O_3), and $FeNiO_4$ which does not have a Fe–Ni bond either. For tetramers ($n = 4$), a competition between three-dimensional and planar structures is found. All pure TM clusters have a tetrahedral structure, but their respective non-oxidized nanoalloys have a planar rhombic structure in some cases, such as Fe_2Co_2, Fe_1Co_3 and all Co–Ni nanoalloys for $n = 3$, Fe_2Ni_2 and Fe_1Ni_3. Although in oxidized TM clusters [39–41], the tetrahedron prevails for $m = 1, 6, 7$, TM_4O_5 (TM_4O_4) oxides show a slightly bent rhombus (planar ring-like structure). Moreover, for Ni_4O_3 and Co_4O_2, an open tetrahedron is obtained, becoming a rhombus for Ni_4O_2. For the rest of cases, the tetrahedral TM subcluster is preserved. We have found that, in general, more planar structures appear for oxidized nanoalloys than for their bare counterparts. Thus, for $m = 2$, all oxidized nanoalloys are planar except $Fe_2Ni_2O_2$ and $Fe_1Ni_3O_2$, for which the tetrahedron found for Fe_4O_2 still remains. In some cases, changes are found with respect to the tetrahedral structure for oxidized nanoalloys: $Fe_1Co_3O_3$, $Co_2Ni_2O_3$ and Co_1Ni_3O present a 3D structure consisting of two triangles sharing the same side, halfway between a bent rhombus and a tetrahedron.

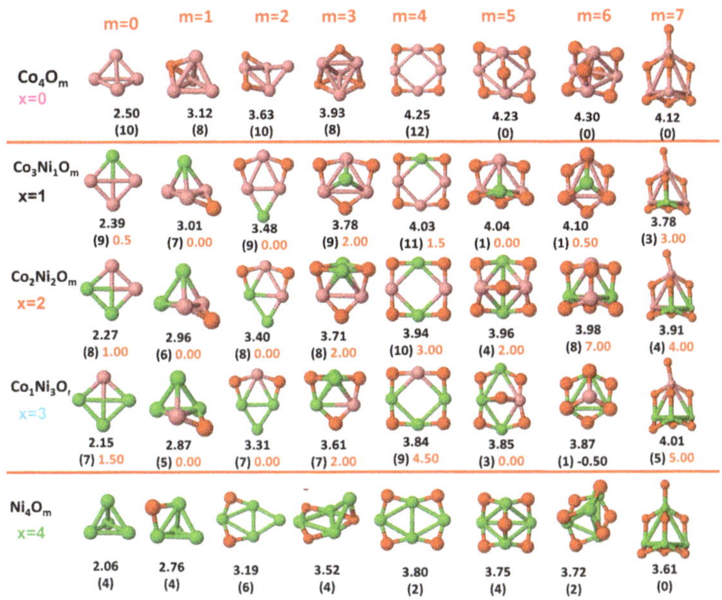

Figure 8. Putative global minimum structures of $Co_{4-x}Ni_xO_m$, x = 0–2, m = 1–6. Numbers below structures are the binding energy per atom (in eV), numbers in parentheses are the magnetic moment (in μ_B) and the third number (when present) is the excess magnetic moment (in μ_B) defined in Equation (2). Co atoms in pink color, Ni atoms in green and oxygen atoms in red.

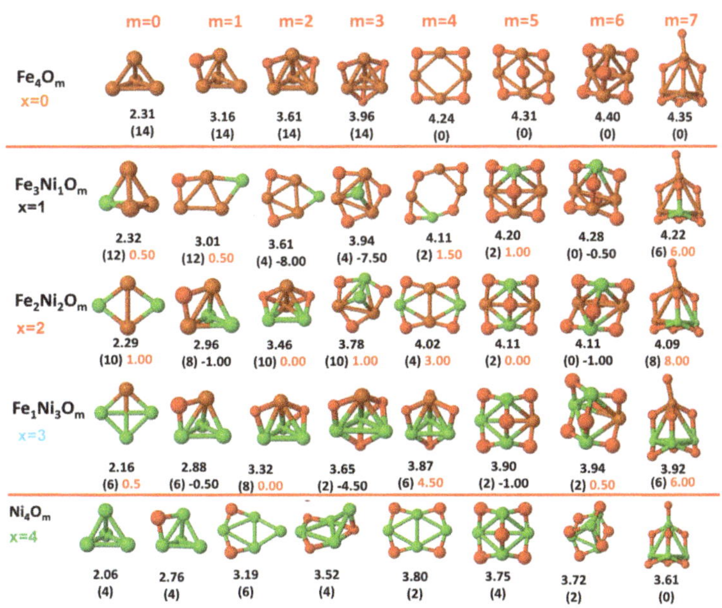

Figure 9. Putative global minimum structures of $Fe_{4-x}Ni_xO_m$, x = 0–2, m = 1–6. Numbers below structures are the binding energy per atom (in eV), numbers in parentheses are the magnetic moment (in μ_B) and the third number (when present) is the excess magnetic moment (in μ_B) defined in Equation (2). Fe atoms in brown color, Ni atoms in green and oxygen atoms in red.

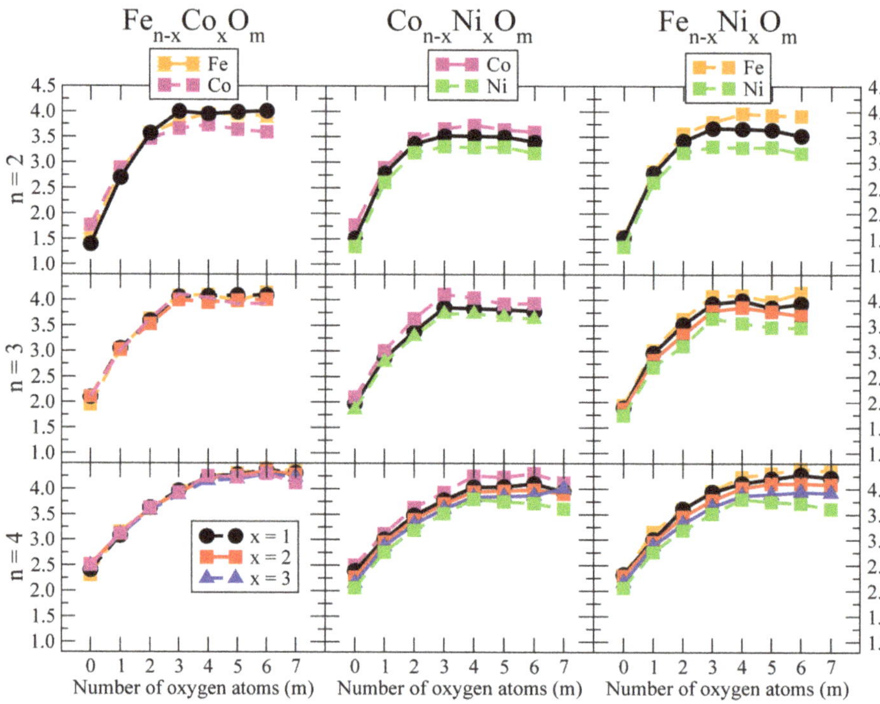

Figure 10. Binding energy per tom $A_{n-x}B_xO_m$ clusters (A, B = Fe, Co, Ni; $0 \leq x \leq n$; m = 1–7).

The following general trends can be also inferred from the results shown in Figures 1–10:

(i) The oxygen atoms tend to occupy bridge sites between the TM atoms ($m \leq n$ for $n = 2, 3$ and $m \leq 6$ for $n = 4$), and when bridge sites are saturated, O atoms tend to occupy top sites ($m = 3, 4$ for $n = 2$, $m = 4, 5, 6$ for $n = 3$ and $m = 7$ for $n = 4$).

(ii) The oxygen binds preferentially with Fe atoms. When the oxygen rate increases, the next preferential coordination is Co and finally Ni.

(iii) The net electronic charge in the O atoms, resulting from the charge transfer from the metal atoms, favors an uniform distribution of the O atoms in the cluster.

(iv) At oxidation rates where the number of O atoms equals the number of metal atoms, ring-like structures are formed for all sizes considered in this work. In this atomic configuration all metal–metal bonds are mediated by oxygen. This trend was already discussed in our previous works for the pure Fe, Co and Ni oxidized clusters [39–41], and it has been also discussed by other groups for other systems [33,51] in this small size range. We now demonstrate that these ring-like structures are preserved in the TM oxidized nanoalloys, except for $Fe_1Ni_3O_4$, for which a tetrahedral structure is obtained, as stated above.

(v) We considered up to m = 6 or 7 oxygen atoms which is enough to analyze the transition from atomic to molecular adsorption in the case of the TM dimers (not yet in the trimers and tetramers). When the oxygen concentration is high enough, oxygen starts to be adsorbed molecularly, preserving short O–O inter-atomic distances. The critical oxygen concentration for this to occur lowers while going from Fe to Co and Ni, as it can be clearly seen in the structures of the oxidized dimers in Figures 1–3: while in Fe_2O_m no molecular adsorption takes place for $m \leq 6$, in Co_2O_m occurs earlier (for $m = 5$) and in Ni_2O_m it starts even earlier (for $m = 4$). Concerning mixed oxidized clusters, in $FeCoO_m$ no molecular adsorption takes place for m = 6 while in $CoNiO_m$

and $FeNiO_m$ it already occurs for $m = 5$. The molecular adsorption is a manifestation of oxygen saturation of the system and its emergence at different oxygen concentrations for the different TM compositions is connected with the relative strength of the respective metal–oxygen bonds. The stronger the metal–oxygen binding is, the more oxygen is bound in an atomic form. In Table 1 we give, as a reference, the binding energies of Fe–O, Co–O and Ni–O, along with those of the different TM dimers. The metal–oxygen binding lowers for TM oxides while going from Fe to Co and Ni, as electronic charge transfer decreases and bond distance slightly increases.

Table 1. Binding energies (E_b) in eV, bond distance (d) in Å and electronic charge transfer (q) in e units, of the TMO oxides (TM = Fe, Co, Ni) and TM dimers.

	FeO	CoO	NiO	Fe_2	Co_2	Ni_2
E_b (eV)	2.75	2.63	2.29	1.56	1.77	1.35
d (Å)	1.67	1.70	1.70	2.04	2.01	2.17
q (e)	0.46	0.45	0.43	–	–	–

(vi) The metal–metal binding also lowers while going from Fe to Ni (see Table 1). However, the Co dimer has the highest binding (and the lowest bond distance). General structural trends are also consistent with the binding energy per atom plotted in Figure 10 for the oxides. Consequently, for TM oxide dimers, and in connection with the discussion of the oxygen saturation, we note a shift to lower m in the maximum of binding energy while going from Fe to Co and Ni which means that adsorption of more oxygen atoms beyond this critical value of m does not increase the binding, because the metal–metal bonding weakens as m increases. The weakening of the metal–metal bonding is manifested in the less compact metal skeleton (lower average metal–metal inter-atomic distance) that results as m increases.

The upper panel of Figure S1 shows the increase on the average metal–metal distance for the $Co_2Ni_1O_m$ ($0 \leq m \leq 6$) nanoalloys. This distance increases until the triangular structure has all the oxygen atoms on the bridge positions ($m = 3$), the increase being higher when both bridge and top positions are occupied by oxygen ($m = 6$), where an open triangular subcluster is found. The average Co–Co and Ni–Co distances are also shown. Both increase also as the oxygen rate increases, with the Co atoms remaining somewhat further apart from each other than the Ni atoms, for a given oxidation rate. The lower panel of Figure S1 displays the average distance between the TM atoms and the O atoms located on bridge sites, which slightly decreases as the amount of oxygen decreases, mainly due to the reduction of the average Co–O distance. As the triangular subcluster opens up, the TM atoms are nearest to the bridge O atoms.

Moreover, the results for the binding energies indicate that the higher the concentration of Fe in the TM oxide clusters, the higher their absolute stability, which may be important for practical purposes. On the other hand, the higher the Ni content, the lower the exothermic character of oxidation, or in other words, the less prone to oxidation the nanoparticle will be, which may also be important for practical purposes.

(vii) The relative differences of electronegativities between the TM atoms, although small, also support this trend. Electronegativity increases while going from Fe to Ni so that the tendency towards electron donation to O should be stronger in Fe atoms. Electronic charge transfers also corroborate this fact. Figure 11 illustrates an example of the correlation between the binding energies and the electronic charge transfer from the TM atoms to the O ones for all pure oxides and oxidized nanoalloys for a given oxidation rate, $A_{4-x}B_xO_6$ (A, B = Fe, Co, Ni, $0 \leq x \leq 4$ and $m = 6$). Moreover, the binding energy per atom and the average electronic charge transfer are plotted in Figure S2 for $Fe_{4-x}Co_xO_6$, $Co_{4-x}Ni_xO_6$, and $Fe_{4-x}Ni_xO_6$ as a function of x. The binding energy decreases for pure TM oxides from Fe_4O_6 to Co_4O_6 and from Co_4O_6 to Ni_4O_6, and it also decreases for any type of oxidized mixture, as the concentration (x) of the B constituent increases. For Fe–Co nanoalloys, the decrease of binding energy is the lowest among all mixtures, the highest being for Fe–Ni nanoalloys. Regarding the electronic charge transfer, Ni atoms are less prone to donate electrons to O due to their higher electronegativity. This is illustrated in the local electronic charges on the atoms of the respective structures shown in the Figure 11. In addition, as the concentration of Ni increases both in Co–Ni and Fe–Ni mixtures, each of theTM atoms (Fe, Co, Ni) increases its electronic donation, and consequently, the total transferred electronic charge becomes higher (see Figure S2); and this fact is reflected in a weakening of the bonding between the TMs, and therefore, in a decrease of the binding energy. However, in the case of FeCo mixtures, the binding energies do not vary significantly as the concentration of Co(x) increases, because the electronic transfer remains almost constant. Since the electronic transfer of Fe is higher than that of Co, and there is a lesser concentration of the former, the total electronic charge transfer from the TM atoms to oxygen decreases slightly instead of increasing as in the rest of the mixtures. Consequently, the binding energies in Fe–Co mixtures are more similar. That is, the oxides scarcely loose stability upon the substitution of Fe by Co. This fact, together with the magnetism that these mixtures present, makes them remarkable. In Figure 11, both average TM–TM and TM–O distances are also provided. The former are almost constant for both Fe–Co and Fe–Ni mixtures, and in the case of Co–Ni mixtures for $x = 2$, the TM subcluster is somewhat more compact. The TM–O distances have little variation.

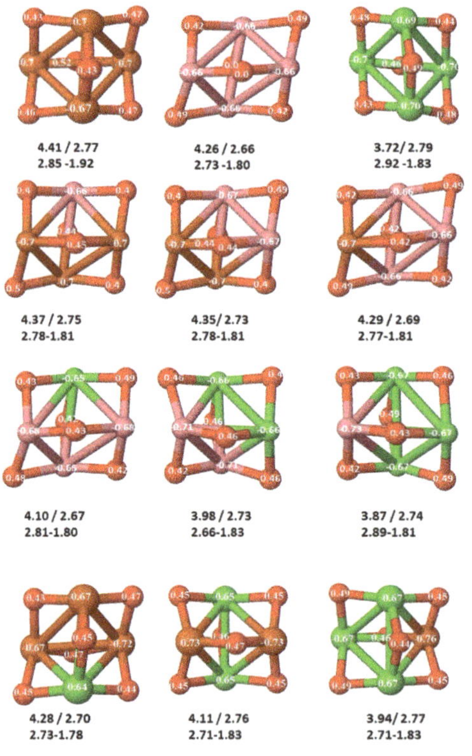

Figure 11. (Color online) Ground states of $A_{4-x}B_xO_6$ (A, B = Fe, Co, Ni; x = 0–4) oxides. The excess (or defect) of the electronic charge for Fe, Co and O is given for each atom. Positive (negative) numbers indicate excess (defect). Below each of the structures, there are two rows of numbers. In the first one, binding energy per atom (in eV) and average electronic charge transfer (in e) are given. In the second one, both average TM–TM and TM–O distances are given. Fe atoms in brown color, Co atoms in pink, Ni atoms in green color and oxygen atoms in red.

Up to which extent those facts have consequences on the magnetic properties of these systems is discussed in the next subsection.

Although several general structural and energetic trends have been found and rationalized in terms of relatively simple arguments, quantum confinement effects and the non scalability of physical and chemical properties inherent to the nanoscale lead to certain anomalies or unexpected structures that depart from the general behavior. This confirms again the need of performing explicit calculations for each individual system at the nanoscale in order to capture the details. Example of some peculiarities are, for instance, for dimers, Ni_2O_6 with three oxygen molecules, and the related to the geometrical changes above discussed. In the case of oxidized trimers: (i) $Fe_1Co_2O_m$ ($m = 4, 5$) and $Fe_2Co_1O_6$ oxides with two bound oxygen atoms on the same bridge, in the same way that they were found in Co_3O_4 oxide [39]; (ii) $Co_2Ni_1O_6$ oxide, with the sixth oxygen atom on hollow site instead of top position after saturating all bridge sites, similar to the fourth one for Ni_3O_4 and Fe_3O_4 oxides. For tetramers, in addition to the geometric changes already described: (i) Fe_3O_4 and Ni_3O_4 with an oxygen atom in hollow site instead of an top after saturating all bridge sites; (ii) Co_3O_4 with two bound oxygen atoms on the same bridge.

3.2. Magnetic Properties

Small clusters of late 3d TM atoms are superparamagnetic with very low magnetic anisotropy energy [52–54]. However, the parallel magnetic couplings and relatively high spin polarization due to electron localization make them have high total magnetic moments which is their magnetic figure of merit. Nanoparticles with high total moments and low magnetic anisotropy are relevant, among others, for applications in which rotation of the nanoparticles should not affect the orientation of the moment when an external field is applied. This behavior is adequate for designing magnetic markers in nanomedicine, where particle aglomeration is also to be avoided. TM oxide clusters are among the most relevant magnetic nanoparticles in this context but, besides the absolute stability already discussed in the previous subsection, it is important to achieve (i) high total magnetic moments, and (ii) robustness of the total magnetic moment (mainly contributed by the TM atoms) against oxidation. The discussion of the present subsection is oriented towards the seeking of those binary TM oxide clusters with particularly high total moments among the investigated ones $A_{n-x}B_xO_m$; the determination of those compositions that lead to more robust magnetic moments against oxidation; and finally, the seeking of general trends that could be valid for larger sizes.

The total magnetic moment of $A_{n-x}B_xO_m$ (A, B = Fe, Co, Ni; $0 \leq x \leq n$) as a function of the oxygen concentration m is shown in the upper panels of Figures 12, 13 and 14 for n = 2, 3 and 4, respectively.

As a simple measure of the benefit of mixing TM atoms from the magnetic point of view, we define the excess magnetic moment or magnetic excess, μ_{exc}, which in our systems is calculated as:

$$\mu_{exc}(A_{n-x}B_xO_m) = \mu(A_{n-x}B_xO_m) - \frac{n-x}{n}\mu(A_nO_m) - \frac{x}{n}\mu(B_nO_m) \tag{2}$$

where $\mu(S)$ is the total magnetic moment of the system S in its ground state. Thus, $\mu(A_nO_m)$ and $\mu(B_nO_m)$ are the total moments of the oxide clusters of TM elements, A and B, in their respective global minimum structures at the oxygen concentration m. The excess magnetic moment is zero for TM oxide clusters of a single TM element by definition. In the rest of the cases, an excess magnetic moment equal to zero means that the mixture is an ideal mixture. The ideal mixture of $A_{n-x}B_xO_m$ would follow a simple Vegard law, according to which the total moment of the oxide nanoalloys follows a linear behavior connecting the magnetic moments of the pure oxide clusters, A_nO_m and B_nO_m, obeying the quantization of the total spin. Positive excess magnetic moments indicate that the formation of the corresponding oxidized nanoalloy is magnetically favorable as compared to an ideal mixture, and often presents a magnetism even higher than that associated with the two pure systems. We note that similar magnitudes are often used in the analysis of energetic trends or the compactness of nanoalloys (excess energy [55] and excess radius [56], respectively), but to our knowledge, the excess magnetic moment was not defined. In the lower panels of Figures 12–14, we plot the excess magnetic moment of the binary TM oxide nanoalloys, as a function of the oxygen concentration m.

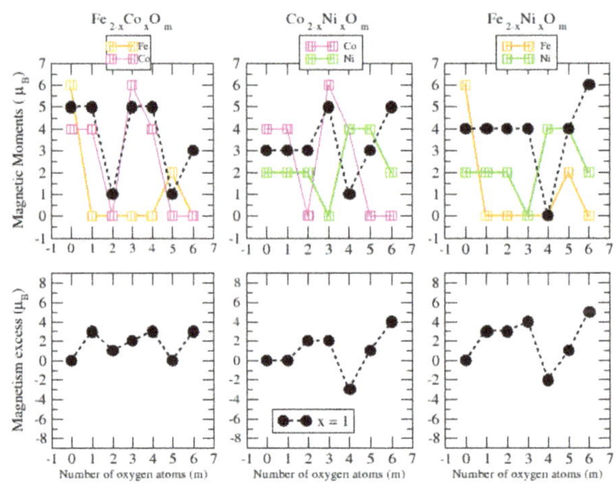

Figure 12. (**upper pannels**) Total magnetic moments of $A_{2-x}B_xO_m$ clusters (A, B = Fe, Co, Ni; $x \leq 2$; m = 1–6). Brown, pink and green curves correspond, respectively, to Fe_nO_m, Co_nO_m and Ni_nO_m oxides. Black curves correspond to nanoalloys with $x = 1$. (**lower panels**) Magnetism excess of ABO_m nanoalloys (A, B = Fe, Co, Ni; m = 1–6).

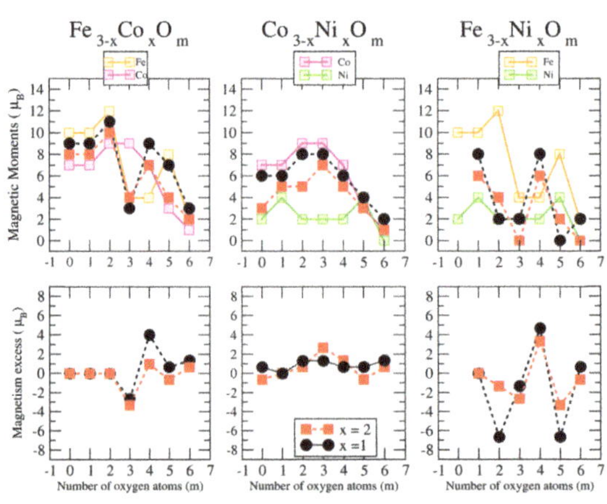

Figure 13. (**upper pannels**) Total magnetic moments of $A_{3-x}B_xO_m$ clusters (A, B = Fe, Co, Ni; $0 \leq x \leq 3$; m = 1–7). Brown, pink and green curves correspond, respectively, to Fe_nO_m, Co_nO_m and Ni_nO_m oxides. Black and red curves correspond to nanoalloys with x = 1, 2, respectively. (**lower panels**) Magnetism excess of $A_{3-x}B_xO_m$ clusters (A, B = Fe, Co, Ni; $1 \leq x \leq 2$; m = 1–7).

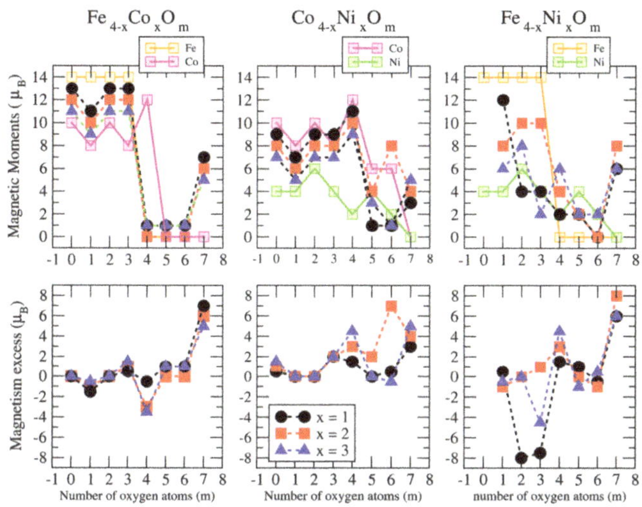

Figure 14. (**upper panels**) Total magnetic moments of $A_{4-x}B_xO_m$ clusters (A, B = Fe, Co, Ni; $0 \leq x \leq 4$; m = 1–7). Brown, pink and green curves correspond, respectively, to Fe_nO_m, Co_nO_m and Ni_nO_m oxides. Black, red and blue curves correspond to nanoalloys with x = 1, 2, 3, respectively. (**lower panels**) Magnetism excess of $A_{4-x}B_xO_m$ clusters (A, B = Fe, Co, Ni; $1 \leq x \leq 3$; m = 1–7).

3.2.1. A-B Non-Oxidized Nanoalloys; A, B = Fe, Co, Ni

Before discussing the results for the oxidized mixtures, $A_{n-x}B_xO_m$, let us analyze the total magnetic moments and the values of the excess magnetic moment when m = 0, that is, for the bare $A_{n-x}B_x$ nanoalloys without the presence of oxygen. The preferred Fe–Fe magnetic coupling is parallel and Fe_n (n = 2–4) clusters show the highest magnetic moments equal to 6 μ_B, 10 μ_B and 14 μ_B, respectively. The Ni_n (n = 2–4) clusters, due to the lowest spin polarization, present the smallest values, equal to 2 μ_B, 2 μ_B and 4 μ_B, respectively, despite their ferromagnetic-like coupling, the Co_n (n = 2–4) clusters being those with the intermediate values, equal to 4 μ_B, 7 μ_B and 10 μ_B, respectively. In the case of non-oxidized Fe–Co alloys, the magnetic moments obtained follow the simple Vegard law with indexes equal to zero for all sizes studied here (n = 2–4). This behavior is also followed by the rest of the dimers ($n = 2$) regardless the transition metal. The Fe–Ni nanoalloys with n = 3–4, although having lower moments than the corresponding Fe–Co ones, present higher excess magnetic moments (see Figures 11–13, for $m = 0$). The Co–Ni nanoalloys (particularly for $n = 4$) exhibit the highest positive excess magnetic moments between 0.5 and 1.5. (Figure 13, for $m = 0$).

3.2.2. Fe–Co Oxided Nanoalloys

Among the late 3d elements, Fe has the most d-holes in its active valence states. This means that it has the largest local magnetic moments among the late TM elements but, at the same time, AP couplings are easier to be promoted in Fe systems than in Co and Ni ones. Electron transfer to O makes Fe electronically approach Mn, which already exhibits AP couplings in certain atomic arrangements and weakens the Fe–Fe interaction and the P magnetic couplings reminiscent of the Fe bulk ferromagnetism. This is the mechanism through which oxidation quenches the moment of the Fe dimer for nearly all oxidation rates m (see Figure 11), and strongly reduces those of the trimer (Figure 12) and tetramer (Figure 13) except for the lowest m in both cases, and also for $m = 5$ in the case of dimers and trimers. This reentrance of μ at relatively high oxidation rates is an unexpected magnetic trend related with the particular structural arrangement and the strengthen of the Fe–O interaction to the point that the magnetic coupling between Fe and O in that particular geometry forces the O mediated Fe–Fe magnetic coupling to be P. This has been discussed in detail in our previous paper, where we show that further oxidation quenches the magnetic moment again [41]. Therefore, a general trend is that the Fe cluster must reach a critical size and metal–metal coordination to retain the P couplings under oxidation. When this happens (Fe_3O_m with $m < 3$, and Fe_4O_m with $m < 4$) we find the TM oxides with the maximum total moment among all oxidized nanoalloys investigated here—that is, those not further enhanced by any mixing with Co or Ni—and also larger than those of any Co–Ni oxide nanoalloy of the same size. However, another general trend is that the magnetic moment of iron oxide clusters is not robust against oxidation.

As compared to iron clusters, those of cobalt approach more the magnetic saturation due to the less d-holes in the active band states, which render them more ferromagnetic-like, although with less spin-polarization than their Fe counterparts. It is for this reason that Co oxide clusters retain, in general, a magnetic moment at oxidation rates at which Fe oxide clusters not, or they have a larger moment despite the less number of d-holes. Examples of this are Co_2O_m with $m = 1,3,4$; Co_3O_m with $m = 3,4$; Co_4O_m with $m = 4$. In particular, ring-like structures with n = 3–4 result in ferromagnets with magnetic moments equal to 9 μ_B and 12 μ_B [39], respectively, whereas their Fe counterparts exhibit an antiferromagnetic coupling [41] with magnetic moments of 4 μ_B and 0 μ_B, respectively. The reentrance or punctual enhancement of the moment at certain high oxidation rates observed in Fe oxide clusters is not found, however, in the Co oxides, at least up to m = 7. It is for this reason that at those particular oxidation rates discussed in the previous paragraph, Fe oxide clusters exhibit a larger magnetic moment than their Co counterparts (Fe_2O_5, Fe_3O_5).

Combining the stronger tendency of Co towards P couplings with the larger spin polarization of Fe is particularly beneficial in the case of dimers (see left panel of Figure 11) where, for $m = 1,2,4$ and 6, $FeCoO_m$ has larger total moment than both Co_2O_m and Fe_2O_m (in these Fe oxides the moment is quenched except for $m = 5$, where Fe–Co mixture show a linear magnetic behavior). This is reflected in the excess magnetic moment which is positive and high in dimers and with maxima in most of those values of m. The moment is also quenched in the cobalt oxide dimer at large oxidation rates (for $m = 5,6$), but Fe–Co mixing leads to magnetic ground states even at those oxidation rates. For $m = 6$, where the magnetic moment of Fe_2O_6 is also quenched, $FeCoO_6$ has a magnetic moment of 3 μ_B, despite the high oxidation rate. Thus, Fe–Co alloying produces a magnetic nanoparticle with magnetic moment close ($m = 3$) to or higher ($m = 1,2,4,5,6$) than that of the cobalt oxide dimers, and therefore, mixing is clearly favorable for the smallest Fe–Co nanoalloys.

For larger sizes, $n = 3$ and $n = 4$ (see left panel of Figures 12 and 13), we find a very well defined trend in the regime of low oxidation rates ($m < 3$ for the trimers and $m < 4$ for the tetramers). In this low oxidation rate, the oxidized nanoalloys have a total magnetic moment that lies in between those of the pure Fe and Co oxide clusters (lower than that of Fe_nO_m and higher than that of Co_nO_m). Besides, the larger the relative concentration of Fe with respect to Co, the larger the moment of the oxidized nanoalloy. This is due to two facts: (i) the magnetic couplings are parallel, like in the non oxidized nanoalloy; (ii) the spin polarization in Fe is larger than in Co due to the larger number of d-holes in the former. Thus, from the magnetic point of view, Fe–Co oxidized nanoalloys outperform Co oxide clusters but not Fe oxide clusters, and since no magnetic quenching takes place, the magnetic excess index is close to zero at low oxidation rates. When $n = m = 3,4$, Co oxides have larger magnetic moments than oxidized mixtures. However, at high oxidation rates is where we find that certain oxidized nanoalloys outperform both the pure oxide clusters—in particular, for $m = 4$, $m = 5$ and $m = 6$ in the case of oxidized trimers, and for $m = 7$ in the case of tetramers with the reentrance of the magnetic moment after its quenching in the intermediate oxidation rates. The magnetic excess index exhibits high positive values at those values of m, reflecting the outperformance of those nanoalloys. It is interesting to analyze this reentrance of magnetism at $m = 7$ in the oxidized tetramers. Figure 15 shows, for this oxidation rate and for oxidized pure Fe and Co clusters and Fe–Co nanoalloys, the total and local magnetic moments in each one of the Fe, Co and O atoms of the corresponding geometrical structure. Both Fe_4O_7 and Co_4O_7 have quenched magnetic moments due to antiparallel couplings. Alloying switches on the reentrance of the magnetic moment, as it changes the magnetic order from antiferromagnetic-like to ferromagnetic-like. Then, once the parallel couplings have been restored, increasing the relative concentration of Fe with respect to Co leads to larger total moments of the oxidized nanoalloy (5 μ_B in $Fe_1Co_3O_7$, 6 μ_B in $Fe_2Co_2O_7$, and 7 μ_B in $Fe_3Co_1O_7$) The ideal, again, is to have as many Fe atoms as possible in the system with P magnetic couplings.

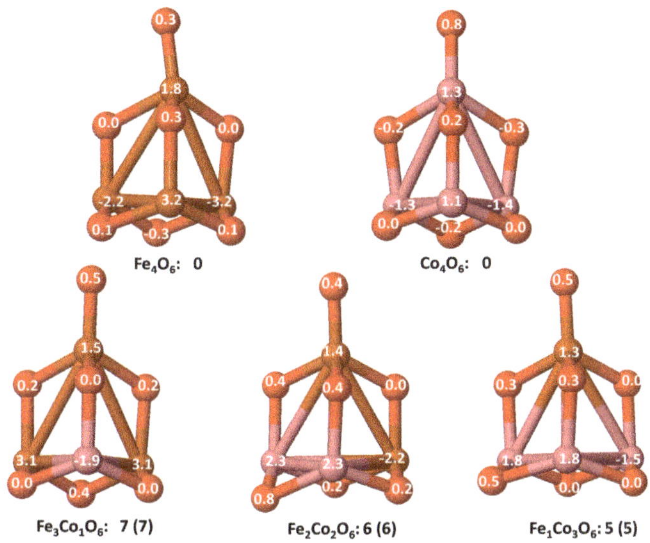

Figure 15. (Color online) Ground states of $Fe_{4-x}Co_xO_6$ (x = 0–4) oxides. Local magnetic moments of Fe, Co and O atoms are indicated. Positive (negative) numbers indicate spin up (down). Below each of the structures, total magnetic moment is given. Numbers between parentheses are the values of the magnetic excess. Both quantities are in μ_B units. Fe atoms in brown color, Co atoms in pink and oxygen atoms in red.

3.2.3. Fe–Ni Oxided Nanoalloys

Ni clusters exhibit magnetic saturation with strong P couplings. Since Ni has even less d-holes than Co in the active valence states, the spin polarization is lower than in Co, which gives rise to smaller local magnetic moments in general. Ni oxide clusters also retain the magnetic moment under oxidation, only being quenched for the largest oxidation rates explored in this work (Ni_3O_6, Ni_4O_7) and in Ni_2O_3, which corresponds to one of the structural anomalies already mentioned in the previous subsection. In fact, this cluster adopts a bent linear structure with exclusively metal–oxygen bonds which imposes indirect Ni–Ni magnetic couplings mediated by O. Nevertheless, the magnetic moments of oxidized Ni dimers exceed or equal those of oxidized Co dimers for $m = 2, 4, 5, 6$, and for larger sizes ($n = 3$ and 4) the same occurs for $m = 5$, and $m = 5, 6$, respectively. The magnetic moment of Ni oxide clusters, although low, can be also qualified as robust against oxidation.

Mixing Fe and Ni has a somewhat similar effect as mixing Fe and Co (right panel of Figures 11–13). In dimers it is beneficial for most oxidation rates, leading to a rather constant moment of 4 μ_B for $m \leq 3$ and $m = 5$, and of 6 μ_B for $m = 6$, values that at least double the moment of the corresponding Ni_2O_m and explain the high positive value of the excess magnetic moment (5 μ_B for $m = 6$). Besides, the magnetic moments of $FeNiO_2$, $FeNiO_5$ and $FeNiO_6$ are 3 μ_B larger than their FeCo counterparts which means that for $m = 2$, $m = 5$ Ni promotes P couplings, but not Co, and for $m = 6$ the promotion of P couplings is larger for Ni. However, for the rest of oxidation rates, $m = 1, 3, 4$, the magnetic moments of FeCo alloys are higher than those of FeNi ones, although $FeNiO_4$ is in a single state associated with the anomalous structure (already mentioned in the previous subsection) in which no direct Fe–Ni bond exists and thus the indirect Fe–Ni magnetic coupling is mediated by O. In trimers, Fe–Ni alloying only increases the moment at the oxygen concentration $m = 4$, and otherwise the pure Fe oxide clusters perform better. The excess magnetic moment reflects this fact with negative values for most oxidation rates and marked positive values (4.6 μ_B and 3.3 μ_B) at $m = 4$ for both Fe–Ni alloy compositions, but they are higher in the case of $Fe_2Ni_1O_4$ for the same reason as in Fe–Co nanoalloys: the higher the Fe content with P magnetic couplings in the system, the higher the total moment. In tetramers, alloying is only beneficial for $m = 4$; at an oxidation rate where the Fe–Co mixture did not increase the magnetic character with respect to the corresponding pure oxides; and particularly for $m = 7$ where the explanation is exactly the same as in the oxidized Fe–Co tetramers at $m = 7$ discussed above and related to the quenching of the moment in both Fe_4O_7 and Ni_4O_7 and the reentrance of magnetic moment promoted by alloying. We note that $m = 7$ is an oxygen rate at which the structures of the pure and binary TM tetramer oxides are similar, but that this is not essential for the alloying to be beneficial. For instance, alloying is beneficial to both Fe–Ni trimer and tetramer nanoalloys at $m = 4$ despite the fact that nanoalloys have a different structural arrangement from that of the pure TM oxides. The magnetic excess reflects the benefit of alloying particularly for those oxygen rates.

It is worth noting that the Fe–Ni mixture with $n = m = 2$ and ring structure has a magnetic moment of 4 μ_B, higher than that of its pure oxides counterparts, leading to a noticeable magnetic excess of 3 μ_B. In fact, this particular Fe–Ni nanoalloy is the most favorable among all the single-compound oxides and nanoalloys. For $n = m = 4$, the Fe–Ni mixtures also present higher magnetic moments (of 4 μ_B at $x = 2$, and 6 μ_B at $x = 3$), with high magnetic excess of 3 μ_B and 4.5 μ_B, respectively.

3.2.4. Co–Ni Oxided Nanoalloys

Finally, in Co–Ni nanoalloys we mixed two TM elements with marked tendencies towards P magnetic couplings. From the point of view of maximizing the total moment, pure Co oxide clusters are better performing than pure Ni oxide clusters for most oxidation rates except at high oxidation rates ($m = 5, 6$ for $n = 2, 4$, and $m = 5$ for $n = 3$).

Regarding Co–Ni oxidized nanoalloys, and in the case of dimers, the magnetic moments are higher than those corresponding to both pure oxides for $m = 2, 6$ and those of Co oxides for $m = 5$. Positive values of excess magnetic moment result in these oxidation rates, and also for $m = 3$, where the Co–Ni mixture substantially improves the magnetic moment of the corresponding pure antiferromagnetic Ni oxide. Nanoalloys have higher magnetic moments than Ni oxides for most m values ($m = 1, 2, 3, 6$). For trimers and tetramers, the nanoalloys present larger magnetic moments than Ni oxides for all oxidation rates. Moreover, Co–Ni nanoalloys for $n = 3$ again present higher magnetic moments than both pure oxides at high oxidation rates, $m = 5, 6$. In addition, the values of excess magnetic moments are high at $m = 3$, particularly for $x = 2$, where the antiferromagnetic character of the Ni_3O_3 ring is lost due to the presence of Co, due to which a net magnetic moment arises. For tetramers, alloying is beneficial at high oxidation rates, giving rise to high magnetic moments for all nanoalloys ($x = 2$), and to high values of excess of magnetism 4–6 μ_B .Therefore, alloying is beneficial for $m = 2, 3, 6$ in dimers, and at high oxidation rates in the larger nanoalloys ($m = 5, 6$ for trimers and $m = 6, 7$ for tetramers). The better performance of the nanoalloys for those oxidation rates is reflected in the magnetic excess, particularly at $m = 6$ in dimers (the moment of the oxide nanoalloy is more than twice that of the pure oxide clusters), and $m = 6, 7$ in tetramers. At $m = 7$, alloying again switches on the reentrance of the magnetic moment in the oxidized tetramers by promoting parallel couplings, like in the other nanoalloys. We note that in $CoNiO_4$ the magnetic excess is negative (-3). The reason is that it is in a low-spin state with 1 μ_B while both its pure counterparts have 4 μ_B. However, there is a nearly degenerated state of $CoNiO_4$ (0.009 eV higher in total energy) in a high-spin state with 5 μ_B and a magnetic excess value of 1 μ_B.

It is noteworthy that Co–Ni alloying has an important effect at the equiatomic composition, $n = m = 2, 3, 4$. These ring structures present an antiferromagnetic character in the cases of pure Fe oxides (0 μ_B, 4 μ_B, 0 μ_B) and pure Ni oxides (2 μ_B), and ferromagnetic behavior for pure Co oxides (0 μ_B, 9 μ_B, 12 μ_B). While Fe–Co alloying only increases the magnetic moment by 1 μ_B in the case of $n = m = 2$, and Fe–Ni alloying promotes a ferromagnetic character in the case of $n = m = 2, 4$, Co–Ni alloys, whose magnetic moment is equal to 3 μ_B for dimers ($n = m = 2$), higher than their pure counterparts, achieve a ferromagnetic character for trimers ($n = m = 3$) and tetramers ($n = m = 4$), resulting in quite high magnetic moments of 8–7 μ_B and 11–9 μ_B, with magnetic excesses of 1.3–2.6 μ_B and 1.5–3–4.5 μ_B, respectively.

4. Conclusions

We have performed an extensive DFT study of the possibility of inducing ferromagnetic-like order to promote a high-spin state in small TMO-NP (TM = Fe, Co, Ni) by their doping with a different TM elements of the list. We investigated all nanoalloys resulting from the different stoichiometries and with different oxidation rates up to saturation. Our calculations were performed using the SIESTA code [44] with the PBE functional of the GGA to treat the exchange and correlation effects [45].

From the energetic point of view, the binding energies indicate that the higher the concentration of Fe in the TMO-NP, the higher their absolute stability. On the other hand, the higher the Ni content, the lower the exothermic character of oxidation, or in other words, the less prone to oxidation the nanoparticle will be. The critical O concentration beyond which molecular adsorption is energetically favorable as compared to atomic adsorption lowers while going from Fe to Co and Ni. Since molecular adsorption is a manifestation of oxygen saturation of the system, those nanoalloys that are rich in the later TM elements admit less oxygen. Moreover, there is an important correlation between binding energies and electronic charge transfer for all mixtures; low binding energies correlate with large electronic charge transfer to oxygen atoms.

From the structural point of view, oxygen atoms tend to occupy bridge sites between the TM atoms followed by top sites, and preferably bind with Fe atoms, followed by Co and finally Ni. The net electronic charge in the O atoms favors their uniform distribution on the nanoparticle. Weakening of the metal–metal binding while increasing the oxidation rate is manifested in a less compact metal skeleton of the nanoparticle. In a particular case, ring-like structures with O-mediated metal–metal bonds are formed when the number of O atoms equals the number of metal ones, except for $Fe_1Ni_3O_4$.

Regarding the magnetic properties, a general trend is that an Fe cluster must reach a critical size and metal–metal coordination to retain the parallel magnetic couplings under oxidation. When this happens, since Fe has the highest spin polarization among the considered TMs, we found the TM oxides with the maximum total moments and the maximum stabilities among all oxidized nanoalloys. Unfortunately, another general trend is that the magnetic moment of these small iron nanoparticles is not robust against oxidation and the moment dramatically drops while increasing the oxidation rate due to antiparallel couplings. We have found that combining the stronger tendency of Co and Ni to parallel couplings and the larger spin polarization of Fe (more d-holes) is particularly beneficial for certain nanoalloys in order to achieve a high total magnetic moment and more robust against oxidation. For instance, the oxidized Fe–Co nanoalloys present magnetic moments that exceed at least that of one of their single-compound counterparts, and those of both in the case of dimers. In addition, the larger the relative concentration of Fe with respect to Co, the larger the moment of the oxidized nanoalloy. On the other hand, Co–Ni nanoalloys wherein two TM elements with marked tendencies towards P magnetic couplings are mixed, are robust magnetic grains, with high moments particularly at the equiatomic metal composition. For larger sizes, at high oxidation rates, certain oxidized nanoalloys also outperform both the pure counterparts. In particular, for $m = 7$ in the case of tetramers, alloying of Fe with Co or Ni even promotes the reentrance of magnetism where the pure oxidized counterparts exhibit quenched magnetic moments. This reentrance is due to a change of magnetic order from antiferromagnetic-like to ferromagnetic-like, and once the parallel couplings have been restored, increasing the relative concentration of Fe with respect to Co or Ni leads to higher total moments and higher stability of the oxidized nanoalloy.

We believe that the results reported here may contribute in a better understanding of the mechanisms through which alloying can improve the magnetic figures of merit in an oxidized TM nanoparticle, and thus to help in the design of magnetic TMO-NP for those applications where superparamagnetism is required.

Supplementary Materials: The following are available online at http://www.mdpi.com/2079-4991/10/9/1814/s1. Figure S1: (Color online) (**a**) Upper panel: average TM–TM distance (green), Co–Co distance (blue) and average Co–Ni distance (red) for $Co_2Ni_1O_m$ oxides as a function of the number of oxygen atoms, m. (**b**) Lower panel: average TM–O distance (green), average Co-O distance (blue),and average Ni-O distance (red) for $Co_2Ni_1O_m$ oxides as a function of the number of oxygen atoms, m. Continuous (dashed) lines correspond to average distances between oxygens on bridge (top) positions and TM atoms. Figure S2: (Color online) Binding energies in eV (**upper panel**) and average electronic charge transfer in *e* (**lower panel**) of $A_{4-x}B_xO_6$ (A, B = Fe, Co, Ni; $x = 0 - 4$) oxides. Blue, green and red curves correspond to Fe–Co, Fe–Ni and Co–Ni nanoalloys, respectively.

Author Contributions: Conceptualization, A.V. and M.B.T.; data curation, M.B.T.; formal analysis: R.H.A.-d.-T. and M.B.T.; funding acquisition: A.V.; investigation, R.H.A.-d.-T., M.B.T., F.A.-G. and A.V.; methodology, M.B.T., R.H.A.-d.-T. and A.V.; project administration: A.V.; supervision, A.V. and M.B.T.; validation, R.H.A.-d.-T., M.B.T., F.A.-G. and A.V.; visualization; R.H.A.-d.-T. and M.B.T.; writing—original draft preparation, A.V.; writing—review and editing, A.V. and M.B.T. All authors have read and agreed to the published version of the manuscript.

Funding: This research was funded by Ministerio de Ciencia, Innovación y Universidades, grant number Project PGC2018-093745-B-I00 and Junta de Castilla y León, Spain, project number VA124G18.

Acknowledgments: R.H.A.T acknowledges financial support from the University of Valladolid (Spain) for the postdoctoral contract (contract number E-47-2019-0197368).

Conflicts of Interest: The authors declare no conflict of interest.

References

1. Lungu, I.I.; Rădulescu, M.; Mogoşanu, G.D.; Grumezescu, A.M. pH sensitive core-shell magnetic nanoparticles for targeted drug delivery in cancer therapy. *Rom. J. Morphol. Embryol.* **2016**, *57*, 23–32. [PubMed]
2. Jadhav, S.; Gaikwad, S.; Nimse, M.; Rajbhoj, A. Copper oxide nanoparticles: Synthesis, characterization and their antibacterial activity. *J. Clust. Sci.* **2011**, *22*, 121–129. [CrossRef]
3. Qiang, Y.; Antony, J.; Sharma, A.; Nutting, J.; Sikes, D.; Meyer, D. Iron/iron oxide core-shell nanoclusters for biomedical applications. *J. Nanopart. Res.* **2006**, *8*, 489–496. [CrossRef]
4. Nie, S.; Xing, Y.; Kim, G.J.; Simons, J.W. Nanotechnology applications in cancer. *Annu. Rev. Biomed. Eng.* **2007**, *9*, 257–288. [CrossRef]
5. Gupta, A.S. Nanotechnology applications in diagnosis and treatment of metastasis. *Nanomedicine* **2014**, *9*, 1517–1529. [CrossRef]
6. Azam, A.; Ahmed, A.S.; Oves, M.; Khan, M.S.; Habib, S.S.; Memic, A. Antimicrobial activity of metal oxide nanoparticles against Gram-positive and Gram-negative bacteria: A comparative study. *Int. J. Nanomed.* **2012**, *7*, 6003. [CrossRef]
7. Laurent, S.; Forge, D.; Port, M.; Roch, A.; Robic, C.; Elst, L.V.; Muller, R.N. Magnetic iron oxide nanoparticles: Synthesis, stabilization, vectorization, physicochemical characterizations, and biological applications. *Chem. Rev.* **2008**, *108*, 2064–2110. [CrossRef]
8. Jones, N.; Ray, B.; Ranjit, K.T.; Manna, A.C. Antibacterial activity of ZnO nanoparticle suspensions on a broad spectrum of microorganisms. *FEMS Microbiol. Lett.* **2008**, *279*, 71–76. [CrossRef]
9. Pankhurst, Q.A.; Connolly, J.; Jones, S.K.; Dobson, J. Applications of magnetic nanoparticles in biomedicine. *J. Phys. D Appl. Phys.* **2003**, *36*, R167. [CrossRef]
10. Wang, H.; Cui, L.-F.; Yang, Y.; Casalongue, H.S.; Robinson, J.T.; Liang, Y.; Cui, Y.; Da, H. Mn_3O_4-graphene hybrid as a high-capacity anode material for lithium ion batteries. *J. Am. Chem. Soc.* **2010**, *132*, 13978–13980. [CrossRef]
11. Koo, B.; Xiong, H.; Slater, M.D.; Prakapenka, V.B.; Balasubramanian, M.; Podsiadlo, P.; Johnson, C.S.; Rajh, T.; Shevchenko, E.V. Hollow iron oxide nanoparticles for application in lithium ion batteries. *Nano Lett.* **2012**, *12*, 2429–2435. [CrossRef] [PubMed]

12. Wu, Z.S.; Ren, W.; Wen, L.; Gao, L.; Zhao, J.; Chen, Z.; Zhou, G.; Li, F.; Cheng, H.M. Graphene anchored with Co_3O_4 nanoparticles as anode of lithium ion batteries with enhanced reversible capacity and cyclic performance. *ACS Nano* **2010**, *4*, 3187–3194. [CrossRef] [PubMed]
13. Stoimenov, P.K.; Klinger, R.L.; Marchin, G.L.; Klabunde, K.J. Metal oxide nanoparticles as bactericidal agents. *Langmuir* **2002**, *18*, 6679–6686. [CrossRef]
14. Tran, M.; Mir, A.; Mallik, D.; Sinha, A.; Nayar, S.; Webster, T.J. Bactericidal effect of iron oxide nanoparticles on Staphylococcus aureus. *Int. J. Nanomed.* **2010**, *5*, 277–283.
15. Sondi, I.; Salopek-Sondi, B. Silver nanoparticles as antimicrobial agent: A case study on *E. coli* as a model for Gram-negative bacteria. *J. Colloid Interface Sci.* **2004**, *275*, 177–182. [CrossRef]
16. Zhang, S.; Zhao, X.; Niu, H.; Shi, Y.; Cai, Y.; Jiang, G. Superparamagnetic Fe_3O_4 nanoparticles as catalysts for the catalytic oxidation of phenolic and aniline compounds. *J. Hazard. Mater.* **2009**, *167*, 560–566. [CrossRef]
17. Jiao, F.; Frei, H. Nanostructured cobalt oxide clusters in mesoporous silica as efficient oxygen evolving catalysts. *Angew. Chem.* **2009**, *121*, 1873–1876. [CrossRef]
18. Teja, A.S.; Koh, P.-Y. Synthesis, properties, and applications of magnetic iron oxide nanoparticles. *Prog. Cryst. Growth Charact. Mater.* **2009**, *55*, 22–45. [CrossRef]
19. Carnes, C.L.; Klabunde, K.J. The catalytic methanol synthesis over nanoparticle metal oxide catalysts. *J. Mol. Catal. A Chem.* **2003**,*194*, 227–236. [CrossRef]
20. Lopez, N.; Norskov, J.K. Catalytic CO oxidation by a gold nanoparticle: A density functional study. *J. Am. Chem. Soc.* **2002**, *124*, 11262–11263. [CrossRef]
21. Johnson, G.E.; Reilly, N.M.; Castleman, A.W., Jr. Effect of charge state and stoichiometry on the structure and reactivity of nickel oxide clusters with CO. *Int. J. Mass Spectrom.* **2009**, *208*, 93–100. [CrossRef]
22. Wan, J.; Yuan, R.; Zhang, C.; Wu, N.; Yan, F.; Yu, S.; Chen, K. Potential application of metal dichalcogenides double-layered heterostructures as anode materials for Li-ion batteries. *J. Phys. Chem. C* **2016**, *120*, 23799–23806. [CrossRef]
23. Casula, M.F.; Conca, E.; Bakaimi, I.; Sathya, A.; Materia, M.E.; Casu, A.; Falqui, A.; Sogne, E.; Pellegrino, T.; Kanaras, A.G. Manganese doped-iron oxide nanoparticle clusters and their potential as agents for magnetic resonance imaging and hyperthermia. *Phys. Chem. Chem. Phys.* **2016**, *18*, 16848–16855 . [CrossRef] [PubMed]
24. Szczerba, W.; Zukrowski, J.; Przybylski, M.; Sikora, M.; Safonova, O.; Shmeliov, A.; Nicolosi, V.; Schneider, M.; Granath, T.; Oppmann, M.; et al. Pushing up the magnetisation values for iron oxide nanoparticles via zinc doping: X-ray studies on the particle's sub-nano structure of different synthesis routes. *Phys. Chem. Chem. Phys.* **2016**, *18*, 25221–25229. [CrossRef]
25. Wang, Y.; Chen, Q.; Wang, J. Ab initio study of structure and magnetism of late transition metal oxide TM_nO_m (TM = Fe, Co, Ni, n = 1, 2, m = 1–6). *J. Nanosci. Nanotechnol.* **2012**, *12*, 6488–6493. [CrossRef] [PubMed]
26. Erlebach, A.; Hunhn, C.; Jana, R.; Sierka, M. Structure and magnetic properties of $(Fe_2O_3)_n$ clusters ($n =$ 1–5). *Phys. Chem. Chem. Phys.* **2014**, *16*, 26421–26426. [CrossRef] [PubMed]
27. Reilly, N.M.; Reveles, J.U.; Johnson, G.E.; Khanna, S.N.; Castelman, A.W. Experimental and theoretical study of the structure and reactivity of $Fe_{1-2}O^-_{\leq 6}$ clusters with CO. *J. Phys. Chem. A* **2007**, *111*, 4158–4166. [CrossRef]
28. Reilly, N.M.; Reveles, J.U.; Johnson, G.E.; del Campo, J.M.; Khanna, S.N.; Köster, A.M.; Castleman, A.W. Experimental and theoretical study of the structure and reactivity of $Fe_mO_n{}^+$ ($m = 1{,}2$; $n =$ 1–5) with CO. *J. Phys. Chem. C* **2007**, *111*, 19086–19097. [CrossRef]
29. Ota, K.; Koyasu, K.; Ohshimo, K.; Misaizu, F. Structures of cobalt oxide cluster cations studied by ion mobility mass spectrometry. *Chem. Phys. Lett.* **2013**, *588*, 63–67. [CrossRef]
30. Tung, N.T.; Tam, N.M.; Nguyen, M.T.; Lievens, P.; Janssens, E. Influence of Cr doping on the stability and structure of small cobalt oxide clusters. *J. Chem. Phys.* **2014**, *141*, 044311. [CrossRef]
31. Johnson, G.E.; Reveles, J.U.; Reilly, N.M.; Tyo, E.C.; Khanna, S.N.; Castelman, A.W., Jr. Influence of stoichiometry and charge state on the structure and reactivity of cobalt oxide clusters with CO. *J. Phys. Chem. A* **2008**, *112*, 11330–11340. [CrossRef]
32. Kumavat, S.; Deshpande, M. Alkali metal doped nickel oxide clusters: A density functional study. *Comput. Theor. Chem.* **2014**, *1035*, 19–27. [CrossRef]
33. Yin, S.; Xue, W.; Ding, X.-L.; Wang, W.-G.; He, S.-G.; Ge, M.-F. Formation, distribution, and structures of oxygen-rich iron and cobalt oxide clusters. *Int. J. Mass Spectrom.* **2009**, *281*, 72–78. [CrossRef]

34. Dible, C.J.; Akin, S.T.; Ard, S.; Fowler, C.P.; Duncan, M.A. Photodissociation of cobalt and nickel oxide cluster cations. *J. Phys. Chem. A* **2012**, *116*, 5398–5404. [CrossRef]
35. Kirilyuk, A.; Fielicke, A.; Demyk, K.; von Helden, G.; Meijer, G.; Rasing, T.H. Ferrimagnetic cagelike Fe_4O_6 cluster: Structure determination from infrared dissociation spectroscopy. *Phys. Rev. B* **2010**, *82*, 020405(R). [CrossRef]
36. Li, S.; Zhai, H.-J.; Wang, L.-S.; Dixon, D.A. Structural and electronic properties of reduced transition metal oxide clusters, M_4O_{10} and $M_4O_{10}^-$ (M = Cr, W), from photoelectron spectroscopy and quantum chemical calculations. *J. Phys. Chem. A* **2012**, *116*, 5256–5271. [CrossRef]
37. Wang, H.-Q.; Li, H.-F. Probing the structural and electronic properties of small vanadium dioxide clusters by density functional theory and comparison with experimental photoelectron spectroscopy. *J. Chem. Phys.* **2012**, *137*, 164304. [CrossRef] [PubMed]
38. Ohshimo, K.; Komukai, T.; Moriyama, R.; Misaizu, F. Isomer separation of iron oxide cluster cations by ion mobility mass spectrometry. *J. Phys. Chem. A* **2014**, *118*, 3899–3905 . [CrossRef]
39. Aguilera-del-Toro, R.H.; Aguilera-Granja, F.; Vega, A.; Balbás, L.C. Structure, fragmentation patterns, and magnetic properties of small cobalt oxide clusters. *Phys. Chem. Chem. Phys.* **2014**, *16*, 21732–21741. [CrossRef]
40. Aguilera-del-Toro, R.H.; Aguilera-Granja, F.; Balbás, L.C. Structure, fragmentation patterns, and magnetic properties of small nickel oxide clusters. *Phys. Chem. Chem. Phys.* **2017**, *19*, 3366–3383. [CrossRef]
41. Aguilera-del-Toro, R.H.; Aguilera-Granja, F.; Torres, M.B.; Vega, A. Relation between structural patterns and magnetism in small iron oxide clusters; reentrance of magnetic moment at hich oxidation rates. to be published.
42. Torres, M.B.; Aguado, A.; Aguilera-Granja, F.; Vega, A.; Balbás, L.C. Structural, vibrational, and magnetic properties of $FeCoO_n^{0/+}$ (n= 1–6) bimetallic oxide clusters. *Phys. Chem. C* **2015**, *119*, 11200–11209. [CrossRef]
43. Borzorth, R.M. *Ferromagnetism*; Wiley-IEEE Press: Piscataway, NJ, USA, 1993; ISBN 978-0-780-31032-2.
44. Soler, J.M.; Artacho, E.; Gale, J.D.; García, A.; Junquera, J.; Ordejón, P.; Sánchez-Portal, D. The SIESTA method for ab initio order-N materials simulation. *J. Phys. Condens. Matter* **2002**, *14*, 2475. [CrossRef]
45. Perdew, J.P.; Burke, K.; Ernzerhof, M. Generalized gradient approximation made simple. *Phys. Rev. Lett.* **1996**, *77*, 3865. [CrossRef] [PubMed]
46. Troullier, N.; Martins, J.L.Efficient pseudopotentials for plane-wave calculations. *Phys. Rev. B* **1991**, *43*, 1993. [CrossRef] [PubMed]
47. Kleinman, L.; Bylander, D.M. Efficacious form for model pseudopotentials. *Phys. Rev. Lett.* **1982**, *48*, 1425. [CrossRef]
48. Louie, S.G.; Froyen, S.; Cohen, M.L. Nonlinear ionic pseudopotentials in spin-density-functional calculations. *Phys. Rev. B* **1982**, *26*, 1738. [CrossRef]
49. Pauling, L. *The Nature of the Chemical Bond*; Cornell University Press: Ithaca, NY, USA, 1960.
50. Aguilera-del-Toro, R.H.; Aguilera-Granja, F.; Balbás, L.C.; Vega, A. Synthesis and characterization of indium oxide nanoparticles. *Theor. Chem. Acc.* **2018**, *137*, 54. [CrossRef]
51. Ohshimo, K.; Azuma, A.; Komukai, T.; Moriyama, R.; Misaizu, F. Structures and CO-adsorption reactivities of nickel oxide cluster cations studied by ion mobility mass spectrometry. *J. Phys. Chem. C* **2015**, *119*, 11014–11021. [CrossRef]
52. Fritsch, D.; Koepernik, K.; Richter, M.; Eschrig, H. Transition metal dimers as potential molecular magnets: A challenge to computational chemistry. *J. Comput. Chem.* **2007**, *29*, 2210–2219. [CrossRef]
53. Błoński, P.; Hafner, J. Magnetic anisotropy of transition-metal dimers: Density functional calculations. *Phys. Rev. B* **2009**, *79*, 224418. [CrossRef]
54. Błoński, P.; Hafner, J. Magneto-structural properties and magnetic anisotropy of small transition-metal clusters: A first-principles study. *J. Phys. Condens. Matter* **2011**, *23*, 136001. [CrossRef]
55. Fortunelli, A.; Velasco, A.M. Structural and electronic properties of Pt/Fe nanoclusters from EHT calculations. *J. Mol. Struct. Theochem* **1999**, *487*, 251–266. [CrossRef]
56. Aguado, A.; López, J.M. Identifying structural and energetic trends in isovalent core-shell nanoalloys as a function of composition and size mismatch. *J. Chem. Phys.* **2011**, *135*, 134305. [CrossRef]

nanomaterials

MDPI

Article

Structures and Properties of the Self-Assembling Diphenylalanine Peptide Nanotubes Containing Water Molecules: Modeling and Data Analysis

Vladimir Bystrov [1,*], Jose Coutinho [2], Pavel Zelenovskiy [3,4], Alla Nuraeva [3], Svitlana Kopyl [5], Olga Zhulyabina [6] and Vsevolod Tverdislov [6]

1 Institute of Mathematical Problems of Biology, Keldysh Institute of Applied Mathematics, RAS, Pushchino, Moscow 142290, Russia

2 Department of Physics & I3N, University of Aveiro, Campus Santiago, 3810-193 Aveiro, Portugal; jose.coutinho@ua.pt

3 School of Natural Sciences and Mathematics, Ural Federal University, Ekaterinburg 620000, Russia; zelenovskiy@urfu.ru (P.Z.); alla.nuraeva@urfu.ru (A.N.)

4 Department of Chemistry & CICECO-Aveiro Institute of Materials, University of Aveiro, 3810-193 Aveiro, Portugal

5 Department of Physics & CICECO-Aveiro Institute of Materials, University of Aveiro, 3810-193 Aveiro, Portugal; svitlanakopyl@ua.pt

6 Faculty of Physics, Lomonosov Moscow State University, Moscow 119991, Russia; zhulyabina.o@yandex.ru (O.Z.); tverdislov@mail.ru (V.T.)

* Correspondence: vsbys@mail.ru

Received: 11 August 2020; Accepted: 27 September 2020; Published: 10 October 2020

Abstract: The structures and properties of the diphenylalanine (FF) peptide nanotubes (PNTs), both L-chiral and D-chiral (L-FF and D-FF) and empty and filled with water/ice clusters, are presented and analyzed. DFT (VASP) and semi-empirical calculations (HyperChem) to study these structural and physical properties of PNTs (including ferroelectric) were used. The results obtained show that after optimization the dipole moment and polarization of both chiral type L-FF and D-FF PNT and embedded water/ice cluster are enhanced; the water/ice cluster acquire the helix-like structure similar as L-FF and D-FF PNT. Ferroelectric properties of tubular water/ice helix-like cluster, obtained after optimization inside L-FF and D-FF PNT, as well of the total L-FF and D-FF PNT with embedded water/ice cluster, are discussed.

Keywords: diphenylalanine; peptide nanotubes; self-assembly; water molecules; DFT; molecular modelling; semi-empirical methods; polarization; chirality

1. Introduction

Self-assembly of complex molecular structures based on various amino acids (AAs) is one of the most important phenomena both in living nature and in nanomaterials development [1–4]. At the same time, it is now known that the chirality of the initial molecules plays an important role in self-assembly processes [5–7]. All this is important both for our understanding of the basic principles of the emergence of life, and for numerous practical applications [5–7], including the new nanomaterials synthesis for nanoelectronics [3,8,9] and biomaterials for usage in biomedicine [10,11], including targeted drug delivery [12,13].

One necessary research approach is computer molecular modeling of the processes of self-organization of molecular systems at different levels and by different methods [14–16]. All AAs have their own dipole moments [17], which interact with each other and self-organize into more complex molecular and crystalline structures, such as peptide nanotubes (PNTs) and similar nanostructures [18].

As results, many of these structures have piezoelectric and ferroelectric properties [19–24]. This has been shown and investigated in detail in our previous works [24–30]. Self-assembly of such PNTs occurs in aqueous media rather quickly and under certain conditions that affect the rate of their growth and the shape of self-organizing structures [31,32]. It is important that their structural and physical properties turn out to be dependent on the chirality of the original molecules of amino acids and dipeptides [29–32]. In many cases, water molecules also appear in the internal hydrophilic cavity of such PNTs [29–36]. They affect the physical properties of PNT and largely determine their changes. However, the experimental detection of water molecules by X-ray diffraction methods is very difficult. In this case, it is computer simulation methods that can help: they play an important role for a clearer identification of these water structures [37] and their study, establishing their structural and physical properties and their effect on the properties of PNT as a whole. In principal, computer molecular modeling allows us to calculate, investigate, and predict the basic physical properties of these nanostructures based on any various AAs, with and without including water molecules.

In this paper, we continue our further study of the structural and physical (including polar and ferroelectric-like) properties of PNTs based on diphenylalanine (FF) (FF PNTs) of different chirality—L-chiral (L-FF) and D-chiral (D-FF)—and focus on the embedded molecular clusters of water in the PNTs' internal hydrophilic cavity. An optimum possible number of water molecules per one unit cell of the D-FF and L-FF hexagonal crystal structures was determined. These structural data and water molecules are also considered in molecular models for D-FF and L-FF PNTs, which have at least two coils of the PNT helix structure [25,28,29] that correspond to the period of the hexagonal unit cell along the c axis. The influences of these water molecules on PNT properties are analyzed, including a change in the dipole moments and polarization of the PNTs, as well as a change in the structure and properties (dipole moment and polarization) of water clusters embedded in a PNT cavity for both chirality types.

In all calculations, the quantum semi-empirical AM1, PM3, RM1 methods in the Hartee-Fock approximations (from the HyperChem package [38]) are used. The initial structural data of D-FF and L-FF from the crystallographic database [39] were taken, and for their calculations and optimization, the density functional theory (DFT) methods (in Vienna Ab initio Simulation Package (VASP) program [40]) are applied, taking into account the Van der Waals interactions (VdW correction by "PBE + D3" method was used, available in VASP). All results obtained in comparison with known and experimental data are analyzed.

2. Models and Computational Details

2.1. Main Methods and Software

The calculations were carried out using DFT methods, as implemented by VASP [40–43]. The exchange-correlation potential was evaluated using the generalized gradient approximation (GGA) according to Perdew, Burke, and Ernzerhof (PBE) functional [43,44]. Core states were described by means of the projector augmented wave method [45], while the Kohn-Sham task (for calculating the energies of the ground states of systems) was calculated using plane waves with kinetic energy up to Ecut = 400 eV to expand the wave functions.

In the calculations of self-assembled systems based on amino acids, it is also necessary to take into account the Van der Waals (VdW) interactions. A useful pragmatic method to work around this problem is to add a correction to the conventional Kohn-Sham DFT energy. Here, we used the "PBE + D3" method for VdW correction developed by Grimme S., Antony J., Ehrlich S., and Krieg S. [46] and named D3. This method is compatible with PBE and is implemented in VASP (PBE + D3).

In this paper, further study of diphenylalanine peptide nanotubes is based on the models constructed from experimental data of their crystallographic structures obtained by X-ray methods and recorded in the crystallographic database of the Cambridge Crystallographic Data Center (CCDC) [39]. (These data correspond to No. CCDC 16337 for L-FF [33] and No. CCDC 1853771 for D-FF [25,29]). These structural data and visual models of both PNT L-FF and D-FF structures are easily reproduced in accordance with their periodic crystallographic cell parameters (see Table 1) in various software systems (such as "Mercury"—https://www.ccdc.cam.ac.uk/solutions/csd-system/components/mercury/— a system for visualization and analysis of crystallographic structures compatible with the CCDC base). Using all these data, the source files were built up for modeling and calculations based on DFT methods in the VASP program [40] (see Figure 1).

Table 1. Lattice cell parameters for L-FF and D-FF PNT (from references [25,29,33] according to CCDC [39]).

	L-FF	D-FF
Space Group	**P6$_1$**	**P6$_5$**
a, Å	24.0709 (13)	23.9468 (14)
b, Å	24.0709 (13)	23.9468 (14)
c, Å	5.4560 (4)	5.4411 (2)
V, Å^3	2737.7 (3)	2702.2 (2)

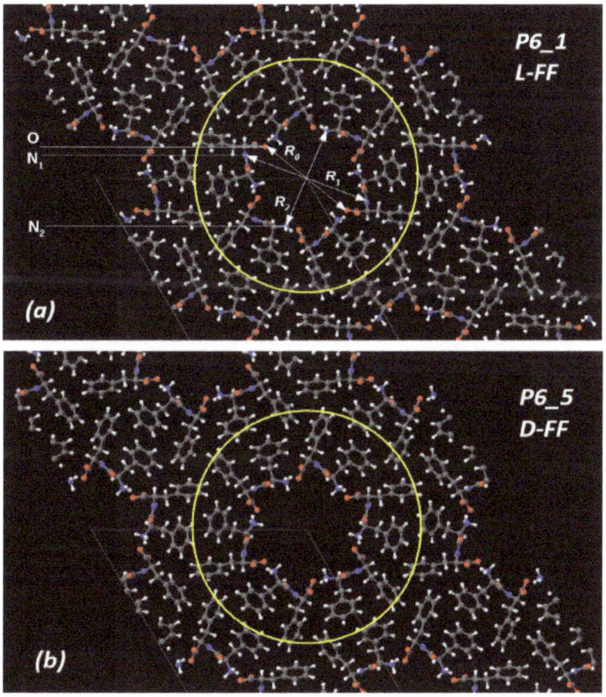

Figure 1. Images of structures based on the Cambridge Crystallographic Data Center (CCDC) for two different diphenylalanine (FF) enantiomers with symmetry elements (obtained using Jmol after performing Vienna Ab initio Simulation Package (VASP) calculations) for cases: (**a**) L-chiral (L-FF) with space group P6$_1$ and (**b**) D-chiral (D-FF) with space groups P6$_5$. Hexagonal cells are marked by thin lines. Selected atoms and molecules for the formation of peptide nanotubes (PNTs) are shown by a circle in yellow lines. Atoms are marked with colors: oxygen—red, nitrogen—blue, carbon—gray, hydrogen—white. The upper figure (**a**) shows the distances in the inner cavity: R$_0$—between the oxygen atoms O ... O; R$_1$—between the nitrogen atoms N$_1$...N$_1$ (larger "far" diameter PNT); R$_2$—between the nitrogen atoms N$_2$...N$_2$ of the opposing NH_3^+ groups (short "near" inner diameter PNT).

Then, these structures were also converted and transferred into files in the HyperChem [38] workspace for their further analysis and calculations of their polar properties with various quantum-mechanical semi-empirical methods.

2.2. *Models of Initial Water-Free Crystal Structures*

The initial and water-free crystal structures of both L-FF and D-FF PNT types are shown in Figure 1. The anhydrous and non-centrosymmetric hexagonal unit cell for both enantiomers crystal structures contains six FF molecules, and it is formed by a total of 258 atoms, while the space groups of these enantiomers are different: $P6_1$ space group for L-FF and $P6_5$ for D-FF. These initial crystal configurations were obtained from X-ray experimental Cambridge Crystallographic Data Center (CCDC) data [39]. In all cases, the Brillouin zone (BZ) was sampled using a Monkhorst-Pack [47] scheme with a $1 \times 1 \times 3$ mesh k-point sampling. The Hartree-Fock exact exchange was evaluated using the same k-point grid computed for the DFT potential. This enables a real-space grid of $120 \times 120 \times 28$ points to keep along a1, a2, and a3 lattice vectors, respectively, which corresponds to the experimental lattice constants a1 = a2 = a = b ~24 Å and a3 = c = ~5.44 Å of the D-FF crystal structures data (Table 1). The relevant grid density appears to be about 5 points/Å along all the three directions.

The relaxation (optimization of the total energy) of both initial structures was carried out, and the same procedure was also performed for all cases of simulated structures with different numbers of water molecules in the cavity of the inner channel of the nanotubes. The relaxation cycle was stopped when the maximum force acting on lattice vectors and ions became less than 10 mV/Å. The main method used here for finding the minimum total energy in VASP [40] is the conjugate gradient algorithm; but in some cases, we used a different algorithm for the case close to a local minimum. For better optimization, in some cases, we varied the maximum acting force limit.

2.3. *Model of Water/Ice Clusters*

The influence of water molecules on D-FF and L-FF PNT properties was studied with the use of the hexagonal ***Ih*** ice cluster model [48], which served as the basis for the construction of the initial model of a water cluster with different numbers n of water molecules (in the case of small n*(H_2O) water clusters with n = 2, 4, 6, etc., only a short part of such ***Ih*** ice cluster was used)). The model clusters constructed were introduced into the cavity of the initial anhydrous nanotubes, as was done in our previous recent work [26]. Then, the whole D-FF and L-FF PNTs structures, filled with this embedded water/ice cluster with n water molecules, was optimized, keeping the lattice parameters a, b, c, of the initial nanotubes constant (Table 1). This is necessary to obtain a correlation with the initial experimental data. The obtained optimized structures with different numbers of water molecules in their cavity were collected and stored for further analysis of their parameters and visual control using different methods.

2.4. *Estimation of Interaction Energy of Water/Ice Cluster and PNT*

To estimate the energy of water molecule interaction E_i with PNT for each optimized structure with different quantities of water molecules n, we calculated a change in the total energy as the number of water molecules (average energy per FF unit cell) as a function of the number of water molecules increased:

$$E_i = E_{tot} - E_{PNT} - n \cdot E(H_2O) \quad (1)$$

where E_{tot} is the total energy per unit cell for the optimized PNT structure with water molecules, E_{PNT} is the energy of the optimized PNT structure without water molecules, $E(H_2O)$ is the energy of a relaxed single water molecule H_2O, and n is the number of H_2O molecules used in the calculation. This approach is similar in general to that of Ref. [36] (but differs in some computational details). The E_i obtained here is the average interaction (or binding) energy that provides information concerning the interaction between water molecules and their surroundings, i.e., the sum interaction between water molecules and the inner wall of the nanotube via hydrogen bonds (HBs) and the intensity of the intermolecular HB interactions formed between water molecules [49–51]. Naturally, with an increase in the number of water molecules in the inner cavity of the nanotube, this energy also changes.

To determine changes in the main distances (R_1, R_2) inside the PNT cavity after relaxation and optimization of all the structures, we use the *Jmol* software tool for visual presentation of all the structures extracted after calculations. For extraction of all atomic files and transformation of their formats, the *OpenBabel* and *Cyberduck* software tools were used too.

2.5. Semi-Empirical Calculations

To obtain the values of the energy, dipole moment, and polarization of the optimized D-FF and L-FF structures with and without water and individual extracted water clusters, the HyperChem package [38] was used with various quantum-mechanical semi-empirical AM1 (Austin Model 1, developed by Dewar M.J.S., et al. [52,53]), PM3 (Parametrization Method No. 3, is a reparameterization of AM1, developed by Stewart J.J.P., et al. [53–57]), RM1 (Recife Model 1, is a reparameterization of AM1, developed by Rocha G.B., et al. [58,59]) methods (the restricted Hartree-Fock (RHF) approximation was used). For this purpose, the optimization of 21 water molecules embedded in D-FF and L-FF cavity with fixed unit cell parameters (corresponding to the experimental data [25,29,33,39]) was performed. Only these water molecules were optimized, whereas atomic positions for all other atoms of FF molecules were kept "frozen". The optimized water cluster structures were extracted from the D-FF and L-FF cavity to be investigated independently using *OpenBabel*, *Cuberduck*, and HyperChem software [38]. Scheme of the main steps of these procedures are presented in Figure 4 (see below and more details in the next section).

Thus, the optimized structures were transformed from periodical crystal structures in VASP to HyperChem workspace as one, two, and more coil helix molecular structures, where all further calculations were carried out.

3. Results and Discussions

3.1. Determination of the Optimal Number of Water Molecules in the PNT Cavity

To investigate the presence of water molecules and find their optimal number, the PNT structures (both D-FF and L-FF) with different numbers of water molecules were calculated and optimized. As result, optimized PNT models containing different number of water molecules in the inner cavity of the PNTs were obtained, and the dependence of PNT properties on the number of water molecules was studied. Note that optimizing L-FF PNTs with water is concerned with difficulties since these structures are less stable as compared to D-FF PNTs. This also agrees with previously obtained data: D-FF nanotubes have a denser and stronger deep packing than L-FF PNTs [25,28,29]; L-FF PNTs have larger cavity sizes, with looser and less uniform surface of the internal cavity compared to D-FF PNTs [28,30]. It was necessary to vary some parameters and methods of the optimization procedure in VASP to obtain the most stable and suitable optimized structure. These calculations required more time to proceed on a computer cluster. Nevertheless, it was possible to obtain good optimized structures for both D-FF and L-FF and choose the better ones for further analysis.

As a result, the dependence of the average interaction energy E_i (1) per one unit cell FF as a function of the number of water molecules n were obtained for both types of structures L-FF and D-FF PNTs, convincingly showing that the minimum of this energy E_i is observed for $n = 21$ (Figure 2). The peculiarity of the behavior of these dependence (for both chirality type L-FF and D-FF), the values and position of the minimum of energies turned out to be similar to the data [36] and are comparable in magnitudes: the value of the energy is in order of ~−14 eV in the minimum position, corresponding for 21 H_2O, which is the same as in our case.

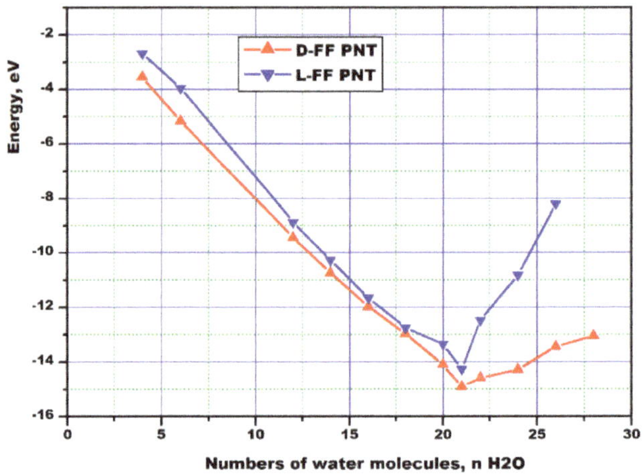

Figure 2. Average interaction energy E_i between water and PNT as a function of number of water molecules confined in the hydrophilic channel of FF PNTs for each chirality types: L-FF and D-FF.

The calculated water-PNT interaction energy E_i is the energy of an interaction between the water molecules and the inner wall of the nanotube through the hydrogen bonds (HBs) [49,50]. The formation of a network of HBs between water molecules and the nearest hydrophilic oxygen and nitrogen atoms at the inner surface of the nanotube cavity was found for all optimized D-FF and L-FF structures (for more details, see discussion below in Section 3.2).

These results are also confirmed by the calculated dependence of the internal diameters of the nanotube cavity, determined by the distances between the main nitrogen atoms N_1, N_2: R_1 (N_1-N_1) and R_2 (N_2-N_2) (these distances were determined in Figure 1a). As the number of water molecules n increases, the size of the inner PNT cavity changes for both types of chirality L-FF and D-FF. Comparing calculated and experimental values of R_1 and R_2, we can see that, when the number of molecules is equal to $n = 21$, the calculated curves intersect with the experimental values, which are according to X-ray structural data for the initial nanotubes (see data in Table 2), for both chirality type L-FF and D-FF PNTs.

Table 2. Parameters of inner hydrophilic cavity of L-FF and D-FF PNTs.

Parameter	L-FF		D-FF	
	Initial	Opt (No Water)	Initial	Opt (No Water)
a, Å	24.0709	23.8308 (284)	23.9468	23.7877 (806)
b, Å	24.0709	23.8308 (284)	23.9468	23.7877 (806)
c, Å	5.456	5.4035 (861)	5.4411	5.4022 (7125)
R_0, Å	12.236	12.091	12.102	12.075
R_1, Å	15.271(698)	15.042 (076)	15.180 (569)	15.030 (688)
R_2, Å	12.218(349)	12.098 (817)	12.135 (396)	12.075 (906)
Etot, eV	−1593.318267	−1657.643468	−1608.735638	−1657.600241

These graphs (Figure 3a,b) show that $n = 21$ and the obtained curves of the dependencies of the internal dimensions of the cavities of the optimized structures of the nanotubes of both chiral types coincide in very close sizes variations with the experimental data.

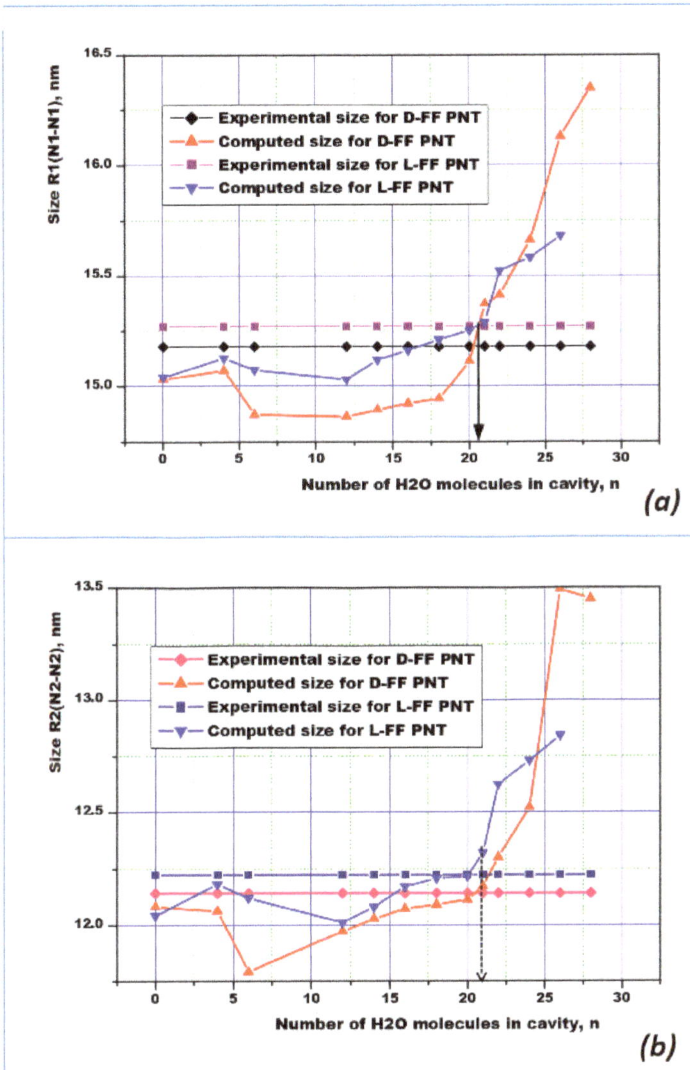

Figure 3. Dependences of the internal cavity sizes of the optimized structures of nanotubes of both types of chirality L-FF and D-FF containing a different number of water molecules on the number n of these water molecules, in comparison with the known sizes of the internal sizes of nanotubes according to experimental data (Table 2) [25,29,33,39]: (**a**) R_1—between the nitrogen atoms $N_1 ... N_1$ (larger "distant" diameter PNT); (**b**) R_2—between the nitrogen atoms $N_2 ... N_2$ of the opposite $NH3^+$ groups (short "near" inner diameter PNT).

As a result, it can be argued that in the L-FF and D-FF cavities of both types of chirality there are about $n = 21$ water molecules per unit cell of these FF periodic molecular crystal structures.

Preliminary analysis showed that, in all cases, water-ice clusters after optimization in the internal cavity of both types L-FF and D-FF PNT change their structure and properties. However, the case with the found amount of 21 water molecules is of greater interest, which corresponds to the minimum total energy of both types of chirality. Extraction of the optimized water cluster (with 21 H_2O molecules) from the inner cavity shows that this cluster has changed as compared to the initial hexagonal ice structure and acquired a helix-like structure, close to the helix, which is typical for L-FF and D-FF PNTs per se (see details below in Section 3.3). Moreover, a water cluster splits into inner and outer parts, and the latter actively forms hydrogen bonds with the atoms of nitrogen and oxygen of the inner surface of the PNT cavity. All these data need more deep and detailed analysis.

3.2. Water Cluster Structures Details

Let us consider the structures of water clusters obtained with 21 H_2O molecules per a unit cell in greater detail. In this case, the initial water cluster based on the hexagonal ***Ih*** ice cluster models and consisting from 21 H_2O molecules per unit cell is presented on Figure 4 for the case of D-FF PNT. A similar initial structure is also used and for L-FF PNT. During the optimization process (using the VASP program, as described above), the structure of water clusters changes—there is a displacement of water molecules inside the cavity under the influence of an electric field inside the cavity (arisen from fixed FF dipoles, which create total strong polarization here and, accordingly, a strong electric field appears along the axis of the nanotube [24–29]).

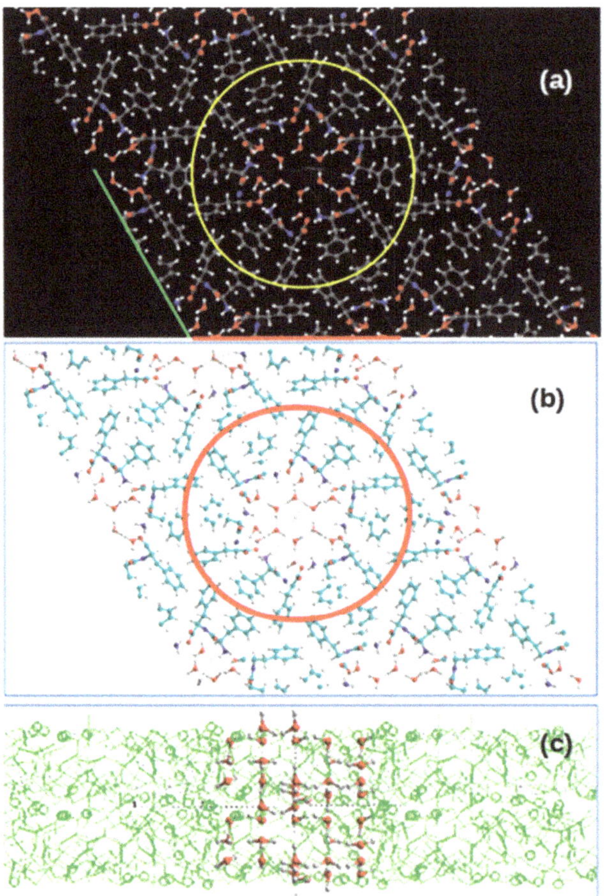

Figure 4. Schematic images of the initial water/ice cluster embedded into the inner hydrophilic cavity of the D-FF PNT: (**a**) top Z-projection images consisting from four unit cells of D-FF crystal structure obtained using *Jmol* from VASP initial data before optimization calculations (green and red lines marked

the unit cell; yellow circle shows the selected atoms and molecules that form the PNT with cavity in the center filled with water/ice cluster molecules; the colors of atoms are the same as in Figure 1); (**b**) the same Z-projection images converted from VASP to HyperChem workspace using *Cyberduck* and *OpenBabel* software (red circle shows the same selected atoms and molecules that form the PNT with cavity in the center filled with water cluster molecules; the colours of atoms here are—Carbon is Cyan, Red is Oxygen, White (or Gray) is Hydrogen); (**c**) side Y-projection and cross-section of D-FF structures with selected water molecules formed initial water cluster.

As noted above, we optimize only the structure of the water cluster inside a PNT cavity, keeping fixed ("frozen") atomic positions of all the atoms of all the FF molecules and the unit cell parameters of the PNT crystal structures. In this case, a rearrangement of hydrogen bonds occurs both between water molecules and between water molecules and FF molecules (in particular, with nitrogen and oxygen atoms of FF molecules on the inner hydrophilic surface of the nanotube's cavity). This happens both in D-FF PNTs and in L-FF PNTs, but in different ways in accordance with their different internal structure and chirality. As a result, we get two altered and significantly different structures of water clusters after their optimization in D-FF and L-FF chirality cases (Figures 5 and 6).

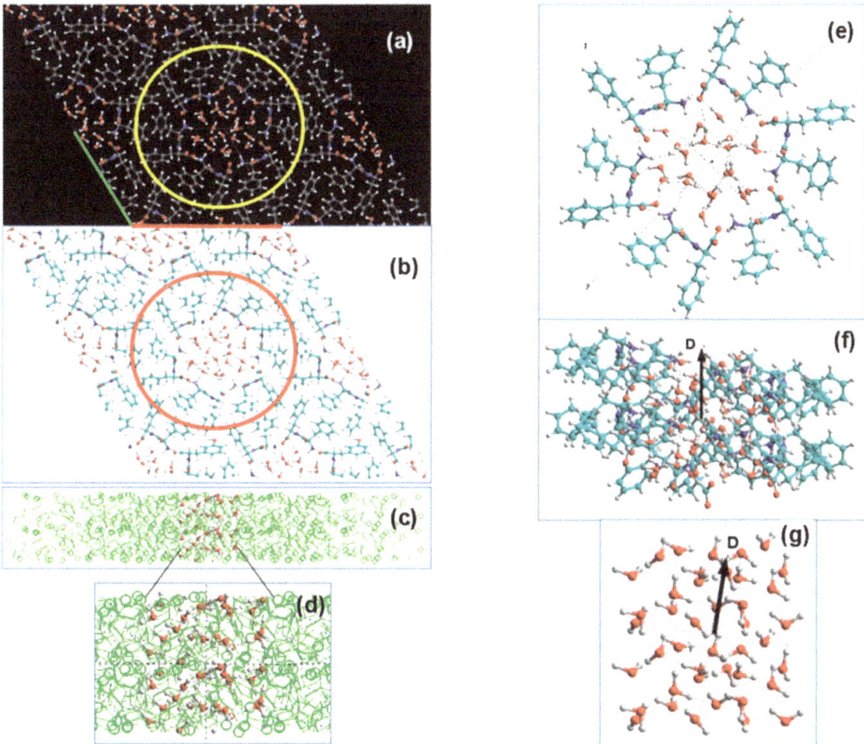

Figure 5. Schematics of the D-FF PNT structures with optimized 21 H_2O per unit cell embedded water cluster (the designations used are the same as in Figure 4): (**a**) Z-projection image from VASP computed data with four unit cells; (**b**) the same image converted into HyperChem workspace; (**c**) side Y-projection and cross-section of D-FF structures with selected water molecules formed initial water cluster; (**d**) inset with biggest image of water cluster; (**e**) Z-projection image of the D-FF 2 coils with 42 H_2O embedded water cluster; (**f**) Y-projection image of the D-FF 2 coils with 42 H_2O embedded water cluster (**D** show the total dipole momentum); (**g**) Y-projection image of the 42 H_2O water cluster extracted from D-FF PNT after optimization (**D** shows the total dipole momentum for water cluster).

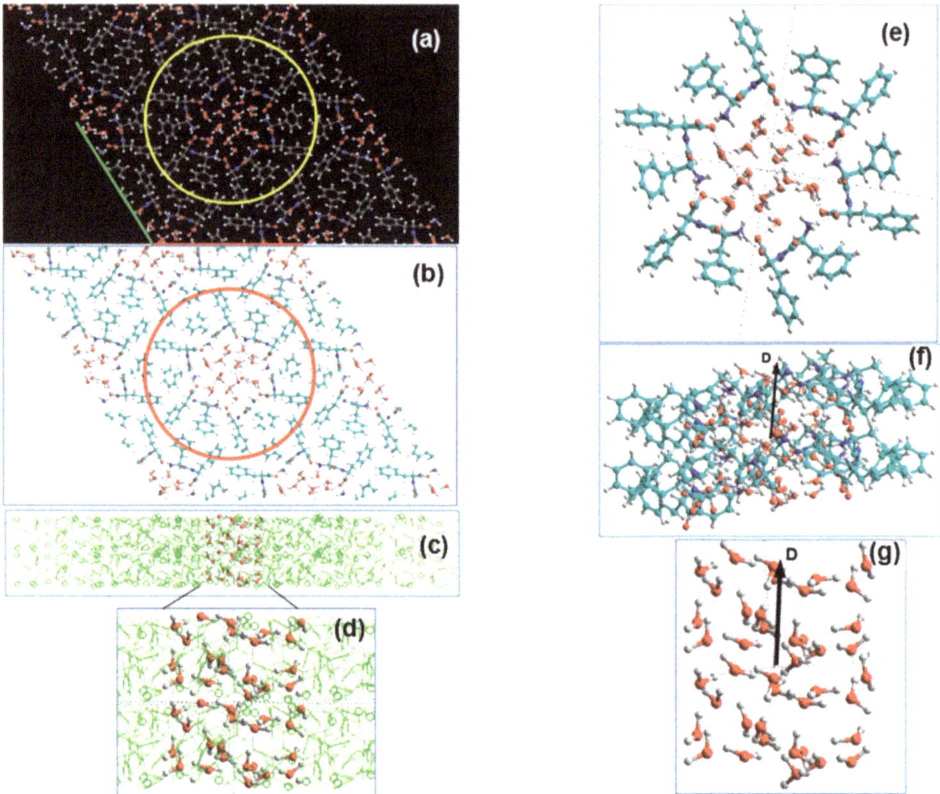

Figure 6. Schematics of the L-FF PNT structures with optimized 21 H_2O per unit cell embedded water cluster (the designations used are the same as in Figure 4): (**a**) Z-projection image from VASP computed data with four unit cells; (**b**) the same image converted into HyperChem workspace; (**c**) side Y-projection and cross-section of L-FF structures with selected water molecules formed initial water cluster; (**d**) inset with biggest image of water cluster; (**e**) Z-projection image of the L-FF 2 coils with 42 H_2O embedded water cluster; (**f**) Y-projection image of the L-FF 2 coils with 42 H_2O embedded water cluster (**D** show the total dipole momentum); (**g**) Y-projection image of the 42 H_2O water cluster extracted from L-FF PNT after optimization (**D** shows the total dipole momentum for water cluster).

The initial water cluster with 21 H_2O molecules constructed on the basis of an ***Ih*** hexagonal ice structure in general had a correct symmetrical organization and a small dipole moment directed on average perpendicular to the axis of the nanotube (see Tables 3 and 4). After optimization of this cluster embedded in the cavity of both PNT, its structure has changed. Moreover, this happened in different ways, depending on the type of the surrounding PNT structure. In both cases (L and D), a distortion of the structure occurs with the formation of a strong dipole moment oriented along the PNT axis.

For a more detailed analysis, we transformed the optimized structures obtained from VASP into HyperChem workspace and distinguished between the models of the structures of both D-FF and L-FF PNTs themselves (consisting of two helix coils) and models of nanotubes containing 42 H_2O molecules each. These models correspond to two layers extracted from the VASP structures with a super-cell of their four initial crystal unit cells in plane (and $8 = 2 \times 2 \times 2$ unit cells total) for each D-FF and L-FF structures. Figures 5 and 6 schematically show the procedures for such transformations. Then, we identified the structures of extracted water clusters themselves from the structures of nanotubes in the HyperChem workspace. This is also shown in Figures 5 and 6.

Table 3. Energy, dipole moment and polarization of D-FF PNTs and water clusters, computed using Austin Model 1 (AM1) restricted Hartree-Fock (RHF) method (HyperChem). Similar and close data were obtained by PM3 and RM1 methods.

Calculated Values		Extracted 42 H_2O Cluster (21 H_2O Per Unit Cell)		Two Coils of D-FF		
		Initial (*Ih*) Structure	After Optimization Inside D-FF	Initial D-FF Structure without H_2O	With 21 H_2O Per u.c. of Initial Structure	With 21 H_2O Per u.c. after Optimization
1		2	3	4	5	6
Total energy, a.u.	E_t	−534.5776	−537.78035	−1739.5256	−2271.06543	−2277.23035
	ΔE_t	−3.20275 (−87.1513 eV)		–	−6.16492 (−167.7561 eV)	
Binding energy, eV	E_b	−312.84856	−399.99734	−2265.27936	−2495.46785	−2663.21967
	ΔE_b	−87.14878		–	−167.75182	
Dipole moment, D	D_t	1.104	29.404	140.385	139.52	158.461
	D_z	−0.876	−28.385	−140.349	−139.447	−158.441
Polarization, C/m^2	P_t	0.00569	0.15075	0.139927	0.119485	0.133218
	P_z	−0.00451	−0.14554	−0.139892	−0.119423	−0.133201
VdW volume, Å^3	V	647.8	650.55	3346.47	3894.86	3967.61

Table 4. Energy, dipole moment and polarization of L-FF PNTs and water clusters, computed using AM1 RHF method (HyperChem). Similar and close data were obtained by PM3 and RM1 methods.

Calculated Values		Extracted 42 H_2O Cluster (21 H_2O Per Unit Cell (u.c.))		Two Coils of L-FF		
		Initial (*Ih*) Structure	After Optimization Inside L-FF	Initial L-FF Structure without H_2O	With 21 H_2O Per u.c. of Initial Structure	With 21 H_2O Per u.c. after Optimization
1		2	3	4	5	6
Total energy, a.u.	E_t	−534.5776	−537.6803	−1739.0274	−2272.7630	−2278.8142
	ΔE_t	−3.10268 (−84.42816 eV)		–	−6.05124 (−164.66275 eV)	
Binding energy, eV	E_b	−312.8486	−397.2743	−2251.7229	−2541.6595	−2706.317
	ΔE_b	−84.426		–	−164.657	
Dipole moment, D	D_t	1.104	28.646	140.757	133.11	157.8331
	D_z	−0.876	−28.386	−140.217	−130.279	−157.035
Polarization, C/m^2	P_t	0.00569	0.14824	0.1395	0.113128	0.13252
	P_z	−0.00451	−0.14690	−0.13897	−0.110722	−0.13185
VdW volume, Å^3	V	647.8	644.55	3365.6	3924.73	3972.63

Figure 7 shows 42 H_2O water clusters extracted from two-coil PNTs (initial and optimized in the D-FF and L-FF PNT). The main feature is that the shift of the individual water molecules is different for D-FF and L-FF PNT—they acquired a helix-like structure with a helix pitch equal to the period of the corresponding D-FF and L-FF PNT helix (that is equal to the period of their periodical crystal structure parameter *c*—see Table 1). Figure 7b,c schematically shows various helix-like structures (in Y-projection) in comparison with the initial water cluster structure (Figure 7a). Figure 7d–f shows a separation of the molecular groups of the water cluster into their internal (or inner) and external (or outer) sub-groups. The outer molecules interact actively with FF atoms on the inner hydrophilic PNT cavity surface, namely with nitrogen and oxygen atoms due to hydrogen bonds.

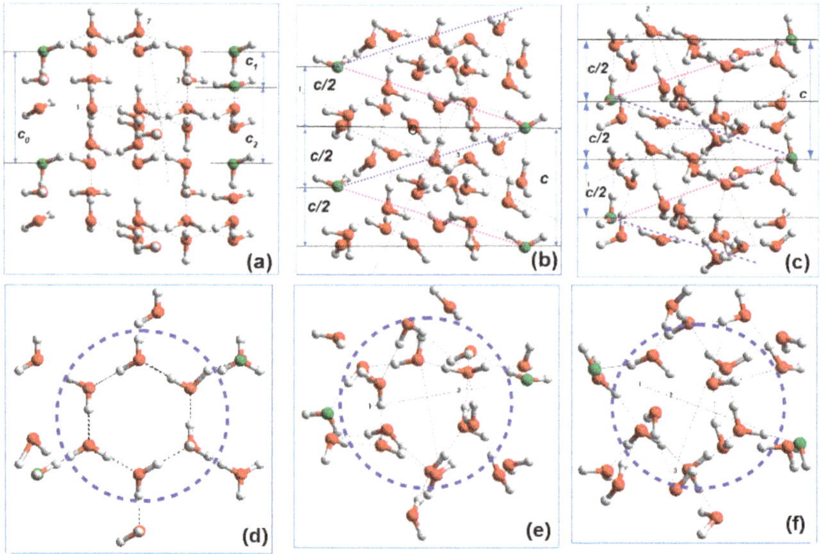

Figure 7. Extracted 42 H_2O water clusters in Y-projection (**a**,**b**,**c**) and Z-projection (**d**,**e**,**f**): (**a**) and (**d**) for initial water cluster based on ***Ih*** structure; (**b**) and (**e**) after optimization within the D-FF PNT; (**c**) and (**f**) after optimization within the L-FF PNT. Dashed lines (with two colors: red and blue) on (**b**) and (**c**) show different direction of the helix-like structure water/ice cluster formation inside the D-FF PNT and L-FF PNT cavity cases. Dashed circles on (**d**), (**e**), and (**f**) show difference between inner and outer water/ice molecules groups. Parameters of periodical water/ice clusters structures are (**a**) c_0 = 5.434 Å, c_1 = 1.670 Å, c_2 = 3.764 Å; (**b**) c = 5.441 Å; (**c**) c = 5.456 Å—these both correspond to the unit cell parameter c in D-FF and L-FF.

A more detailed analysis of the optimized water structure (particularly, for the D-FF PNT structure with 21 water molecules in the cavity) showed that water molecules near the cavity wall arrange approximately halfway between the layers of PNT. These water molecules have the strongest HBs with COO^- and $NH3^+$ groups of PNT. This result was confirmed recently by a dielectric spectroscopy study [51]. These results are also confirmed by visual-differential analysis [30], which are to be published in a separate article.

3.3. Polarization Details

Let us now consider in greater detail the polar properties of both nanotubes with integrated water clusters and the water clusters themselves. Energetic and polar properties (dipole moments and polarization) of empty and filled with 21 water molecules per a unit cell D-FF and L-FF PNTs as well as the extracted water clusters separately were studied using quantum-chemical calculations based on semi-empirical quantum methods AM1, RM1, and PM3 in the restricted Hartree–Fock (RHF) approximation. The data obtained (energies, dipole momentum, polarization) as well the volumes of all molecular structures were presented in Tables 3 and 4.

Accurate calculations are possible with HyperChem using methods that neglect some, but not all, of the electron-electron interactions. These methods are called neglect of differential overlap (NDO) methods. These methods were further improved and developed by Stewart et al. [52–59], and they are based on the so called "neglect of diatomic differential overlap (NDDO)" approximation, with several modifications and with the choice of a wide number of parameters enables one to reproduce experimental quantities. NDDO retains all one-center differential overlap terms when Coulomb and exchange integrals are computed. The NDDO approximation is the basis for the Modified

Neglect of Diatomic Overlap (MNDO), Austin Model 1 (AM1), and Parametrization No. 3 (PM3) methods. AM1 is a modified MNDO method proposed and developed by Dewar et al. at the University of Texas at Austin [52–54]. AM1 is generally the most accurate computational method and is often the best method for collecting quantitative information [52–54]. PM3 differs from AM1 only in the values of the parameters [55–57]. The parameters for PM3 were obtained by comparing a large number of experiments with calculation results. As a rule, non-covalent interactions in the PM3 method are less repulsive than in AM1. Recife Model 1 (RM1) is another semiempirical method that was parameterized to calculate dipole moments and enthalpies of formation, with errors smaller than those for AM1 and PM3 [58,59]. Energetic properties of organic compounds can be calculated in both forms: isolated, and in solvent medium. All methods are available and provide close data for the biomolecular systems studied in this article based on C, O, N, and H atoms.

The total energy per a unit cell, E_t (or energy of two-coils PNT models, shown in Figure 5 for D-FF and on Figure 6 for L-FF models), and the binding energy, E_b (calculated here as a difference of total energies of the structure at the optimal distances between all elements E_t (r ~ r_{opt}) and energy of all its elements removed away at the infinite distance one from another E_t (r ~ ∞), were calculated before and after water optimization (Table 3 for D-FF and Table 4 for L-FF PNT models).

These energies were calculated automatically by HyperChem software when we calculate any molecular structure by quantum semi-empirical methods in the single point (SP) option or in the geometry optimization option.

For the extracted water clusters before and after the optimization, the variations of the total ΔE_t and binding ΔE_b energies are the same (Table 3 for D-FF it is $\Delta E_t \sim \Delta E_b \sim -87.15$ eV and Table 4 for L-FF it is $\Delta E_t \sim \Delta E_b \sim 84.43$ eV, see in columns 2 and 3), thus demonstrating that changes in energy are only due to the realignment of water molecules in the cluster and changes in VdW interaction between water molecules and reorganization of the HB network structure. At the same time, for both D-FF and L-FF PNT with water cluster inside, the E_t and E_b energy decreased: (1) it is at around ~ −167.75 eV after the optimization for D-FF that is twice higher than that for the extracted water cluster ~ −87.15 eV (see Table 3, columns 5 and 6); (2) and for L-FF from ~ −164.66 eV up to ~ −84.43 eV (see Table 4, columns 5 and 6). This can mean that the formation of HBs between the water cluster and inner surface of a nanotube takes twice as much energy as merely a rearrangement of bonds inside the water cluster. In principle, this corresponds to the approaches in references [49,50], but we obtain here some new important quantitative data.

The main results obtained were (1) after optimization the embedded water molecular cluster has a big own dipole moment strongly oriented along D-FF and L-FF PNT channel, while before it did not have such a defined orientation and a very small dipole moment (Figure 5, Figure 6f,g, Figures 8 and 9 as well as Tables 3 and 4); (2) the structure organization of both these water clusters after optimization inside D-FF and L-FF cavity has changed greatly—both have acquired helix properties, with the same helix step as D-FF and L-FF PNT (Figure 7); (3) the total dipole moment and polarization of both D-FF and L-FF PNT after optimization with embedded 21 H_2O water/ice clusters enhance in the direction of main c-axis of each PNTs (Tables 3–5 and Figures 8 and 9). Important feature is that after optimization inside D-FF and L-FF PNTs cavity the water/ice clusters acquired different directions of rotation of a helix. This property of helix lines as known is called chirality [6,7]: "right chirality" (D—from the Latin "dextra") and "left chirality" (L from the Latin "laeva").

This last result (showing in Figures 7–9), concerned with the very pronounced realignment of water cluster structure, is confirmed by some other important computed data. The energy changes show that the main energy shift is concerned with changes of the VdW and HB energies of water molecule structures. It is clearly seen from changes in the extracted water molecules clusters obtained after optimization and their energies ΔE_t and ΔE_b, as was discussed above.

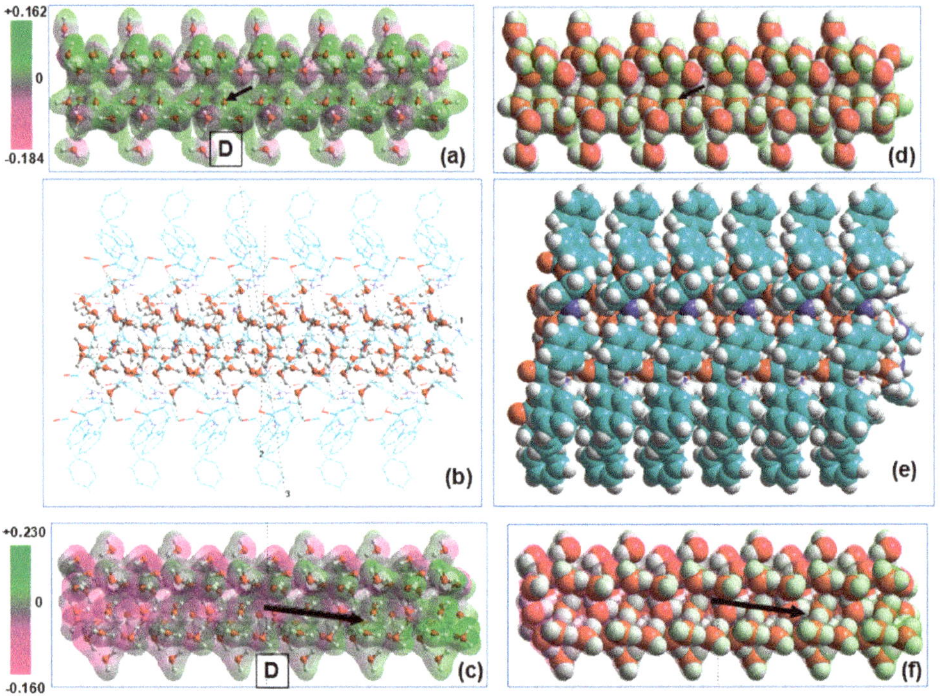

Figure 8. Model images of the water/ice tubular nanostructures and D-FF PNT with filled hydrophilic cavity, consisting from six repeated coils along the c-axis (in X-projection): (**a**) initial tubular water/ice structure, (**b**) D-FF PNT with embedded water/ice structures under optimization process, (**c**) water/ice tubular helix structure after optimization inside D-FF PNT, (**d**), (**e**), and (**f**) the same in the VdW surface presentation. Vector **D** show the direction of the total dipole momentum in the initial (D ~ 1 Debye) and optimized water/ice cluster (D ~ 97 Debye). Translucent 3D-mapped isosurface on (**a**) and (**c**) (as well as on (**d**) and (**f**)) illustrates the re-distribution of the electrostatic potential of water/ice cluster before and after optimization within D-FF PNT.

It is interesting to note that water molecules embedded into a carbon nanotube under the influence of high pressure and temperature lead to a formation of a similar helix structure [60]. It is known that the water confined to nanopores is investigated not only in carbon nanotubes, but in other nanoporous structures, for example, in boron nitride nanotubes (BNNTs) using first-principles calculations [61]. Another study of the polar property, polarization, and even the ferroelectricity in the ice-type (or water-type) nanostructures has recently been performed in nanoporous silicate materials, which have an ordered system of narrow cylindrical pores [62]. It is assumed that, in such filamentous pores of the studied materials, ferroelectric ice XI is formed.

We do not yet insist directly on the occurrence of ferroelectricity of water and ice structures inside peptide nanotubes, since reliable phase transitions between the polar (ferroelectric) and nonpolar (paraelectric) phases, which have pronounced changes in the dielectric constant according to the well-known Curie-Weiss law [19], have not yet been detected. Meanwhile, ferroelectricity was observed by Bdikin I., Bystrov V., Kopyl S., et al. in [63], though in other β-sheet FF PNT structure [17,24], not in α-helix, as is studied now [25–29]. Besides, water was not known to occur in the internal cavity of such PNT. This is to be clarified in the future, and possible results will be achieved with more detailed dielectric measurements [51]. Otherwise, the presence of such a significant and strictly oriented polarization clearly indicates the possibility of the existence of the ferroelectric-like phenomena here.

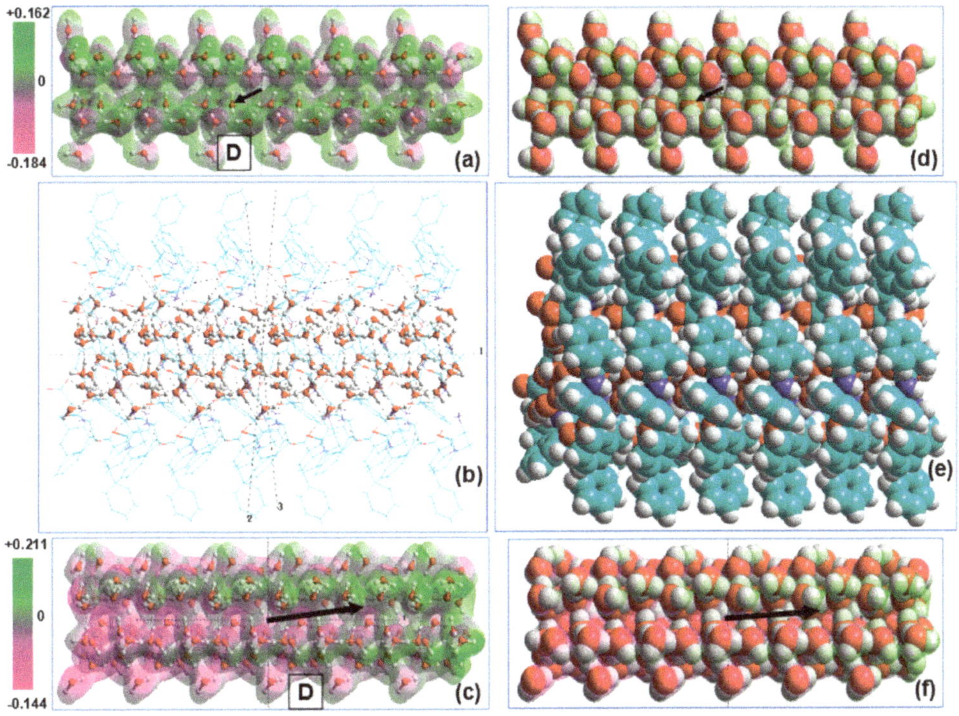

Figure 9. Model images of the water/ice tubular nanostructures and L-FF PNT with filled hydrophilic cavity, consisting from six repeated coils along c-axis (in X-projection): (**a**) initial tubular water/ice structure; (**b**) L-FF PNT with embedded water/ice structures under optimization process; (**c**) water/ice tubular helix structure after optimization inside l-FF PNT; (**d–f**) the same in the VdW surface presentation Vector **D** show the direction of the total dipole momentum in the initial (D ~1 Debye) and optimized water/ice cluster (D ~97 Debye). Translucent 3D-mapped isosurface on (**a**) and (**c**) (as well as on (**d**) and (**f**)) illustrates the re-distribution of the electrostatic potential of water/ice cluster before and after optimization within L-FF PNT.

Table 5. Dipole moments and polarization for water/ice cluster after optimization in D-FF and L-FF PNT, consisting from 126 H_2O molecules (see Figures 8 and 9), computed by various methods.

<table>
<tr><th rowspan="3" colspan="2">126 H_2O Water/Ice Cluster after Optimization</th><th colspan="6">Method Used (In RHF)</th></tr>
<tr><th colspan="3">From D-FF</th><th colspan="3">From L-FF</th></tr>
<tr><th>AM1</th><th>PM3</th><th>RM1</th><th>AM1</th><th>PM3</th><th>RM1</th></tr>
<tr><td rowspan="4">Dipole moment, Debye</td><td>Dt</td><td>95.92</td><td>97.355</td><td>100.575</td><td>94.982</td><td>97.226</td><td>99.533</td></tr>
<tr><td>Dz</td><td>−93.496</td><td>−95.058</td><td>−98.130</td><td>−94.653</td><td>−96.787</td><td>−99.154</td></tr>
<tr><td>Dy</td><td>−15.324</td><td>−14.816</td><td>−15.481</td><td>7.304</td><td>8.528</td><td>8.068</td></tr>
<tr><td>Dx</td><td>14.978</td><td>14.912</td><td>15.695</td><td>2.991</td><td>3.521</td><td>3.188</td></tr>
<tr><td rowspan="4">Polarization, C/m^2</td><td>Pt</td><td>0.166</td><td>0.1685</td><td>0.1741</td><td>0.1655</td><td>0.1698</td><td>0.1738</td></tr>
<tr><td>Pz</td><td>−01618</td><td>−0.1645</td><td>−0.1699</td><td>−0.1653</td><td>−0.1690</td><td>−0.1731</td></tr>
<tr><td>Py</td><td>−0.0265</td><td>−0.0256</td><td>−0.0265</td><td>0.0128</td><td>0.0149</td><td>0.0141</td></tr>
<tr><td>Px</td><td>0.0259</td><td>0.0258</td><td>0.0259</td><td>0.0052</td><td>0.0061</td><td>0.0056</td></tr>
<tr><td colspan="2">VdW Volume, Å^3</td><td>1927.21</td><td></td><td></td><td>1910.28</td><td></td><td></td></tr>
</table>

Returning to the first important result, it should be noted, that the extracted water clusters after optimization have the helix nanotube structure similar D-FF and L-FF PNT and possesses high of dipole moments and corresponding polarization value directed along nanotube OZ axis. The total dipole moment D_t =|**D**| increases from ~1 D for the initial structure up to an order of ~29 D after the optimization for both chirality D-FF and L-FF (see Tables 3 and 4, columns 2 and 3). The D_t =|**D**| of the water-filled D-FF nanotube also increases after the optimization from ~139.5 D up to ~158.5 D (Table 3, columns 5 and 6), and, similarly, from ~133.1 D up to ~157.8 D for L-FF PNT (Table 4, columns 5 and 6). The dipole moment found for empty PNT is about ~140.3 D for D-FF and ~140.8 D for L-FF, which are close to the values obtained in previous works for similar empty PNT [22–29]. For the filled PNT, the absolute value of the projection of the dipole moment at the nanotube axis, D_z, is almost the same as D_t, thus demonstrating that the dipole moment is oriented strongly along the D-FF and L-FF PNT axis. At the same time, the dipole moment of the extracted water cluster is shifted at about 15° from down the axis for D-FF PNT (Figure 8c), while for L-FF it is shifted at about 10–15° up from the L-FF PNT axis (Figure 9c). A water nanotube acquires such polar properties under the influence of D-FF and L-FF PNT—it is self-consistent and self-organized process modulated and induced by an electric field from the strongly oriented dipole moments of D-FF and L-FF. However, this effect is different for each chirality type.

The total polarization, P_t, of D-FF and L-FF PNT with water molecules in a cavity (as well polarization of water clusters) can be calculated as a dipole moment per volume of the unit cell (in SI units P_t [C/m^2] = 3.33556255·D_t [D]/V [Å^3]). The values of the VdW volume of studied water clusters, D-FF and L-FF PNT cells, as well corresponding P_t values are presented in Tables 3–5. Since water inside the nanochannel increases the volume of the unit cell (as was demonstrated by DFT calculations), the value of P_t for filled D-FF PNT after the optimization may be somewhat less than that of the empty PNT. So, for empty D-FF nanotube P_t is around ~0.14 C/m^2, whereas for water-filled PNTs, after optimization, it is 0.133 C/m^2 (see Table 3, columns 4 and 6); and for empty L-FF nanotubes, P_t is around ~0.139 C/m^2, whereas for water-filled PNTs, after optimization, it is 0.132 C/m^2 only (see Table 4, columns 4 and 6). At the same time, the P_t for separate water cluster after optimization is higher than that for filled D-FF and L-FF PNT and reaches ~0.15 C/m^2 (Table 3, column 3), and ~0.148 C/m^2 for L-FF (Table 4, column 3). This mismatch also demonstrates the strong effect of the electric field produced by the PNT. The maximum value of the corresponding electric field for both L-FF and D-FF (with polarization P_t ~0.14 C/m^2) can be estimated as $E = P_t/(\varepsilon\varepsilon_0) \approx 4.0$ GV/m (for $\varepsilon = 4$ [18,24]—dielectric permittivity of the PNT, and ε_0—dielectric vacuum constant). Such strong electric field oriented inside a cavity along the PNT's *c*-axis induces the orientation of each water molecules dipole along this direction (Figure 5, Figure 6f,g, Figures 8 and 9).

In this connection, it should be mentioned here that, in a carbon nanotube, water molecules form a solid-like structure under an applied electric field of 1 V/nm = 1 GV/m for simulation temperatures up to 350 K [64]. The authors Winarto W., Yamamoto E. Yasuoka K. of Ref. [64] suggested that the electric field induces a phase transition from liquid to ice-nanotube at temperatures as high as 350 K, and the electrostatic interaction within the ice-nanotube under an electric field is stronger than that in the absence of an electric field.

Another important point is that water cluster inside D-FF PNT acquiring per se strongly oriented polar properties and can reveal ferroelectric properties. Moreover, if we build longer models of optimized water ice clusters, we get extended needle-shaped strongly polarized ferroelectric-like elements that may have promising potential applications in various fields. As an example, Table 5 shows the results of calculations for more extended water clusters having even higher values of dipole moments and polarization. At the same time, here, we presented the calculation data using different methods, and we see that RM1 gives the highest values of the dipole moments (up ~100 Debye along PNT c-axis) and polarization (up ~0.174 C/m^2). An important feature is that the value of the perpendicular components Dy and Py has different signs for D-FF and L-FF, corresponding to various orientations of the polar vector in accordance with chirality of PNT (see on Figures 8c and 9c).

Some manifestations of ferroelectric properties in water structures are known, especially in their nanostructured form in various nanotube types, and under different influences [60–62,64,65]. In addition, it should be noted that the straightforward study of ferroelectric phenomena in water and ice tubular and one-dimensionally ordered nanostructures is actively carried out by some research teams. For example, Mikami F., Matsuda K., Kataura H. and Maniwa Y. in Ref. [65], novel ferroelectric properties of a new form of ice inside single-walled carbon nanotubes was investigated by molecular dynamics simulation. The authors called them "ice nanotubes" (ice NTs) and found them to consist of polygonal water rings stacked one-dimensionally along the nanotube axis. Mikami F., et al. in Ref. [65], ice NTs were revealed to show stepwise polarization with a significant hysteresis loop as a function of the external field strength. In particular, pentagonal and heptagonal ice NTs are found to be the world's smallest ferroelectrics, with spontaneous polarization of around 1 $\mu C/cm^2 \sim 0.01\ C/m^2$. In our case, for example, the P_t value of water cluster in D-FF PNT is more than order of magnitude higher than this value (0.133 C/m^2), and just greater up to ~0.15 C/m^2 in the case of a such ice nanotube alone (and it is also higher for more long extended clusters; see Table 5). Thus, the presence of ferroelectric properties can be assumed as well. However, additional investigations are necessary.

All these findings show potential applications of nanotube encapsulating dielectric materials for the fabrication of the smallest ferroelectric devices. In turn, we assume that such and similar polar systems based on the tubular ice nanostructures, formed inside peptide nanotubes, have a great future for a variety of applications in many nanodevices.

Ferroelectric properties, in addition to polarization, also represent such a property as piezoelectricity, which has numerous practical applications [19–32]. Recently, it was shown by Bystrov V.S., Bdikin I.K., Singh B. in Ref. [18] that FF PNTs and also some AAs have rather high polarization values and, as a consequence, piezoelectric coefficients. Thus, the results obtained above not only mean that the presence of water/ice structures inside any AA and PNT increases and strengthens their polar properties, but also has very significant potential opportunities for the development of new types of nanostructures with enhanced piezoelectric properties.

4. Conclusions

The results obtained allow us to conclude that, on average, after optimizing the water/ice cluster enclosed in the inner hydrophilic cavity of the nanotubes, they acquire strongly anisotropic electrical properties, with significant magnitudes with dipole moments oriented mainly along the axis of the nanotubes. In this case, the average polarization values reach 0.16–0.17 C/m^2 when calculated by any methods (the highest values are given by RM1, see Table 5) along the surface of the VdW surface.

In conclusion it must be stated, that the main results obtained include (1) after optimization, the embedded water molecular clusters have significantly larger dipole moments that are strongly oriented along D-FF and L-FF PNTs channel, while before optimization, it did not have such a defined orientation and had a very small dipole moments; (2) the structure organization of this water cluster has very significant and high changes—this water/ice cluster acquires helix properties, with the same helix step as D-FF PNT; (3) the dipole moment and polarization of the initial empty D-FF and L-FF PNT was increased and enhanced after optimization with the presence of the embedded water/ice cluster inside their cavity, particularly with the most optimal 21 H_2O molecular cluster.

Such an obvious and drastic change modulated by the influence of D-FF properties and structural features was established here for the first time using direct modeling and calculations using DFT (VASP) and various semi-empirical quantum methods.

The whole set of the findings allowed us to assume the ferroelectric properties of such water clusters. However, the experimental confirmation of this assumption is still to be done. In this connection it must be noted that it is necessary to develop the investigation of the stability and the evolution of such clusters, especially, with the temperature, namely by doing MD simulations (similarly to that demonstrated by Andrade-Filho T. et al. in Ref. [36]), develop and perform the dielectric measurements of these structures (similarly to that reported by Bystrov V.S. et al. in Ref. [51]). In positive cases, these discoveries may lead in future to the development of unexpected applications and may also serve for further development in nanotechnological, medical and biological breakthroughs.

Author Contributions: V.B. and P.Z. wrote the manuscript. J.C. supervised and supported this study and advised on cluster computing with the VASP program. V.T. developed the concept of chirality and critically revised the manuscript. S.K., A.N., and O.Z. participated in the processing of experimental data and analysis of calculated data. All authors have read and agreed to the published version of the manuscript.

Funding: This research received no external funding.

Acknowledgments: This work was partially supported by the Fundacão para a Ciência e a Tecnologia (FCT, Portugal) through project UID/CTM/50025/2013 and UIDB/50011/2020 & UIDP/50011/2020. P.Z. and S.K. are grateful to the FCT (Portugal) through the project "BioPiezo," PTDC/CTM–CTM/31679/2017 (CENTRO-01-0145-FEDER-031679). The theoretical and computational parts of the study was completed within the framework of the non-commercial Agreement on scientific and technical cooperation between Institute of Mathematical Problems of Biology (IMPB) of KIAM RAS and Department of Physics and I3N institution of the University of Aveiro, Portugal. Part of this work was funded by national funds (OE), through FCT (Portugal), in the scope of the framework contract foreseen in the numbers 4, 5, and 6 of the article 23, of the Decree-Law 57/2016, of August 29, changed by Law 57/2017, of July 19.

Conflicts of Interest: We declare no potential conflicts of interest in this article.

References

1. Calvin, M. *Chemical Evolution. Molecular Evolution, towards the Origin of Living System on the Earth and Elsewhere*; Claredon: Oxford, UK, 1969.
2. Pachahara, S.K.; Subbalakshmi, C.; Nagaraj, R. Formation of nanostructures by peptides. *Curr. Protein Pept. Sci.* **2017**, *18*, 1–19. [CrossRef] [PubMed]
3. Aryaa, S.K.; Solankia, P.R.; Dattab, M.; Malhotra, B.D. Recent advances in self-assembled monolayers based biomolecular electronic devices. *J. Biosens. Bioelectron.* **2009**, *24*, 2810–2817. [CrossRef] [PubMed]
4. Mendes, A.C.; Baran, E.T.; Reis, R.L.; Azevedo, H.S. Self-assembly in nature: Using the principles of nature to create complex nanobiomaterials. *Wiley Interdiscip. Rev. Nanomed. Nanobiotechnol.* **2013**, *5*, 582–612. [CrossRef]
5. Yashima, E.; Ousaka, N.; Taura, D.; Shimomura, K.; Ikai, T.; Maeda, K. Supramolecular helical systems: Helical assemblies of small molecules, foldamers, and polymers with chiral amplification and their functions. *Chem. Rev.* **2016**, *116*, 13752–13990. [CrossRef] [PubMed]
6. Tverdislov, V.A. Chirality as a primary switch of hierarchical levels in molecular biological systems. *Biophysics* **2013**, *58*, 128–132. [CrossRef]
7. Tverdislov, V.A.; Malyshko, E.V.; Il'chenko, S.A.; Zhulyabina, O.A.; Yakovenko, L.V. A periodic system of chiral structures in molecular biology. *Biophysics* **2017**, *62*, 331–341. [CrossRef]
8. Naaman, R.; Waldeck, D.H. Spintronics and chirality: Spin selectivity in electron transport through chiral molecules. *Annu. Rev. Phys. Chem.* **2015**, *66*, 263–281. [CrossRef]
9. Burgess, N.C.; Sharp, T.H.; Thomas, F.; Wood, C.W.; Thomson, A.R.; Zaccai, N.R.; Brady, R.L.; Serpell, L.C.; Woolfson, D.N. Modular Design of Self-Assembling Peptide-Based Nanotubes. *J. Am. Chem. Soc.* **2015**, *137*, 10554–10562. [CrossRef]
10. Fleming, S.; Ulijn, R.V. Design of nanostructures based on aromatic peptide amphiphiles. *Chem. Soc. Rev.* **2014**, *43*, 8150–8177. [CrossRef]
11. Smith, K.H.; Tejeda-Montes, E.; Poch, M.; Mata, A. Integrating top-down and self-assembly in the fabrication of peptide and protein-based biomedical materials. *Chem. Soc. Rev.* **2011**, *40*, 4563–4577. [CrossRef]
12. Silva, R.F.; Araújo, D.R.; Silva, E.R.; Ando, R.A.; Alves, W.A. L-diphenylalanine microtubes as apotential drug-delivery system: Characterization, release kinetics, and cytotoxicity. *Langmuir* **2013**, *29*, 10205–10212. [CrossRef] [PubMed]
13. Emtiazi, G.; Zohrabi, T.; Lee, L.Y.; Habibi, N.; Zarrabi, A. Covalent diphenylalanine peptide nanotube conjugated to folic acid/magnetic nanoparticles for anti-cancer drug delivery. *J. Drug Deliv. Sci. Technol.* **2017**, *41*, 90–98. [CrossRef]
14. Orsi, M. Molecular simulation of self-assembly. In *Self-Assembling Biomaterials. Molecular Design, Characterization and Application in Biology and Medicine*, 1st ed.; Azevedo, H.S., da Silva, R.M.P., Eds.; Woodhead Publishing Series in Biomaterials; Elsevier Ltd.: Amsterdam, The Netherlands, 2018; pp. 305–318.

15. Lee, O.S.; Stupp, S.I.; Schatz, G.C. Atomistic molecular dynamics simulations of peptide amphiphile self-assembly into cylindrical nanofibers. *J. Am. Chem. Soc.* **2011**, *133*, 3677–3683. [CrossRef]
16. Brandon, C.J.; Martin, B.P.; McGee, K.J.; Stewart, J.J.P.; Braun-Sand, S.B. An approach to creating a more realistic working model from a protein data bank entry. *J. Mol. Model.* **2015**, *21*, 1–11. [CrossRef] [PubMed]
17. Lehninger, A.L. *Biochemistry. The Molecular Basis of Cell Structure and Function*; Worth: New York, NY, USA, 1972.
18. Bystrov, V.S.; Bdikin, I.K.; Singh, B. Piezoelectric and ferroelectric properties of various amino acids and tubular dipeptide nanostructures: Molecular modeling. *Nanomater. Sci. Eng.* **2020**, *2*, 11–24.
19. Lines, M.E.; Glass, A.M. *Principles and Applications of Ferroelectrics and Related Materials*; Clarendon Press: Oxford, UK, 1977.
20. Bystrov, V.S.; Bdikin, I.; Heredia, A.; Pullar, R.C.; Mishina, E.; Sigov, A.; Kholkin, A.L. Piezoelectricity and Ferroelectricity in biomaterials: From proteins to self-assembled peptide nanotubes. In *Piezoelectric Nanomaterials for Biomedical Applications*; Ciofani, G., Menciassi, A., Eds.; Springer: Berlin/Heidelberg, Germany, 2012; pp. 187–211.
21. Bystrov, V.S.; Seyedhosseini, E.; Kopyl, S.; Bdikin, I.K.; Kholkin, A.L. Piezoelectricity and ferroelectricity in biomaterials: Molecular modeling and piezoresponse force microscopy measurements. *J. Appl. Phys.* **2014**, *116*, 066803. [CrossRef]
22. Kholkin, A.; Amdursky, N.; Bdikin, I.; Gazit, E.; Rosenman, G. Strong piezoelectricity in bioinspired peptide nanotubes. *ACS Nano* **2010**, *4*, 610–614. [CrossRef]
23. Nguyen, V.; Zhu, R.; Jenkins, K.; Yang, R. Self-assembly of diphenylalanine peptide with controlled polarization for power generation. *Nat. Commun.* **2016**, *7*, 13566. [CrossRef]
24. Bystrov, V.S.; Paramonova, E.V.; Bdikin, I.K.; Kopyl, S.; Heredia, A.; Pullar, R.C.; Kholkin, A.L. Bioferroelectricity: Diphenylalanine peptide nanotubes computational modeling and ferroelectric properties at the nanoscale. *Ferroelectrics* **2012**, *440*, 3–24. [CrossRef]
25. Bystrov, V.S.; Zelenovskiy, P.S.; Nuraeva, A.S.; Kopyl, S.; Zhulyabina, O.A.; Tverdislov, V.A. Molecular modeling and computational study of the chiral-dependent structures and properties of the self-assembling diphenylalanine peptide nanotubes. *J. Mol. Model.* **2019**, *25*, 199. [CrossRef]
26. Bystrov, V.S.; Kopyl, S.A.; Zelenovskiy, P.; Zhulyabina, O.A.; Tverdislov, V.A.; Salehli, F.; Ghermani, N.E.; Shur, V.Y.; Kholkin, A.L. Investigation of physical properties of diphenylalanine peptide nanotubes having different chiralities and embedded water molecules. *Ferroelectrics* **2018**, *525*, 168–177. [CrossRef]
27. Bdikin, I.; Bystrov, V.S.; Delgadillo, I.; Gracio, J.; Kopyl, S.; Wojtas, M.; Mishina, E.; Sigov, A.; Kholkin, A.L. Polarization switching and patterning in self-assembled peptide tubular structures. *J. Appl. Phys.* **2012**, *111*, 074104. [CrossRef]
28. Bystrov, V.S.; Zelenovskiy, P.S.; Nuraeva, A.S.; Kopyl, S.; Zhulyabina, O.A.; Tverdislov, V.A. Chiral peculiar properties of self-organization of diphenylalanine peptide nanotubes: Modeling of structure and properties. *Math. Biol. Bioinform.* **2019**, *14*, 94–124. [CrossRef]
29. Zelenovskiy, P.S.; Nuraeva, A.S.; Kopyl, S.; Arkhipov, S.G.; Vasilev, S.G.; Bystrov, V.S.; Gruzdev, D.A.; Waliszek, M.; Svitlyk, V.; Shur, V.Y.; et al. Chirality-dependent growth of self-assembled diphenylalanine microtubes. *Cryst. Growth Des.* **2019**, *19*, 6414–6421. [CrossRef]
30. Filippov, S.V.; Bystrov, V.S. Visual-differential analysis of structural features of internal cavities of two chiral forms of diphenylalanine nanotubes. *Biophysics* **2020**, *65*, 1–8. [CrossRef]
31. Zelenovskiy, P.S.; Shur, V.Y.; Nuraeva, A.S.; Vasilev, S.G.; Vasileva, D.S.; Alikin, D.O.; Chezganov, D.S.; Krasnov, V.P.; Kholkin, A.L. Morphology and piezoelectric properties of diphenylalanine microcrystals grown from methanol-water solution. *Ferroelectrics* **2015**, *475*, 127–134. [CrossRef]
32. Zelenovskiy, P.; Kornev, I.; Vasilev, S.; Kholkin, A. On the origin of the great rigidity of self-assembled diphenylalanine nanotubes. *Phys. Chem. Chem. Phys.* **2016**, *18*, 29681–29685. [CrossRef]
33. Gorbitz, C.H. Nanotube formation by hydrophobic dipeptides. *Chem. Eur. J.* **2001**, *7*, 5153–5159. [CrossRef]
34. Gorbitz, C.H. Hydrophobic dipeptides: The final piece in the puzzle. *Acta Cryst.* **2018**, *B74*, 311–318. [CrossRef]
35. Kim, J.; Han, T.E.; Kim, Y.; Park, J.S.; Choi, J.; Churchill, D.G.; Kim, S.O.; Ihee, H. Role of water in directing diphenylalanine assembly into nanotubes and nanowires. *Adv. Mater.* **2010**, *22*, 583–587. [CrossRef]
36. Andrade-Filho, T.; Martins, T.C.; Ferreira, F.F.; Alves, W.A.; Rocha, A.R. Water-driven stabilization of diphenylalanine nanotube structures. *Theor. Chem. Acc.* **2016**, *135*, 185. [CrossRef]

37. Ryan, H.; Carter, M.; Stenmark, P.; Stewart, J.J.P.; Braun-Sand, S.B. A comparison of X-ray and calculated structures of the enzyme MTH1. *J. Mol. Model.* **2016**, *22*, 1–18. [CrossRef] [PubMed]
38. Hypercube Inc. *HyperChem*, versions 7.51 and 8.0; Hypercube Inc.: Gainesville, FL, USA, 2002. Available online: http://www.hyper.com/?tabid=360 (accessed on 27 July 2020).
39. The Cambridge Crystallographic Data Centre (CCDC). Available online: https://www.ccdc.cam.ac.uk/ (accessed on 27 July 2020).
40. VASP (Vienna Ab initio Simulation Package). Available online: https://www.vasp.at/ (accessed on 27 July 2020).
41. Kresse, G.; Furthmuller, J. Efficient iterative schemes for ab initio total-energy calculations using a plane-wave basis set. *Phys. Rev. B Condens. Matter Mater. Phys.* **1996**, *54*, 11169–11186. [CrossRef] [PubMed]
42. Kresse, G.; Joubert, D. From ultrasoft pseudopotentials to the projector augmented-wave method. *Phys. Rev. B Condens. Matter Mater. Phys.* **1999**, *59*, 1758–1775. [CrossRef]
43. Perdew, J.P.; Burke, K.; Ernzerhof, M. Generalized Gradient Approximation Made Simple. *Phys. Rev. Lett.* **1996**, *77*, 3865–3868. [CrossRef]
44. Paier, J.; Hirschl, R.; Marsman, M.; Kresse, G. The Perdew-Burke-Ernzerhof exchange-correlation functional applied to the G2-1 test set using a plane-wave basis set. *J. Chem. Phys.* **2005**, *122*, 234102. [CrossRef]
45. Blochl, P.E. Projector augmented-wave method. *Phys. Rev. B Condens. Matter Mater. Phys.* **1994**, *50*, 17953–17979. [CrossRef]
46. Grimme, S.; Antony, J.; Ehrlich, S.; Krieg, S. A consistent and accurate ab initio parametrization of density functional dispersion correction (dft-d) for the 94 elements H-Pu. *J. Chem. Phys.* **2010**, *132*, 154104. [CrossRef]
47. Monkhorst, H.J.; Pack, J.D. Special points for Brillouin-zone integrations. *Phys. Rev.* **1976**, *13*, 5188–5192. [CrossRef]
48. Morrison, I.; Li, J.-C.; Jenkins, S.; Xantheas, S.S.; Payne, M.C. Ab-initio total energy studies of the static and dynamical properties of ice Ih. *J. Phys. Chem. B* **1997**, *101*, 6146–6150. [CrossRef]
49. Wang, L.; Zhao, J.; Li, F.; Fang, H.; Lu, J.P. First-principles study of water chains encapsulated in single-walled carbon nanotube. *J. Phys. Chem. C* **2009**, *113*, 5368–5375. [CrossRef]
50. Yang, R.; Hilder, T.A.; Chung, S.H.; Rendell, A. First-principles study of water confined in single-walled silicon carbide nanotubes. *J. Phys. Chem. C* **2011**, *115*, 255–264. [CrossRef]
51. Bystrov, V.S.; Zhulyabina, O.A.; Kopy, S.A.; Zelenovskiy, P.S.; Nuraeva, A.S.; Tverdislov, V.A.; Filippov, S.V.; Salehli, F.; Kholkin, A.L.; Shur, V.Y. Modeling and computer study of diphenylalanine peptide nanotubes, containing the water molecules. In *Abstract Book of the International Online Conference "Research Ferroelectric Materials by Russian Scientists. Centenary of Discovery Ferroelectricity "(SE-100) (Yekaterinburg, 17–19 August 2020)*; Ural Federal University: Yekaterinburg, Russia, 2020; pp. 48–50.
52. Dewar, M.J.S.; Zoebisch, E.G.; Healy, E.F.; Stewart, J.J.P. Development and use of quantum-mechanical molecular models. 76. AM1: A new general purpose quantum mechanical molecular model. *J. Am. Chem. Soc.* **1985**, *107*, 3902–3909. [CrossRef]
53. Dewar, M.J.S.; Dieter, K.M. Evaluation of AM1 calculated proton affinities and deprotonation enthalpies. *J. Am. Chem. Soc.* **1986**, *108*, 8075–8086. [CrossRef]
54. Stewart, J.J.P. MOPAC: A semiempirical molecular orbital program. *J. Comp. Aided Mol. Des.* **1990**, *4*, 1–103. [CrossRef]
55. Stewart, J.J.P. Optimization of Parameters for Semiempirical Methods. I. Method. *J. Comput. Chem.* **1989**, *10*, 209–220. [CrossRef]
56. Stewart, J.J.P. Optimization of parameters for semiempirical methods. *II. Applications. J. Comput. Chem.* **1989**, *10*, 221–264. [CrossRef]
57. Stewart, J.J.P. Optimization of parameters for semiempirical methods V: Modification of NDDO approximations and application to 70 elements. *J. Mol. Mod.* **2007**, *13*, 1173–1213. [CrossRef]
58. Rocha, G.B.; Freire, R.O.; Simas, A.M.; Stewart, J.J.P. RM1: A Reparameterization of AM1 for Y, C, N, O, P, S, F, Cl, Br, and I. *J. Comput. Chem.* **2006**, *27*, 1101–1111. [CrossRef]
59. Lima, N.B.D.; Rocha, G.B.; Freire, R.O.; Simas, A.M. RM1 Semiempirical Model: Chemistry, Pharmaceutical Research, Molecular Biology and Materials Science. *J. Braz. Chem. Soc.* **2019**, *30*, 683–716. [CrossRef]
60. Bai, J.; Wang, J.; Zeng, X.C. Multiwalled ice helixes and ice nanotubes. *Proc. Nat. Acad. Sci. USA* **2006**, *103*, 19664–19667. [CrossRef] [PubMed]

61. Shayeganfar, F.; Beheshtian, J.; Shahsavari, R. First-Principles Study of Water Nanotubes Captured Inside Carbon/Boron Nitride Nanotubes. *Langmuir* **2018**, *34*, 11176–11187. [CrossRef] [PubMed]
62. Bordonskiy, G.S.; Orlov, A.O. The study of ferroelectric phase transitions of water in nanoporous silicates with joint electrical noise and calorimetric measurements. *Phys. Solid State* **2014**, *56*, 1575–1582. (In Russian) [CrossRef]
63. Bdikin, I.; Bystrov, V.; Kopyl, S.; Lopes, R.P.G.; Delgadillo, I.; Gracio, J.; Mishina, E.; Sigov, A.; Kholkin, A.L. Evidence of ferroelectricity and phase transition in pressed diphenylalanine. *Appl. Phys. Lett.* **2012**, *100*, 043702. [CrossRef]
64. Winarto, W.; Yamamoto, E.; Yasuoka, K. Water Molecules in a Carbon Nanotube under an Applied Electric Field at Various Temperatures and Pressures. *Water* **2017**, *9*, 473. [CrossRef]
65. Mikami, F.; Matsuda, K.; Kataura, H.; Maniwa, Y. Dielectric Properties of Water inside Single-Walled Carbon Nanotubes. *ACS Nano* **2009**, *3*, 1279–1287. [CrossRef]

nanomaterials

Article

The Adsorption and Sensing Performances of Ir-modified MoS_2 Monolayer toward SF_6 Decomposition Products: A DFT Study

Hongcheng Liu, Feipeng Wang *, Kelin Hu, Tao Li, Yuyang Yan and Jian Li *

State Key Laboratory of Power Transmission Equipment & System Security and New Technology, Chongqing University, Chongqing 400044, China; swulhcx@163.com (H.L.); 20140802026@cqu.edu.cn (K.H.); litao199811@outlook.com (T.L.); 20143232@cqu.edu.cn (Y.Y.)

* Correspondence: fpwang@cqu.edu.cn (F.W.); lijian@cqu.edu.cn (J.L.); Tel.: +86-185-807-68887 (F.W. & J.L.)

Abstract: In this paper, the Ir-modified MoS_2 monolayer is suggested as a novel gas sensor alternative for detecting the characteristic decomposition products of SF_6, including H_2S, SO_2, and SOF_2. The corresponding adsorption properties and sensing behaviors were systematically studied using the density functional theory (DFT) method. The theoretical calculation indicates that Ir modification can enhance the surface activity and improve the conductivity of the intrinsic MoS_2. The physical structure formation, the density of states (DOS), deformation charge density (DCD), molecular orbital theory analysis, and work function (WF) were used to reveal the gas adsorption and sensing mechanism. These analyses demonstrated that the Ir-modified MoS_2 monolayer used as sensing material displays high sensitivity to the target gases, especially for H_2S gas. The gas sensitivity order and the recovery time of the sensing material to decomposition products were reasonably predicted. This contribution indicates the theoretical possibility of developing Ir-modified MoS_2 as a gas sensor to detect characteristic decomposition gases of SF_6.

Keywords: Ir-modified MoS_2; decomposition components of SF_6; adsorption and sensing; DFT

Citation: Liu, H.; Wang, F.; Hu, K.; Li, T.; Yan, Y.; Li, J. The Adsorption and Sensing Performances of Ir-modified MoS_2 Monolayer toward SF_6 Decomposition Products: A DFT Study. *Nanomaterials* **2021**, *11*, 100. https://doi.org/10.3390/nano11010100

Received: 3 September 2020
Accepted: 21 September 2020
Published: 4 January 2021

1. Introduction

Due to the excellent insulation and arc-extinguishing ability of SF_6 gas, it has been widely used in gas-insulated switchgear (GIS) [1,2]. During long-term operations, SF_6 in GIS decomposes into different characteristic gases, such as H_2S, SO_2, SOF_2, etc., when discharge faults occur [3,4]. Research showed that detecting these characteristic decomposition gases can reflect the operating status and potential faults of the gas-insulated equipment [5–8]. Therefore, the accurate detection of these gases is of important for the online monitoring of GIS and the safe operation of equipment.

Among various detection methods, the metal-oxides-based gas sensor was considered a convenient and effective approach [9]. However, these traditional gas sensors have defects, such as low sensitivity and instability, limiting their further development [10,11]. Two-dimensional (2D) materials are widely used in gas sensing, catalysis, energy storage, etc., due to their large specific surface areas and distinctive physical and chemical properties [12–14]. The MoS_2 monolayer was studied and reported as a potential gas-sensing material [15]. Rathi et al. studied the gas-sensing performance of La-MoS_2 hybrid-heterostructure-based sensor of NO_2 gas. They found that the sensing response of the La-decorated MoS_2 gas sensor was six times more than the pristine MoS_2, indicating that the fabricated sensor was suitable for NO_2 detection [16]. Urs et al. reported the sensitivity of MoS_2 modified with alloy nanoparticles to H_2 at temperatures in the range of 30 to 100 °C. The results showed that the composite nanomaterials can enhance the response to H_2, and this phenomenon was explained using the density functional theory (DFT) simulation [17]. Based on these studies, we deduced that the gas sensitivity of intrinsic MoS_2 can be further effectively improved by introducing appropriate dopants, such as transition metal atoms or metal oxides [18]. In particular, iridium (Ir), which has excellent physicochemical stability, has been proven to effectively improve the gas sensitivity of 2D materials [19,20]. Thus, we speculated that the reasonable combination of Ir and MoS_2 has potential for gas sensing. Considering the huge challenges faced by current sensor development and the importance of detecting SF_6 decomposition components, the adsorption and sensing performance of Ir-modified MoS_2 for these gases should be studied based on DFT for guiding the experiments.

In this paper, we predict the adsorption and sensing properties of Ir-modified MoS_2 monolayer to SF_6 decomposition components (H_2S, SO_2, and SOF_2) based on the DFT calculation. The most stable adsorption structure of Ir-modified MoS_2 for different gases is proposed, and its electronic properties are also systematically explored. The gas-sensing response of the Ir-modified MoS_2 nanomaterial-based sensor to the target gases was reasonably predicted. These results indicated that the Ir-modified MoS_2-based gas sensor is a promising alternative for detecting the decomposition components of SF_6.

2. Computational Detail

In this work, the DMol[3] package based on the DFT method was used to study the electronic and adsorption characteristics [21]. The 4 × 4 MoS_2 monolayer with the lattice constants of 12.664 × 12.664 Å was established, and its size proved to be large enough for gas adsorption [22]. To avoid influence between the adjacent layers of MoS_2, we set the vacuum thickness to 20 Å. The electron exchange and correlation energy were treated by the generalized gradient approximation (GGA), Perdew–Burke–Ernzerhof (PBE), and the DFT-D method. The double numerical plus polarization (DNP) basis set was applied to deal with the relativistic effect of the molecular structure [23]. The k-point was set as 5 × 5 × 1 Monkhorst. To ensure the reasonableness and accuracy of the calculation, the energy convergence accuracy, maximum force, and maximum displacement in this study were selected as 10^{-5} Ha, 0.002 Ha/Å, and 0.005 Å, respectively [24]. The adsorption energy (E_{ad}), charge transfer (Q_t), energy gap (E_g), and recovery time (T_r) for each system were analyzed using the following equations:

$$E_{ad} = E_{gas/Ir\text{-}modified\ MoS2} - E_{gas} - E_{Ir\text{-}modified\ MoS2} \tag{1}$$

$$Q_t = Q_a - Q_b \tag{2}$$

$$E_g = | E_{LUMO} - E_{HOMO} | \tag{3}$$

$$T_r = F_0{}^{-1}\exp(-E_{ad}/K_B T_w) \tag{4}$$

where $E_{gas/Ir\text{-}modified\ MoS2}$, E_{gas}, and $E_{Ir\text{-}modified\ MoS2}$ denote the energy of the Ir-modified MoS_2 structure adsorbed gas, the bare Ir-modified MoS_2 substrate, and the isolated gas, respectively; Q_a and Q_b in represent the charge of gas molecules after and before adsorption, respectively. To study the conductivity change of various adsorption structures, we defined the energy gap (E_g) depicted in Equation (3). HOMO and LUMO represent the highest occupied molecular orbital and the lowest occupied molecular orbital, respectively; the lower the E_g, the higher the conductivity [25]. T_r, F_0 (10^{12} s^{-1}), K_B (8.62×10^{-5} eV K^{-1}), and T_w represent recovery time, attempt frequency, Boltzmann constant, and working temperature, respectively [26].

3. Results and Discussions

3.1. Structure and Properties of the Ir-modified MoS_2 Monolayer

The optimized structures of Ir-modified MoS_2 and target gases are shown in Figure 1. Figure 1a illustrates that the Ir atom forms an optimal configuration with MoS_2 at a distance of 2.742 Å directly above the Mo atom. This doping mode (the S vacancy) proves to be the most stable based on the formation energy analysis [27]. The three optimized Ir-S bond lengths of the Ir-modified MoS_2 structure are 2.250, 2.262, and 2.247 Å. Figure 1b depicts the optimized configurations of three characteristic decomposition components of SF_6. The bond length and bond angle of each gas molecule are marked in the figure, which is consistent with the literature [28,29].

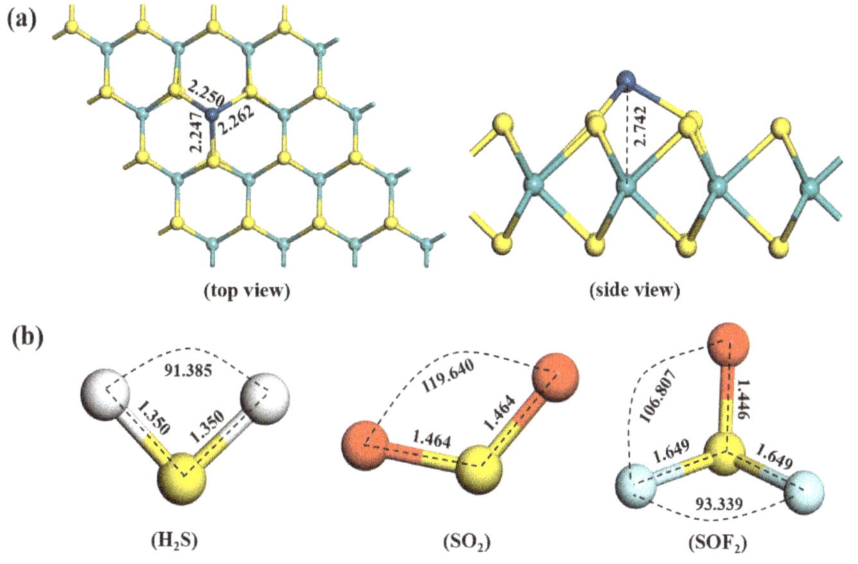

Figure 1. Structures of (**a**) Ir-modified MoS_2 and (**b**) gas molecules.

To further analyze the electronic properties of Ir-modified MoS_2, the density of states (DOS) distributions and band structure are depicted in Figure 2. Figure 2a compares the total density of states (TDOS) of Ir-modified MoS_2 shifted to the left with the intrinsic MoS_2. This phenomenon is mainly attributed to the hybridization of the orbitals of introduced Ir atom with the orbitals of Mo and S atoms in MoS_2. Specifically, the Ir-5d orbital overlaps with Mo-4d and S-3p orbitals in the range of −7.0 to 2.0 eV as observed in the distribution of partial density of states (PDOS), which indicated that the structure formed by the introduced Ir atom and MoS_2 is stable. The Ir-5d peak appears near the Fermi level, indicating that electrons are more easily transferred to the conduction band after Ir atom was introduced, and the conductivity of the gas-sensing material increases. Figure 2b depicts the band structures of intrinsic and Ir-modified MoS_2. We noticed that the bandgap of the composite system reduced from 2.088 to 0.398 eV after introducing an Ir atom due to the introduction of impurity levels near the Fermi level after doping with Ir atom. The electrons are more easily excited from the valence band to the conduction band.

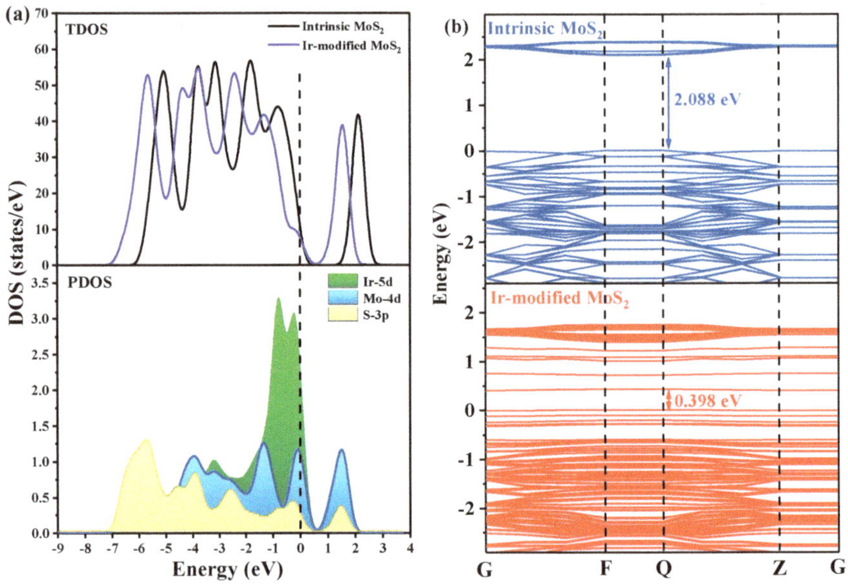

Figure 2. The (**a**) DOS distributions and (**b**) band structure of the intrinsic and Ir-modified MoS_2 systems.

Figure 3a displays the deformation charge density (DCD) of Ir-modified MoS_2 monolayer. The red and blue regions in the figure represent electron accumulation and depletion, respectively [22]. The electrons are mainly concentrated around the Ir atom, which illustrates that the bonds formed by Ir atom and surrounding S atoms have a strong binding force [30]. The Ir atom acts as an electron acceptor and obtains 0.274 e from MoS_2 monolayer, and the S mainly behaves as an electron donator. The HOMO and LUMO distributions of the Ir-modified MoS_2 system are depicted in Figure 3b. We found that a large part of HOMO is distributed near the Ir atom, which indicates that the addition of an Ir atom provides more active sites on the surface of MoS_2 and enhances the sensitivity of the material.

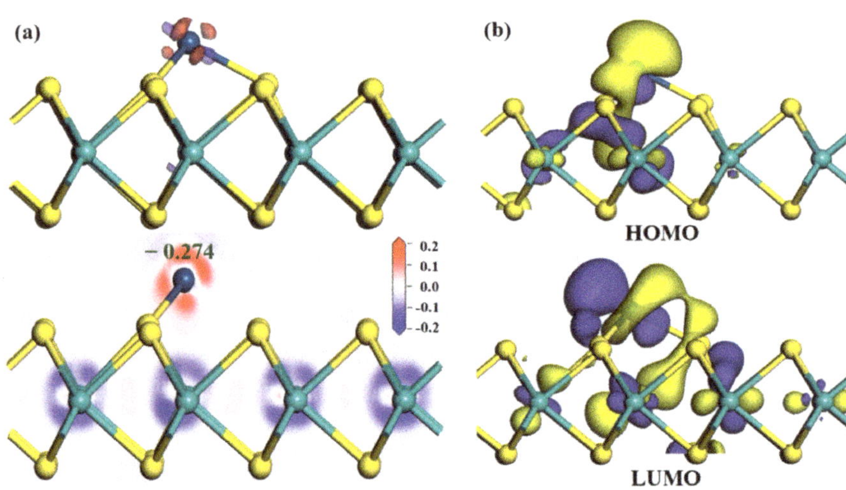

Figure 3. The (**a**) DCD, (**b**) HOMO, and LUMO distributions of Ir-modified MoS_2 system.

3.2. *Gas Molecules Adsorption on the Ir-modified MoS_2 Surface*

3.2.1. Structures of Different Adsorption Systems

To obtain the most stable adsorption structure for each gas, we established various adsorption models in which the gases were close to the Ir-modified MoS_2 surface at different positions. The optimized H_2S-Ir-MoS_2, SO_2-Ir-dopd-MoS_2, and SOF_2-Ir-dopd-MoS_2 structures in various positions are depicted in Figures 4–6, respectively. Based on the figures, we observed that all gas molecules were adsorbed on the surface of the Ir-modified MoS_2 surface in different spatial positions. In particular, in the H_2S adsorption systems of P2 and P3 depicted in Figure 4b,c, the H–S bond in the H_2S gas broke, and the H atom formed a new chemical bond with the Ir atom, indicating that the adsorption of H_2S gas is a chemical adsorption process. The adsorption parameters of Ir-modified MoS_2 for various gases in different positions are depicted in Table 1.We observed from Table 1 that all the adsorption energies were negative, indicating that the adsorption of the target gases is an exothermic process. The absolute value of E_{ad} in all systems was the largest for P3 structure, indicating that the adsorption structure is the most stable compared to the other two configurations, and the gas is most likely to be adsorbed on the surface of the material in this state during the adsorption reaction. The gas in each system was stably adsorbed on the Ir-modified MoS_2 surface at different distances, and the specific adsorption distance values are listed in Table 1. For Q_t, we found that the amount of charge transfer between H_2S and the sensing material was the largest compared with SO_2 and SOF_2 gas adsorption systems, which illustrates that the H_2S gas causes a greater change in the conductivity of the material during the adsorption process.

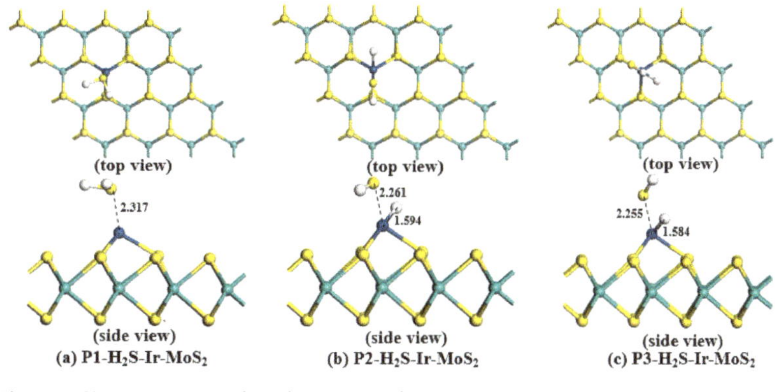

Figure 4. Various optimized configurations of H_2S-Ir-MoS_2 adsorption systems.

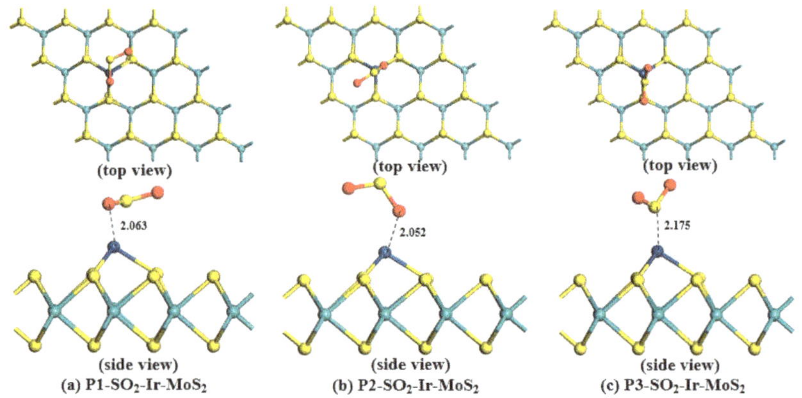

Figure 5. Different optimized structures of SO_2 adsorption on the Ir-modified MoS_2 surface.

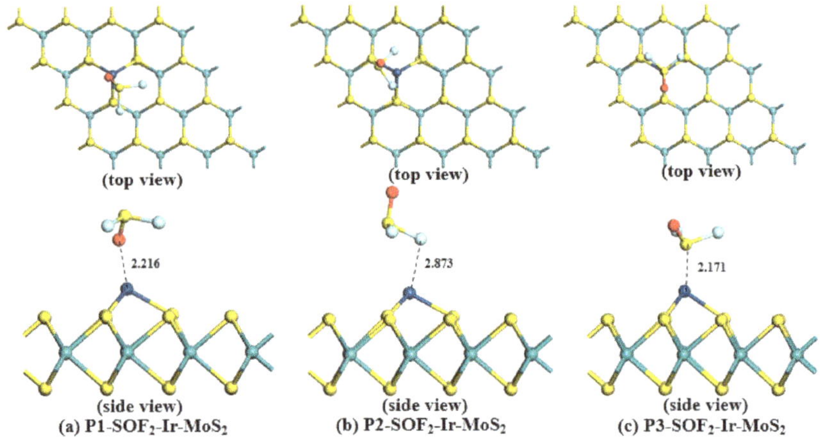

Figure 6. Adsorption systems of SOF_2-Ir-MoS_2 in different positions.

Table 1. The adsorption parameters of Ir-modified MoS_2 for various gases in different positions.

System	Adsorption Position	E_{ad} (eV)	D (Å)	Q_t (e)
	Position 1 (Figure 4a)	−1.578	2.317	0.341
H_2S-Ir-MoS_2	Position 2 (Figure 4b)	−2.310	1.594	0.292
	Position 3 (Figure 4c)	−2.323	1.584	0.286
	Position 1 (Figure 5a)	−1.656	2.063	−0.158
SO_2-Ir-MoS_2	Position 2 (Figure 5b)	−1.053	2.052	−0.190
	Position 3 (Figure 5c)	−1.757	2.175	0.114
	Position 1 (Figure 6a)	−0.411	2.216	0.069
SOF_2-Ir-MoS_2	Position 2 (Figure 6b)	−0.104	2.873	0.048
	Position 3 (Figure 6c)	−1.492	2.171	0.154

To further compare and analyze the different characteristics of Ir-modified MoS_2 sensing material for different SF_6 decomposition components, we selected the most stable structures (Position 3) for each gas adsorption system. The optimized configurations and adsorption parameters of target gases adsorbed on the Ir-modified MoS_2 monolayer are depicted in Figure 7 and Table 2, respectively. We suggest, based on Figure 7a, that the S–H bond in H_2S gas was broken due to the strong metallicity of Ir atom during the adsorption process of H_2S gas, which led to the formation of a new H2–Ir bond between H2 atom and Ir atom with a bond length of 1.584 Å. The H1–S1 bond length is 1.357 Å, which is slightly elongated compared to the bond length of the gas before adsorption (1.350 Å). For the SO_2-Ir-MoS_2 and SOF_2-Ir-MoS_2 adsorption systems depicted in Figure 7b,c, respectively, we observed that the S atom in the gas molecules is close to the Ir atom to form the stable structures. The bond lengths of S2–O1 and S2–O2 in SO_2 gas are 1.467 and 1.468 Å, respectively. The bond lengths of S3–O3, S3–F2, and S3–F1 in SOF_2 gas are 1.449, 1.642, and 1.643 Å, respectively. All of the bond lengths of SO_2 and SOF_2 gases undergo small changes during the adsorption progress, suggesting that gases have interactions with the sensing material.

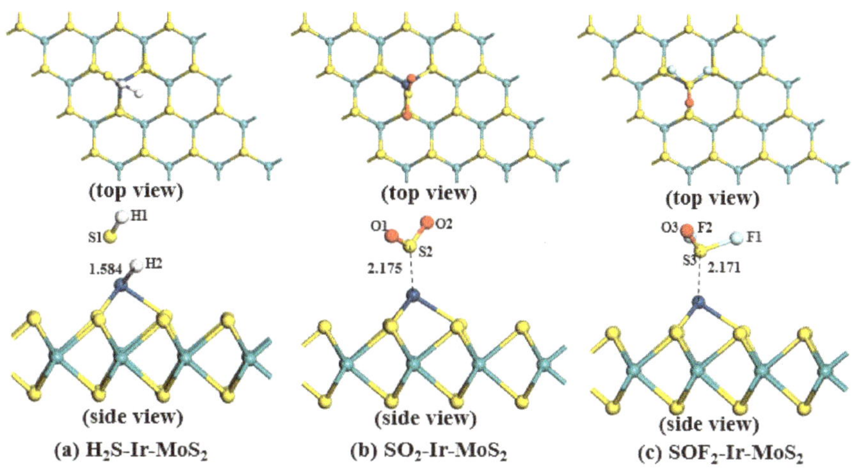

Figure 7. The most stable structures of SF_6 decomposition components adsorbed on the Ir-modified MoS_2 monolayer.

Table 2. The parameters of different Ir-modified MoS_2 adsorption systems.

System	E_{ad} (eV)	D (Å)	Q_t (e)
H_2S-Ir-MoS_2	−2.323	1.584	0.286
SO_2-Ir-MoS_2	−1.757	2.175	0.114
SOF_2-Ir-MoS_2	−1.492	2.171	0.154

Adsorption systems are displayed in Table 2. The E_{ad} of H_2S-Ir-MoS_2 was calculated to be −2.323 eV, which is smaller than the SO_2-Ir-MoS_2 (−1.757 eV) and SOF_2-Ir-MoS_2 (−1.492 eV) systems. This result suggested that all adsorption reactions can proceed spontaneously, the adsorption process of the H_2S gas is the strongest, and the formed structure is the most stable among the three systems. According to the definition of adsorption distance, the D of H_2S-Ir-MoS_2, SO_2-Ir-MoS_2, and SOF_2-Ir-MoS_2 is 1.583, 2.175, and 2.171 Å, respectively. Based on the Mulliken population analysis, we calculated the Q_t of the three adsorption systems as 0.286, 0.114, and 0.154 e for H_2S-Ir-MoS_2, SO_2-Ir-MoS_2, and SOF_2-Ir-MoS_2 configurations, respectively. The largest Q_t of H_2S-Ir-MoS_2 indicates that the H_2S gas interacts strongly with the Ir-modified MoS_2 system, which is consistent with the calculated results of adsorption energy.

3.2.2. DOS Analysis of Different Adsorption Systems

The DOS distributions were used to study the electronic properties and gas sensitivity of SF_6 decomposition components adsorbed on the Ir-modified MoS_2 monolayer, and the results are depicted in Figure 8. Figure 8a depicts the overall TDOS of the H_2S-Ir-MoS_2

system moving to the left after adsorbing H_2S gas, indicating that the electrons easily fill in the conduction band. The huge change in TDOS that happened near the Fermi level indicated that the Ir-modified MoS_2 gas-sensing material has strong interactions with H_2S gas, and the gas sensitivity is improved. The peaks of the PDOS spectrum overlap among Ir, S, and H atoms, suggesting that the adsorption structure has good stability. The hybridization of Ir-3p and S-3p orbitals should be responsible for the huge changes in TDOS near the Fermi level. In the SO_2-Ir-MoS_2 structure displayed in Figure 8b, the TDOS decreases around the Fermi level after SO_2 gas adsorption. This phenomenon illustrates that the number of free electrons decreases after gas adsorption and the resistance of the material increases. We observed the PDOS distributions of Ir-5d orbital near the Fermi level, which may be the main reason for the decrease in TDOS. The hybridization of Ir-5d and S-3p orbitals increases the TDOS near the energy level of −7.0 eV after gas adsorption. For the SOF_2-Ir-MoS_2 system depicted in Figure 8c, the TDOS of the SOF_2-Ir-MoS_2 adsorption system slightly changes near the Fermi level compared with the DOS of Ir-MoS_2, indicating that the material has weak gas-sensitivity to SOF_2 gas. From the PDOS distributions, we found that the novel peaks appearing in DOS at the range of −9.0 to −8.0 eV are mainly caused by the hybridization of the S-3p and F-2p orbitals.

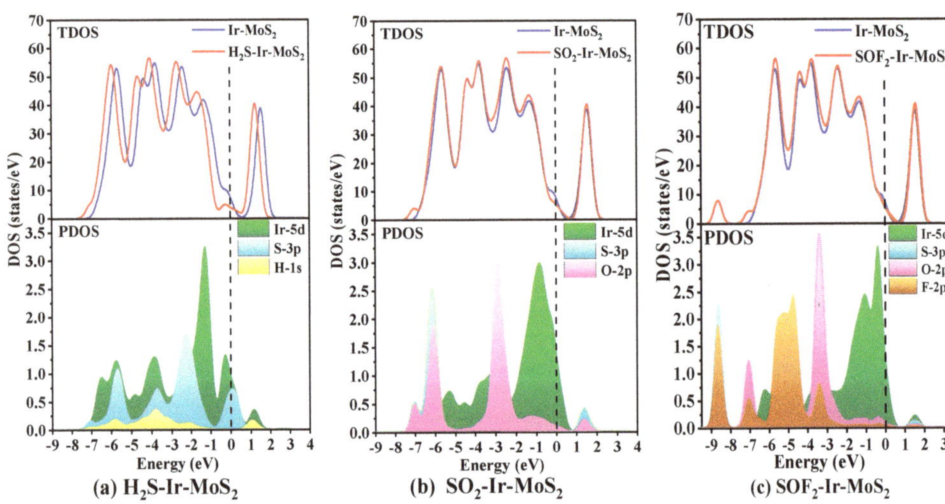

Figure 8. The DOS distributions of different adsorption systems.

3.2.3. DCD Analysis of Various Adsorption Systems

The interactions between SF_6 decomposition components and Ir-modified sensitive materials were studied via the DCD analysis. Figure 9 depicts the corresponding DCD results of various adsorption systems. The charge accumulation in all structures is mainly concentrated on the gas molecules, and charge depletion is distributed around the Ir atom. This distribution suggests that the gases have interactions with the Ir-modified MoS_2 material, which is also consistent with the DOS analysis results. In these adsorption systems, the SF_6 decomposition components act as electron donors, while the Ir-modified MoS_2 material acts as an electron acceptor. Compared with SO_2-Ir-MoS_2 and SOF_2-Ir-MoS_2 adsorption configurations, there is a large amount of charge accumulation and dissipation between the H_2S gas and the Ir-modified MoS_2 layer, which is caused by the strong reaction in the adsorption process [31].

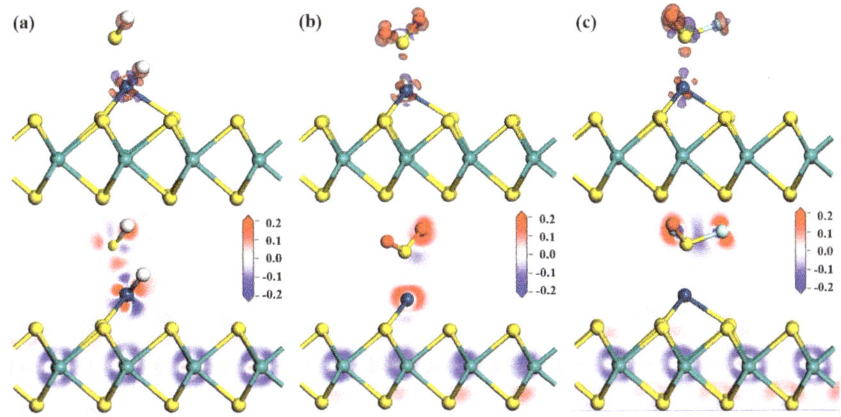

Figure 9. The DCD of (**a**) H_2S-Ir-MoS_2, (**b**) SO_2-Ir-MoS_2, and (**c**) SOF_2-Ir-MoS_2 adsorption systems.

3.2.4. Frontier Molecular Orbital Analysis of Different Systems

The HOMO and LUMO distributions were used to study the electronic behavior of various adsorption systems. According to the frontier molecular orbital analysis, we analyzed the possible change trends of material conductivity to predict the gas-sensing performance of materials [32]. After the target gas is adsorbed, the electron clouds of Ir-modified MoS_2 material is redistributed, causing changes in the energy values of HOMO and LUMO, as depicted in Figure 10. A large amount of HOMO is distributed on and around the gas, especially in the H_2S-Ir-MoS_2 system, indicating that these electrons are not bound and can undergo charge transfer during the adsorption reaction. The Ir-modified MoS_2 sensitive material has obvious electron transfer behavior for the adsorption of these three target gases; thus, we speculate that Ir-modified MoS_2 gas-sensing material can be used as a resistive gas sensor to detect H_2S, SO_2, and SOF_2 gases.

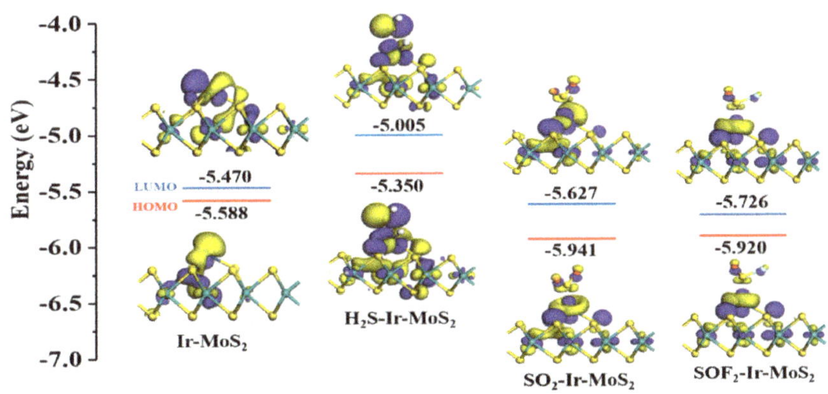

Figure 10. The HOMO and LUMO distributions of various systems.

3.3. *Gas-Sensing Prediction of Ir-modified MoS_2 to SF_6 Decomposition Products*

After gas adsorption, a large change in E_g indicates that the conductivity of the sensing material increases based on the definition of resistive gas sensor sensitivity [33]. The comparative analysis of E_g for different optimized systems are proposed and displayed in Figure 11a. The E_g increased to varying degrees after gas adsorption due to the different changes in the energy levels of HOMO and LUMO (marked in Figure 10). Compared with the E_g of the Ir-modified MoS_2, the degree of E_g change in these systems is as follows:

H_2S-Ir-MoS_2 > SO_2-Ir-MoS_2 > SOF_2-Ir-MoS_2. Combining the above results, we suggest that the gas sensitivity order of the Ir-modified MoS_2 to these SF_6 decomposition products is H_2S > SO_2 > SOF_2. Work function (WF) refers to the minimum energy required for electrons to release from the surface. It represents the contact barrier between the target gas and the material during the gas adsorption process [31]. Figure 11b shows the calculated values of WF for various optimized structures. We observed that WF decreases to 5.252 eV after the adsorption of H_2S gas, but increases to 5.905 and 5.878 eV for SO_2 and SOF_2, respectively, compared with the Ir-modified MoS_2 system (5.469 eV). In other words, the smaller work function value of H_2S-Ir-MoS_2 indicates that H_2S more easily adsorbs on the Ir-modified MoS_2 material compared with the other two systems.

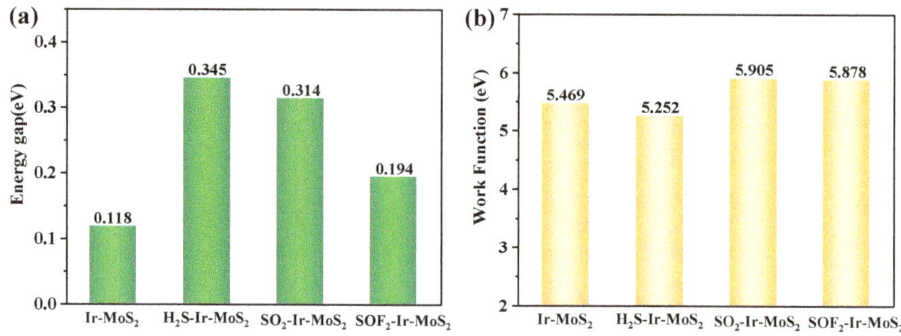

Figure 11. The (**a**) energy gap and (**b**) working function for different optimized systems.

The predicted recovery time for the Ir-modified MoS_2 based sensor is displayed in Figure 12. The recovery time decreases with the increase in temperature due to the rapid desorption of gas molecules at high temperatures. The sequence of the recovery time for these gases at the same temperature is as follows: SOF_2 < SO_2 < H_2S. Although the recovery time of H_2S is very long due to the strong adsorption capacity, the time can be less than 2 min by appropriately increasing the working temperature during the experimental test. This result can provide a theoretical basis for guiding the gas-sensing performance test experiment.

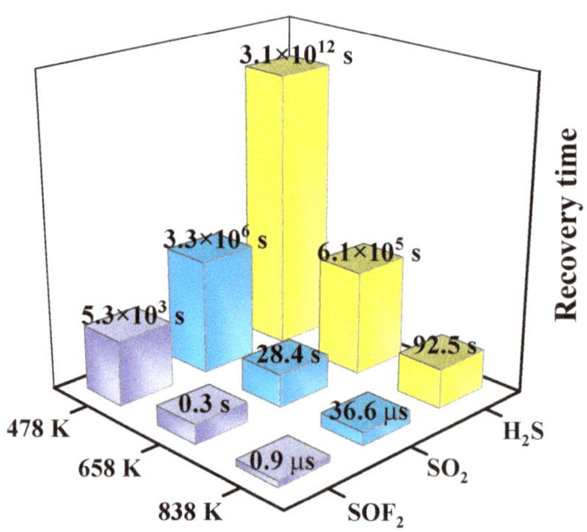

Figure 12. The predicted recovery time of various optimized systems.

4. Conclusions

In this work, we used the theoretical calculation based on the DFT method to investigate the adsorption characteristic and gas-sensing mechanism of Ir-modified MoS_2 to decomposition components of SF_6, including H_2S, SO_2, and SOF_2. We optimized and analyzed the geometric parameters and electronic properties of the Ir-modified MoS_2 system. The results indicated that the introduction of the Ir atom enhances the surface activity of the material and reduces the bandgap of intrinsic MoS_2 from 2.088 to 0.398 eV as well as increases the conductivity. The most stable adsorption structure of Ir-modified MoS_2 for different gases was proposed, and their electronic properties were systematically explored via analyzing the DOS, DCD, molecular orbital theory, and WF. The gas-sensing mechanism study demonstrated that Ir-modified MoS_2 monolayer can adsorb the target gases and cause microscopic electron behavior, especially for H_2S gas. The gas sensitivity order of SF_6 decomposition products was predicted as follows: $H_2S > SO_2 > SOF_2$. The predicted recovery time of the sensor to all target gases can be less than 2 min by appropriately increasing the temperature. Based on these results, the Ir-modified MoS_2 is suggested as a potential candidate for detecting decomposition components of SF_6.

Author Contributions: Conceptualization, H.L., F.W. and J.L.; methodology, H.L., K.H. and F.W.; investigation, H.L. and T.L.; resources, T.L. and Y.Y.; writing—original draft preparation, H.L.; writing—review and editing, H.L., T.L. and Y.Y.; supervision, K.H.; project administration, F.W. and J.L. All authors have read and agreed to the published version of the manuscript.

Funding: This work was supported in part by the National Key Research and Development Program of China (No.2018YFB2100100), the Graduate Research and Innovation Foundation of Chongqing, China (No.CYS20008), the Chongqing Municipality Human Resources and Social Security Bureau (No.cx2017041), and the National "111" Project of the Ministry of Education of China (No.B08036).

Conflicts of Interest: The authors declare no conflict of interest.

References

1. Beslija, D.; Gorenc, D.; Muratovic, M.; Kapetanovic, M. Enhanced Method for Pressure Rise Calculation in SF_6 GIS Due to Fault Arcs. *IEEE Trans. Power Deliv.* **2020**, *35*, 1619–1624. [CrossRef]
2. Wu, S.Y.; Zeng, F.P.; Tang, J.; Yao, Q.; Miao, Y.L. Triangle Fault Diagnosis Method for SF_6 Gas-Insulated Equipment. *IEEE Trans. Power Deliv.* **2019**, *34*, 1470–1477. [CrossRef]
3. Chen, D.C.; Zhang, X.X.; Tang, J.; Cui, Z.L.; Cui, H.; Pi, S.M. Theoretical Study of Monolayer $PtSe_2$ as Outstanding Gas Sensor to Detect SF_6 Decompositions. *IEEE Electron Device Lett.* **2018**, *39*, 1405–1408. [CrossRef]
4. Pepi, F.; Ricci, A.; Di Stefano, M.; Rosi, M. Sulfur hexafluoride corona discharge decomposition: Gas-phase ion chemistry of SOFx+ (x = 1–3) ions. *Chem. Phys. Lett.* **2003**, *381*, 168–176. [CrossRef]
5. Liu, D.K.; Gui, Y.G.; Ji, C.; Tang, C.; Zhou, Q.; Li, J.; Zhang, X.X. Adsorption of SF_6 decomposition components over Pd (111): A density functional theory study. *Appl. Surf. Sci.* **2019**, *465*, 172–179. [CrossRef]
6. Wang, D.W.; Lan, T.S.; Yang, A.J.; Pan, J.B.; Chu, J.F.; Yuan, H.; Li, Y.J.; Wang, X.H.; Rong, M.Z. SF_6 Decomposition Gas Sensor Based on GeP Monolayer: A First-Principle Study. *IEEE Sens. J.* **2020**, *20*, 8997–9003. [CrossRef]
7. Cui, H.P.; Zheng, K.; Tao, L.Q.; Yu, J.B.; Zhu, X.Y.; Li, X.D.; Chen, X.P. Monolayer Tellurene-Based Gas Sensor to Detect SF_6 Decompositions: A First Principles Study. *IEEE Electron Device Lett.* **2019**, *40*, 1522–1525. [CrossRef]
8. Qian, H.; Deng, J.; Xie, Z.C.; Pan, Z.C.; Zhang, J.Y.; Zhou, H.B. Adsorption and Gas Sensing Properties of the Pt_3-$MoSe_2$ Monolayer to SOF_2 and SO_2F_2. *ACS Omega* **2020**, *5*, 7722–7728. [CrossRef]
9. Lu, Z.R.; Zhou, Q.; Wang, C.S.; Wei, Z.J.; Xu, L.N.; Gui, Y.G. Electrospun ZnO-SnO_2 Composite Nanofibers and Enhanced Sensing Properties to SF_6 Decomposition Byproduct H_2S. *Front. Chem.* **2018**, *6*, 9. [CrossRef]
10. Zeng, W.W.; Liu, Y.Z.; Mei, J.; Tang, C.Y.; Luo, K.; Li, S.M.; Zhan, H.R.; He, Z.K. Hierarchical SnO_2-Sn_3O_4 heterostructural gas sensor with high sensitivity and selectivity to NO_2. *Sens. Actuator B Chem.* **2019**, *301*, 9. [CrossRef]
11. Constantinoiu, I.; Viespe, C. Development of Pd/TiO_2 Porous Layers by Pulsed Laser Deposition for Surface Acoustic Wave H_2 Gas Sensor. *Nanomaterials* **2020**, *10*, 760. [CrossRef] [PubMed]
12. Roy, P.K.; Luxa, J.; Sofer, Z. Emerging pnictogen-based 2D semiconductors: Sensing and electronic devices. *Nanoscale* **2020**, *12*, 10430–10446. [CrossRef] [PubMed]
13. Rui, H.S.; Li, L.; Zhang, N.; Lin, X.; Hua, Y.L.; Wu, X.M.; Wang, D.; Yin, S.G. Tunable Deep-Red Electroluminescence From Flexible Quasi-2D Perovskites Light-Emitting Diodes. *IEEE Electron Device Lett.* **2019**, *40*, 59–62. [CrossRef]
14. Gui, Y.G.; Li, T.; He, X.; Ding, Z.Y.; Yang, P.G. Pt Cluster Modified h-BN for Gas Sensing and Adsorption of Dissolved Gases in Transformer Oil: A Density Functional Theory Study. *Nanomaterials* **2019**, *9*, 1746. [CrossRef]

15. Yang, Z.M.; Zhang, D.Z.; Chen, H.N. MOF-derived indium oxide hollow microtubes/MoS_2 nanoparticles for NO_2 gas sensing. *Sens. Actuator B Chem.* **2019**, *300*, 10. [CrossRef]
16. Rathi, K.; Kumar, A.N.; Pal, K. Fabrication of flexible La-MoS_2 hybrid-heterostructure based sensor for NO_2 gas sensing at room temperature. *Nanotechnology* **2020**, *31*, 13. [CrossRef]
17. Urs, K.M.B.; Katiyar, N.K.; Kumar, R.; Biswas, K.; Singh, A.K.; Tiwary, C.S.; Kamble, V. Multi-component (Ag-Au-Cu-Pd-Pt) alloy nanoparticle-decorated p-type 2D-molybdenum disulfide (MoS_2) for enhanced hydrogen sensing. *Nanoscale* **2020**. [CrossRef]
18. Wang, J.X.; Zhou, Q.; Lu, Z.R.; Gui, Y.G.; Zeng, W. Adsorption of H_2O molecule on TM (Au, Ag) doped-MoS_2 monolayer: A first-principles study. *Physica E* **2019**, *113*, 72–78. [CrossRef]
19. Lin, L.; Shi, Z.G.; Huang, J.T.; Wang, P.T.; Yu, W.Y.; He, C.Z.; Zhang, Z.Y. Molecular adsorption properties of CH_4 with noble metals doped onto oxygen vacancy defect of anatase TiO_2(101) surface: First-principles calculations. *Appl. Surf. Sci.* **2020**, *514*, 7. [CrossRef]
20. Basharnavaz, H.; Habibi-Yangjeh, A.; Kamali, S.H. A first-principle investigation of NO_2 adsorption behavior on Co, Rh, and Ir-embedded graphitic carbon nitride: Looking for highly sensitive gas sensor. *Phys. Lett. A* **2020**, *384*, 8. [CrossRef]
21. Zhang, X.X.; Chen, Z.W.; Chen, D.C.; Cui, H.; Tang, J. Adsorption behaviour of SO_2 and SOF_2 gas on Rh-doped BNNT: A DFT study. *Mol. Phys.* **2020**, *118*, 9. [CrossRef]
22. Gui, Y.G.; Liu, D.K.; Li, X.D.; Tang, C.; Zhou, Q. DFT-based study on H_2S and SOF_2 adsorption on Si-MoS_2 monolayer. *Results Phys.* **2019**, *13*, 8. [CrossRef]
23. Wang, J.X.; Zhou, Q.; Lu, Z.R.; Wei, Z.J.; Zeng, W. Gas sensing performances and mechanism at atomic level of Au-MoS_2 microspheres. *Appl. Surf. Sci.* **2019**, *490*, 124–136. [CrossRef]
24. Leenaerts, O.; Partoens, B.; Peeters, F.M. Adsorption of H_2O, NH_3, CO, NO_2, and NO on graphene: A first-principles study. *Phys. Rev. B* **2008**, *77*, 6. [CrossRef]
25. Zhang, X.X.; Fang, R.X.; Chen, D.C.; Zhang, G.Z. Using Pd-Doped gamma-Graphyne to Detect Dissolved Gases in Transformer Oil: A Density Functional Theory Investigation. *Nanomaterials* **2019**, *9*, 1490. [CrossRef]
26. Peng, S.; Cho, K.J.; Qi, P.F.; Dai, H.J. Ab initio study of CNT NO_2 gas sensor. *Chem. Phys. Lett.* **2004**, *387*, 271–276. [CrossRef]
27. Cui, H.; Zhang, X.X.; Zhang, G.Z.; Tang, J. Pd-doped MoS_2 monolayer: A promising candidate for DGA in transformer oil based on DFT method. *Appl. Surf. Sci.* **2019**, *470*, 1035–1042. [CrossRef]
28. Zhang, X.X.; Yu, L.; Wu, X.Q.; Hu, W.H. Experimental Sensing and Density Functional Theory Study of H_2S and SOF_2 Adsorption on Au-Modified Graphene. *Adv. Sci.* **2015**, *2*, 10. [CrossRef]
29. Li, J.; Pang, L.; Cai, F.; Yuan, X.; Kong, F. Adsorption Properties of Pd_3-Modified Double-Vacancy Defect Graphene toward SF_6 Decomposition Products. *Sensors* **2020**, *20*, 4188. [CrossRef]
30. Ma, S.X.; Li, D.J.; Rao, X.J.; Xia, X.F.; Su, Y.; Lu, Y.F. Pd-doped h-BN monolayer: A promising gas scavenger for SF_6 insulation devices. *Adsorpt. J. Int. Adsorpt. Soc.* **2020**, *26*, 619–626. [CrossRef]
31. Cui, H.; Yan, C.; Jia, P.F.; Cao, W. Adsorption and sensing behaviors of SF_6 decomposed species on Ni-doped C_3N monolayer: A first-principles study. *Appl. Surf. Sci.* **2020**, *512*, 7. [CrossRef]
32. Cui, H.; Zhang, X.X.; Chen, D.C.; Tang, J. Adsorption mechanism of SF_6 decomposed species on pyridine-like PtN_3 embedded CNT: A DFT study. *Appl. Surf. Sci.* **2018**, *447*, 594–598. [CrossRef]
33. Wang, F.P.; Liu, H.C.; Hu, K.L.; Li, Y.Q.; Zeng, W.; Zeng, L. Hierarchical composites of MoS_2 nanoflower anchored on SnO_2 nanofiber for methane sensing. *Ceram. Int.* **2019**, *45*, 22981–22986. [CrossRef]

nanomaterials

MDPI

Article

Magnetic Simulations of Core–Shell Ferromagnetic Bi-Magnetic Nanoparticles: The Influence of Antiferromagnetic Interfacial Exchange

Juan A. Ramos-Guivar [1,*], Carlo A. Tamanaha-Vegas [1], Fred Jochen Litterst [2,3] and Edson C. Passamani [4]

1 Grupo de Investigación de Nanotecnología Aplicada para Biorremediación Ambiental, Energía, Biomedicina y Agricultura (NANOTECH), Facultad de Ciencias Físicas, Universidad Nacional Mayor de San Marcos, Av. Venezuela Cdra 34 S/N, Ciudad Universitaria, Lima 15081, Peru; carlo.levano@unmsm.edu.pe

2 Institut für Physik der Kondensierten Materie, Technische Universität Braunschweig, 38106 Braunschweig, Germany; j.litterst@tu-braunschweig.de

3 Centro Brasileiro de Pesquisas Físicas, Rio de Janeiro 22290-180, Brazil

4 Department of Physics, Federal University of Espirito Santo-UFES, Vitória 29075-910, Brazil; edson.caetano@ufes.br

* Correspondence: juan.ramos5@unmsm.edu.pe; Tel.: +51-1-914728212

Abstract: Magnetic properties of ferromagnetic nanostructures were studied by atomistic simulations following Monte Carlo and Landau–Lifshitz–Gilbert approaches. First, we investigated the influence of particle size and shape on the temperature dependence of magnetization for single cobalt and gadolinium nanoparticles and also in bi-magnetic Co@Gd core–shell nanoparticles with different sizes. The Landau–Lifshitz–Gilbert approach was subsequently applied for inspecting the magnetic hysteresis behavior of 2 and 4 nm Co@Gd core–shell nanoparticles with negative, positive, and zero values of interfacial magnetic exchange. We were able to demonstrate the influence of finite-size effect on the dependence of the Curie temperature of Co and Gd nanoparticles. In the Co@Gd core–shell framework, it was possible to handle the critical temperature of the hybrid system by adjusting the Co core size. In addition, we found an improvement in the coercive field values for a negative interfacial exchange energy and for a different core size, suggesting an exchange spring behavior, while positive and zero values of interfacial exchange constant showed no strong influence on the hysteresis behavior.

Keywords: atomistic simulation; core–shell bi-magnetic nanoparticles; Monte Carlo simulation; interfacial exchange

Citation: Ramos-Guivar, J.A.; Tamanaha-Vegas, C.A.; Litterst, F.J.; Passamani, E.C. Magnetic Simulations of Core–Shell Ferromagnetic Bi-Magnetic Nanoparticles: The Influence of Antiferromagnetic Interfacial Exchange. *Nanomaterials* **2021**, *11*, 1381. https://doi.org/10.3390/nano11061381

Academic Editor: Vladimir S. Bystrov

Received: 17 April 2021
Accepted: 20 May 2021
Published: 24 May 2021

1. Introduction

The thermal dependence of the magnetization, *M(T)*, of low dimensional ferromagnets (FM) has been studied and simulated by different methods using the Hinzke–Nowak Monte Carlo (MC) algorithm [1], the magnetic dynamic Landau–Lifshitz–Gilbert (LLG) equation [2], and very recently by the Heisenberg mean-field theory [3]. The Curie temperature (T_C) of FM in the nanoscale regime is still an open issue. It is basically due to the fact that the T_C value will depend on the finite-size effects, when long-range atomic ordering is lost due to material sizes and surface morphologies [4].

A magnetic material with a low T_C value (near to room temperature) is often required for applications, for instance, in magnetic hyperthermia [5]. Some good examples of these materials are doped nanoparticulate systems, such as $La_{1-x}Sr_xMnO_3$ [6], $Mn_{1-x}Zn_xFe_2O_4$ nanoparticles (NPs) [7], and hard–soft mixed ferrites [5]. In addition to this peculiarity, these magnetic nanosystems can be coated with gadolinium (a rare earth metal with a T_C of 293 K that is relatively low as compared to ordinary 3D-FM [8]) to be used as a contrast agent in Magnetic Resonance Imaging (MRI) recording [9,10]. Thus, the understanding of the magnetic interactions and the *M(T)* behavior, which govern the core–shell magnetic nanosystems, are crucial, at least for future biomedical applications. In fact, the *M(T)* behavior in bi-magnetic NPs has been theoretically and experimentally studied in some systems [11,12] but not in cobalt and gadolinium (Co@Gd) core–shell structures having different geometries.

On the other hand, several models have theoretically been employed to study the superparamagnetism in ensembles of magnetically non-interacting and interacting NPs. For instance, Serantes et al. [13] simulated the behavior of an ensemble of non-interacting identical particles contained in a 3D box. The implementation of the effect of superparamagnetism was done by taking into account the classical formulation of Stoner–Wohlfarth using the MC method. Another formulation was considered by Arantes et al. [14], who implemented the classical Heisenberg model also for non-interacting frozen magnetic particles. They traced the transition from the magnetically blocked to the superparamagnetic regime using the concept of blocking temperature that showed an intrinsic dependency with the finite-size effect. Obviously, when the system's complexity is enhanced for nanostructures (including different magnetic exchange types), the simulation times get longer (i.e., costlier), and therefore, new simulation methods need to be developed and implemented to analyze the temperature dependence of magnetic properties of core–shell-like nanostructures. Especially for a combination of magnetically hard and soft magnets, one may expect a magnetic exchange interaction between these two FM phases, resulting in an exchange spring (ES)-like phenomenon. Thus, the hybrid materials will have enhanced magnetic properties such as high coercivity and also high saturation magnetization, as reported by Coehoorn et al. [15].

Considering the above points, we simulated magnetic properties of single Co and Gd NPs and also the nanohybrid Co@Gd core–shell structure by atomistic simulations using VAMPIRE software. We performed several studies applying the MC method in magnetic nanosystems for the following cases: (i) the *M(T)* behavior of individual FM Co and Gd NPs with different geometries (spherical, cubic, and cylindrical) in the range from 1 to 9 nm, (ii) core–shell spherical systems of Co@Gd, where the thermal dependence of the critical temperature was studied using different interfacial exchange values with an insignificant change in the T_C values, suggesting that the bi-magnetic core–shell system can be employed in biomedical applications for imaging contrast with high magnetic response, and finally (iii) the hysteretic behavior of the Co@Gd core–shell system was studied by changing the Co core size and the interfacial exchange constant. An improvement in the coercive (H_C) field was observed for a negative interfacial exchange, suggesting the existence of the ES effect. Thus, from our results it is possible to significantly increase the coercive values and handle the presence of ES behavior at the core–shell bi-magnetic interfaces.

2. Theoretical Background, Models, and Methodology

Atomistic Simulations Using VAMPIRE

In the framework of the atomistic system generation, VAMPIRE 5.0 software (York, England) [2] was used to generate particles with certain geometrical shapes (spheres, cylinders, and cubes). The first step was to create a crystalline network with a size large enough to contain the simulated shape. The exchange interaction of the unit cell is known if one considers the first neighbors' approach, given by Equation (1). Once the unit cell is replicated until the dimensions agree with the first step, atoms are removed from the crystalline structure until a desired geometrical shape is reached [2]. The atoms within this already molded crystalline structure are then assigned to one or more materials. This also enables the formation of core–shell structures, but it will depend on the core structure, shell, and interfacial exchange interactions.

Two protocols were established for simulations with the VAMPIRE software:

(i) To reach equilibrium for a subsequent study of the *M(T)* properties, a Hinzke–Nowak MC algorithm was employed, using 20,000 equilibrium steps. After 20,000 averaging steps, the system was then heated in steps of 5 K till the complete *M(T)* curve was obtained. To investigate the finite-size effects on T_C, the equilibrium magnetization for different sizes of spherical, cylindrical, and cubic NPs were calculated as a function of temperature, $M(T)$. The Hamiltonian for the simulated system is [2]:

$$H = -\sum_{i \neq j} J_{ij} S_i . S_j \tag{1}$$

where J_{ij} is the FM exchange constant, and i,j, refer to FM sites. For the simulations, we fixed J_{ij} = 6.064 × 10^{-21} J/link (=joule/binding) and Co atomic spin moment of 1.72 μ_B (=bulk value of moment). In the case of Gd, J_{ij} = 1.28 × 10^{-21} J/link and Gd atomic spin moment of 7.63 μ_B (bulk value). The exchange integral (J) is defined, based on mean field theory, as [2]

$$J = \frac{3k_B T_c}{\epsilon z} \tag{2}$$

where k_B is the Boltzmann constant, z is the number of nearest neighbors, and ϵ is a correction factor obtained from the mean-field expression [2].

For the core–shell model (e.g., negative interfacial configuration), the Hamiltonian Equation (1) turns to

$$\begin{aligned} H_{exc} = H_{FM1} + H_{AFM} + H_{FM2} = \sum\nolimits_{i \neq j} J_{FM1} S_i . S_j + \sum\nolimits_{i,v} J_{afm} S_i . S_v + \\ \sum\nolimits_{v,i} J_{afm} S_v . S_i + \sum\nolimits_{i \neq j} J_{FM2} S_i . S_j \end{aligned} \tag{3}$$

where S corresponds to the spin unit vector. In Equation (3), the first term refers to FM Co, the last one to the shell of Gd, and the second and third terms are the interfacial exchange energies that, in the VAMPIRE software, are taken with the same value. In other words, Equation (3) describes only exchange coupling between the two FM structures, and other interactions (dipole–dipole, anisotropy, and Zeeman, e.g.) to the total magnetic energy are considered negligible. The dipolar was discarded because we assumed non-interacting NPs, while the magnetic anisotropy was due to the fact that the temperature-dependent magnetic properties of the nanohybrid Co@Gd core–shell system were mainly associated with the strong relation between T_C and J given by Equation (2). The Zeeman contribution was neglected since it is still not implemented in the VAMPIRE software using the MC approach. We only took into consideration the interfacial exchange and direct exchange in the FM core and shell, respectively. In brief, the above model assumes the fact that NPs are monodisperse, and the effect of magnetic interactions is expected to be negligible.

(ii) The Landau–Lifshitz–Gilbert (LLG)–Heun method, implemented in the software VAMPIRE [2], was used to simulate the hysteresis curves [$M(H)$], using Equation (4) as suggested in Ref. [2]:

$$\frac{\partial S_i}{\partial t} = -\frac{\gamma}{(1+\lambda^2)}\left[S_i \times H^i_{eff} + \lambda S_i \times \left(S_i \times H^i_{eff}\right)\right] \tag{4}$$

where S_i is the unit vector that represents the direction of the magnetic spin moment i, λ is the microscopic damping parameter, γ is the gyromagnetic ratio, and H^i_{eff} is the net effective magnetic field on each spin [2] (see Figure 1 for a sketch of the spin precession model). In Equation (4), the H^i_{eff} is a sum of two contribution as given by Equation (5):

$$H^i_{eff} = -\frac{1}{\mu_S}\frac{\partial H}{\partial S_i} + H^i_{th} \tag{5}$$

where the first term is the Hamiltonian used for the LLG method as represented by Equation (6), and the second term is the thermal field, given by Equation (7), as follows:

$$H = H_{exc} + H^{uni}_{ani} + H_{app} \tag{6}$$

$$H_{th}^{i} = \Gamma(t)\sqrt{\frac{2\lambda k_B T}{\gamma \mu_S \Delta t}} \tag{7}$$

where H_{ani}^{uni} is the uniaxial magnetic anisotropy and external applied magnetic field contribution, $\Gamma(t)$ is a Gaussian distribution that represents the thermal fluctuations of each spin i, and μ_S is the magnitude of the atomic magnetic moment.

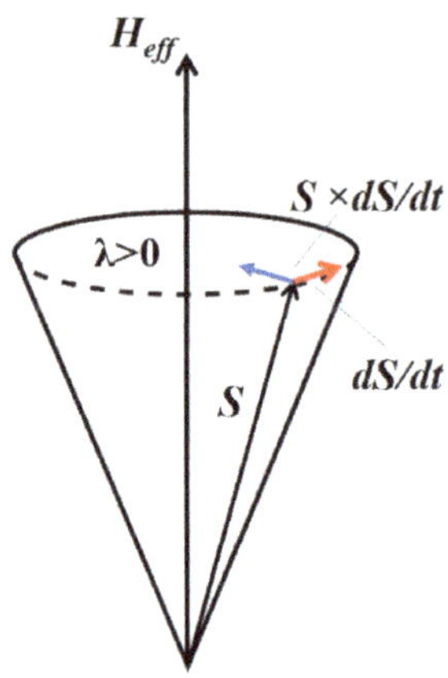

Figure 1. Sketch of spin precession.

For core–shell models, the same input and mat files used for *M(T)* curves were used as templates, but the magnetic field was varied from −3T to +3T with an applied field strength increment of 0.01 T. The λ value was taken equal to 1.0 for an elemental FM as implemented by [2] for all simulations, and the same Hamiltonian given in Equation (4) was used. Additionally, it is important to note that all magnetization data were presented as normalized magnetization (M/M_s), where M_S is the saturation magnetization. For all simulations performed in this work, a simple laptop HP Intel Core i3, serial number 5CB4211908, was used.

3. Results and Discussion

We start this section pointing out that for magnetic configurations (e.g., core–shell structures), a single MC program for non-interacting NPs needs to be modified. Otherwise, it will increase the MC step to a much higher value, as those employed by Nehme et al. [16], where up to 100,000 MCs were used to stabilize the physical parameters of a non-interacting system. This, of course, will increase the computation time significantly to some days or even weeks depending on the Random Access Memory (RAM) computer memory and multi-core processor. On the other hand, the VAMPIRE software allows one to perform atomistic simulations in a single graphics processing unit (GPU) for a variety of magnetic configurations using MC and LLG approaches in much shorter computational times. The software runs easily in serial mode with optimized computational time. Based on that, we applied the VAMPIRE software to study FM bi-magnetic core–shell structures, and the results are shown in the next sections.

3.1. T_c Finite-Size Effects for Single Gd and Co FM NPs with Different Geometries

Before studying the core–shell structures, the *M(T)* behavior was systematically studied independently for single Co and Gd nanostructures with three geometries (sphere, cylinder, and cubic). For the spheres, the diameters were in the range from 1 to 9 nm. For the cube geometry, the cubic length for each face was the same in the rectangular (x,y,z) dimensions and increased from 1 to 9 nm, while for the case of cylinder geometry, the side length increased from 1 to 9 nm, and the base diameter of the cylinder was fixed to 2 nm for all simulations. The results are respectively shown in Figures 2 and 3. A clear distinction in the M/M_S (hereafter called as $M(T)$) profiles is observed with decreasing size. For 1 to 6 nm, the $M(T)$ curves have a tendency to increase continuously above T_C, as would be expected for small FM NP systems (not behaving as an ordinary FM that would show a Brillouin-like curve).

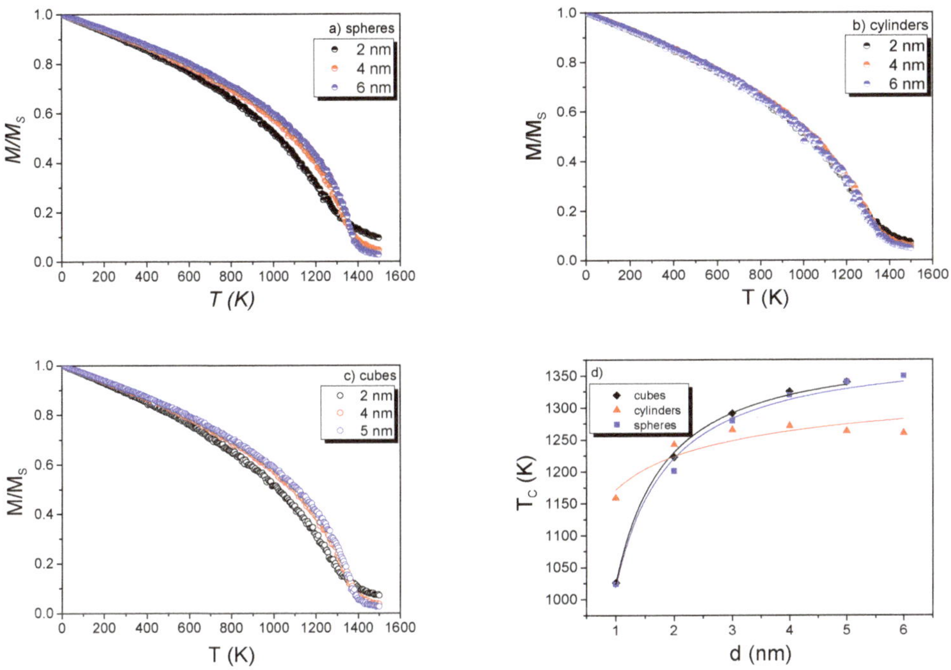

Figure 2. Normalized $M(T)/M_s$ curves for an ensemble of non-interacting Co NPs with sphere (**a**), cylinder (**b**), and cube (**c**) geometries. (**d**) Size dependence of the T_C for the three studied geometries. (in (**d**) the full lines are guides for eyes).

For the determination of T_C, three approximations were employed [1]: (i) Curie–Bloch (Equation (8)), (ii) Kuzmin relation (Equation (9)), and iii) derivative criteria and subsequent polynomial fitting. A first fitting was performed using the Equations (8) and (9):

$$m(T) = (1 - T/T_c)^{\beta} \tag{8}$$

$$m(\tau) = \left[1 - s\tau^{\frac{3}{2}} - (1-s)\tau^{p}\right]^{\beta} \tag{9}$$

where β is suggested to be a universal exponent equal to 0.34 for ideal FM and $\tau = \frac{T}{T_C}$, $s = 1$, and $p = 3/2$ for a pure Bloch FM [1]. However, the Equations (8) and (9) cannot be used to fit the simulated $M(T)$ data when particle sizes are below 9 nm. Values of β ranging from 0.45 to 0.34 were obtained for sizes in the range of 1 to 9 nm, respectively, and consequently no accurate determination of T_C could be obtained. Hence, the third approximation was employed, and the obtained T_C values vs d(nm) are plotted in Figures 2d and 3d for Co and Gd NPs, respectively. T_C values decreased for individual Co and Gd NPs for sizes smaller than 5 nm. As is well-known from the literature [17], the T_C has a finite-size dependence for which critical parameters can be determined using Equation (10):

$$\in(d) = \frac{T_c^{\infty} - T_c(d)}{T_c^{\infty}} = \left(\frac{d_0}{d}\right)^{z} \tag{10}$$

where z is the phenomenological shift exponent, d (nm) is the length of the NPs geometries (diameters for spheres, and cubic and side cylinder lengths), and d_0 is the microscopic length close in value to the single unit cell in the lattice structure of the FM material.

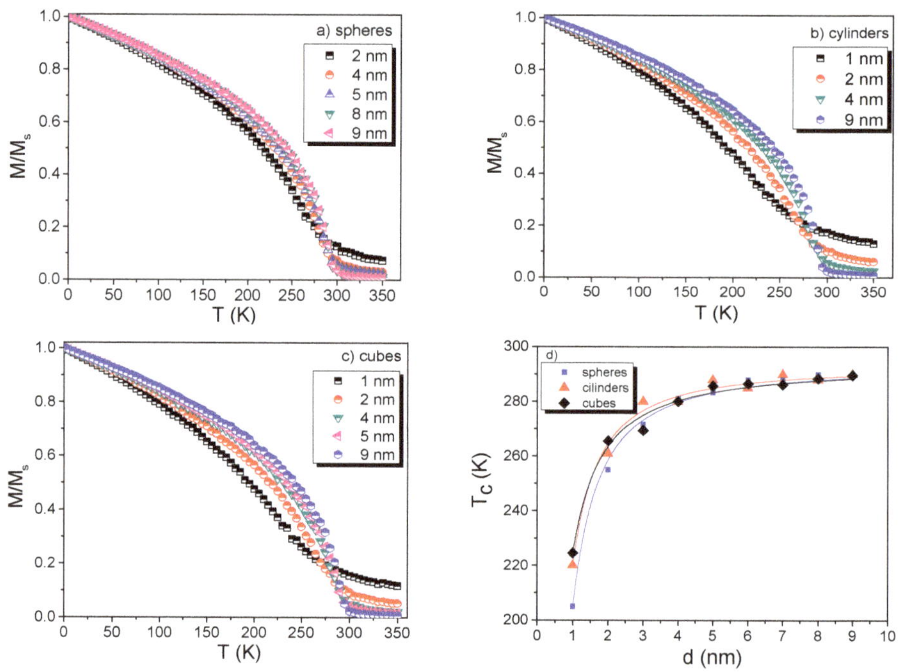

Figure 3. Normalized $M(T)/M_S$ curves for the Gd NPs with sphere (**a**), cylinder (**b**), and cube (**c**) geometries. (**d**) Size dependence of the T_C for the three studied geometries. (in (**d**) the full lines are guides for eyes).

For smaller particle sizes, the magnetic behavior, near T_C, loses its critical magnetic behavior, and T_C can hardly be determined accurately [2]. Commonly, T_C is taken from the minimum of dM/dT. However, at nanoscale, this significantly underestimates the actual temperature in which the magnetic order is lost. We resolved this issue by fitting the dM/dT curve with a polynomial fitting that covers the total curve, including all points and not only the minimum value. The estimated values are individually reported in Table S1 for Co and Gd NPs (they are not much different from their bulk counterparts). Another visible effect, for very small NPs, is the slightly different behavior of the magnetization when the temperature is increasing (not a pure Brillouin-like curve as expected for an ordinary FM, i.e., the increasing tendency of magnetization above T_C is different from bulk FM). This effect on the magnetization has been pointed out to occur from the correlations of local moments that are presented above T_C [2,17]. It only plays a role in NPs when the size of the system is close to the length of the magnetic correlation, ξ [2,17]. Moreover, the critical exponent for Co NPs (Table S1) is close to the value previously reported for Co films, that is, $z = 1.34$ [18]. In Table S1, there are also shown z values for the different geometries. In a first approximation, the z values (Table S1) are similar to that found also by Fisher [19], i.e., $z = 1$. Observing the T_C values (Table S1), one can notice that only the cylinder NPs geometry has shown a change in the T_C value [4], but the values are, in general, not substantially modified when compared among them. In principle, the demagnetization field can hardly change the T_C values of FM described by mean-field theory. Thus, since there were no significant changes in the T_C values upon changing the geometries, especially for spherical geometries in both Co and Gd NPs, we tested the Co@Gd core–shell simulations with spherical-like morphology, because (i) experimentally this is a geometry well accepted to describe real system, (ii) it is a geometry that helps the simulation regarding the spin configuration, and (iii) it is used in the major theoretical, experimental, and application approaches.

3.2. Core Shell Model Co@Gd (Effect on Tc Behavior with Size and Interfacial Exchange)

In the present section, different exchange couplings were used to describe the nanohybrid Co@Gd core–shell properties. For J > 0 (parallel spin configuration of the two FM phases), an artificial ferromagnetic-like heterostructure was achieved, whereas for J < 0 (antiparallel spin configuration), the system behaved as an artificial ferrimagnet-like (FI). We then worked with a spherical-like configuration where the Co core size was varied with a total core–shell size of 2, 4, and 6 nm, using values of interfacial exchange of $J_{int} = -1.38 \times 10^{-21}$ J/link This is a value that was taken from the literature for an ensemble of core–shell bi-magnetic NPs [11]. As shown from Figure 4a, the *M(T)* curves have different behaviors, depending on the Co core sizes. Consequently, it has a strong influence in the critical temperature (T_C^*) value of the nanohybrid core–shell system.

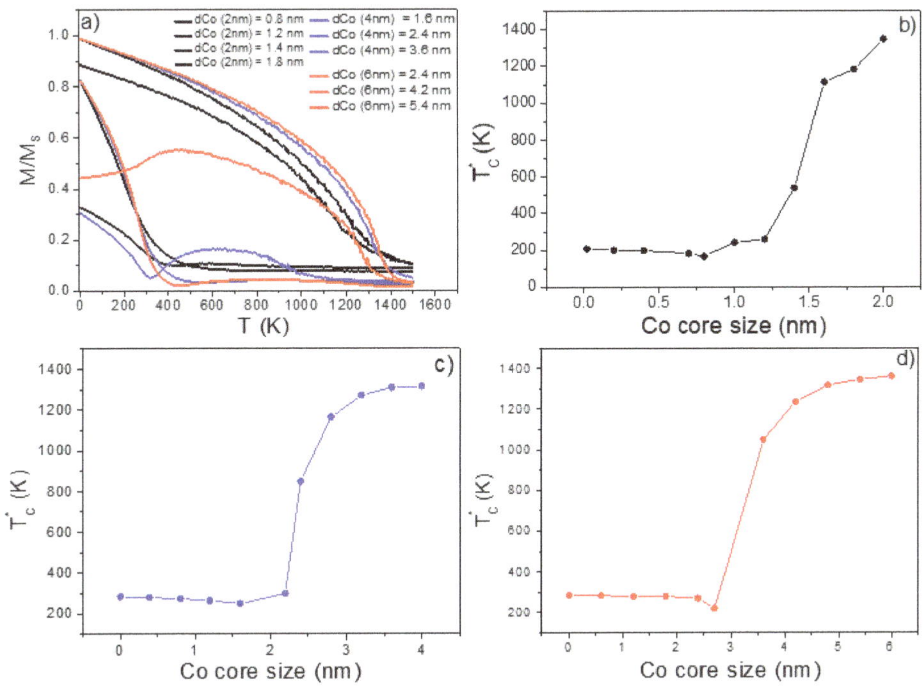

Figure 4. Normalized $M(T)/M_s$ curves for the core–shell bi-magnetic Co@Gd NP with total diameter of 2 (black lines), 4 (blue lines), and 6 nm (red lines) (**a**) for a negative interfacial exchange. Dependence of the T_C^* on the Co core size for a total particle size: 2 nm (**b**), 4 nm (**c**), and 6 nm (**d**). Full lines in (**b–d**) are guides for eyes.

We noticed that for core sizes smaller than 2 nm with total variable size (Figure 4b–d), the T_C^* values of the Co@Gd core–shell decreased significantly. Especially, it reached a value close to that found for pure spherical Gd NPs (~290 K), as can be seen comparing with data shown in Figure 3a. Thus, this T_C^* reduction with decreasing of Co size results in a nanohybrid core–shell material with a T_C^* close to RT (Gd-based materials are shown to present a giant magnetocaloric effect [20,21]).

It is important to highlight the occurrence of a compensation-like temperature in several *M(T)* curves, but not in all of them (see Figure 4a). In principle, this phenomenon only happens due to the temperature dependence of the magnetization of the antiferromagnetically coupled FM Co and Gd sublattices [22,23]. In other words, considering that both sublattices (Gd and Co) have different temperature-dependent behaviors, as the temperature increases, the Gd sublattice will go first to the paramagnetic (PM) state above 293 K, while the Co sublattice will enter into a PM state at higher temperatures. Thus, in a specific temperature of the sample, there will be total magnetization cancellation in case of an ordinary compensation temperature effect.

However, according to our findings, the compensation-like temperature occurs in all nanohybrid Co@Gd core shell systems independent on the sign of exchange interaction, as also displayed in Figures S1 and S2. Thus, we can point out that this effect should be dependent on the fraction (and magnetic state) of each FM structure that forms the nanohybrid Co@Gd core–shell system (the value of the Gd moment is much higher than that of Co). However, for the nanohybrid Co@Gd core–shell with $J_{int} < 0$, one can assume the ordinary compensation temperature effect, and therefore the simulated bi-magnetic core–shell systems behave as a total ferrimagnet with two antiferromagnetically coupled FMs [24]. This interesting effect has been studied in many magnetic systems based on magnetic garnets [22,23] and alloys of ferrimagnetic CoGd [25], but not yet in core–shell Co@Gd NPs. In addition, it can be mentioned that the compensation-like effect physically occurs in a state of high magnetic entropy, consequently favoring the magnetocaloric phenomenon in all nanohybrids here studied, and hence would allow the use of the bi-magnetic core–shell system in biomedical applications. This finding is significantly important since this Co@Gd core–shell system may be applied in magnetic hyperthermia, where systems with T_C close to room temperature (RT) are required.

3.3. *Hysteresis Curves Simulation*

3.3.1. Determination of the Computational Time Step

To simulate the *M(H)* curves for bimetallic FM core–shell Co@Gd, we made use of Equation (4), using a λ value equal to 1.0 for all the simulations. We first evaluated the suitable computational time step Δt, which represents the numerical parameter employed to resolve the LLG equation with higher precision, as suggested by Aurélio et al. [26]. As can be noticed from Figure S3a, the *M(H)* curves have the expected physical behavior for orders of $\Delta t = 10^{-15}$ s. In our case, we took a Δt value equal to 2×10^{-15} s at 0 K for a spherical 4 nm Co NP. For other orders of Δt, the system does not even reach the saturation regime and no physically reasonable values are obtained, which can be significantly affected when evaluating the H_C field values. Thus, we took a fixed value of 2×10^{-15} s for all *M(H)* simulated experiments as taken in [26].

It is important to highlight that this parameter will decide how long the atomistic simulations will take, recalling that the smaller the values, the more expensive will be the computational time and, of course, it will be limited by the particle size (number of atoms). Figure S1b also depicts the Δt dependence for H_C and the M/M_s ratio, where higher values equal to 2×10^{-15} s favor the stabilization of both parameters and hence their correct determination.

3.3.2. Hysteresis Curves for Pure 4 nm Co NP

Having defined and fixed the Δt values, we first proceed with the estimation of the H_C values for single Co NP at 0 K and several temperatures, as plotted in Figure 5a,b. We would also like to give a reminder that the LLG differential equation is only applicable at 0 K. Therefore, the Langevin dynamics concept must be employed [2] using the $\Gamma(t)$ concept, as represented by Equation (7). Then, to perform the atomistic simulations at other temperatures, the VAMPIRE software used the concept of the thermal field. Figure 5c shows that the H_C field starts to decrease significantly with temperature for simulations up to 700 K, in agreement Kneller's law [27]. Moreover, the M/M_s linearly with temperature (Figure 5d), and consequently a linear fit with R^2 equal to 0.999, support this finding.

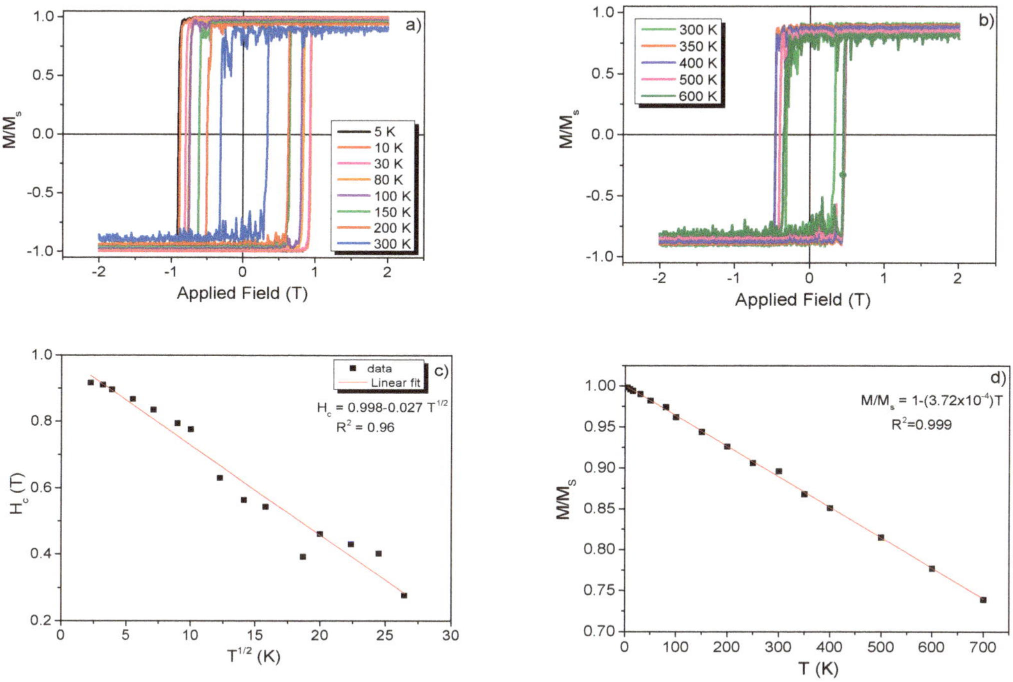

Figure 5. Temperature dependence of the normalized $M(H)/M_S$ curves for the 4 nm Co NPs (**a**,**b**). Thermal dependence of H_C and M/M_s parameters. In (**d**), the ratio M/M_s depends on the temperature and external magnetic field. In (**c**) and in (**d**) the full lines are result from the fits.

3.3.3. Hysteresis for Bi-Magnetic Core–Shell NP Varying the Interfacial Exchange Constant

Sizes of 2 and 4 nm for the bi-magnetic FM Co@Gd core–shell NPs were taken for simulations. The $M(H)$ loops for each size were obtained for negative, positive, and zero interfacial exchange (J_{int}) constants (see Figure 6, Figure S4 and S5). In the case of $J_{int} < 0$, the bi-magnetic core–shell NP system exhibited an improvement in the H_C values for both sizes, as can be observed by broad $M(H)$ loops displayed in Figure 6c,d. Interestingly, it occurred when the Co core size ranged from 1.2 to 1.5 nm. This means that under the present magnetic configuration (antiparallel spin interfacial exchange) the core–shell system will have an ES-like behavior. This kind of magnetic exchange occurs in systems composed of hard and soft magnetic materials with comparable or small exchange length [28,29]. Therefore, we can assume that the finite-size effect (in this case NP of 2 and 4 nm) will have a strong influence on the appearance of the ES phenomenon. In our system, the magnetic hard core of Co was expected to retain the uniaxial anisotropy of the Gd soft shell, causing an increment of the H_C values. In case of parallel magnetic configuration ($J_{int} > 0$), the core–shell $M(H)$ loops were not too broad, as seen for $J_{int} < 0$, but an enhancement in the H_C values was also observed (Figure S4c,d) in the range from 1.2 to 1.5 nm (for 2 and 4 nm of core–shell size). On the other hand, the H_C value reached 38% of the maximum value (Hc = 2.3 T) estimated for the antiparallel interfacial configuration (it became constant with the Co core size). For a zero interfacial exchange configuration, there was no magnetic interaction between the core and the shell. In this case, the magnetic exchange response happened individually for core and shell, and we could clearly observe two superposed magnetic hysteresis loops in the simulated $M(H)$, as displayed in Figure S5a,b. This could also be observed from the Co core diameter dependence of the H_C and M/M_s values that exhibited two independent responses due to Gd and Co, as plotted in Figure S5c,d.

Additionally, the M/M_s values were constant for individual non-coupled FM for 2 and 4 nm. This was expected because no temperature effect was taken into account.

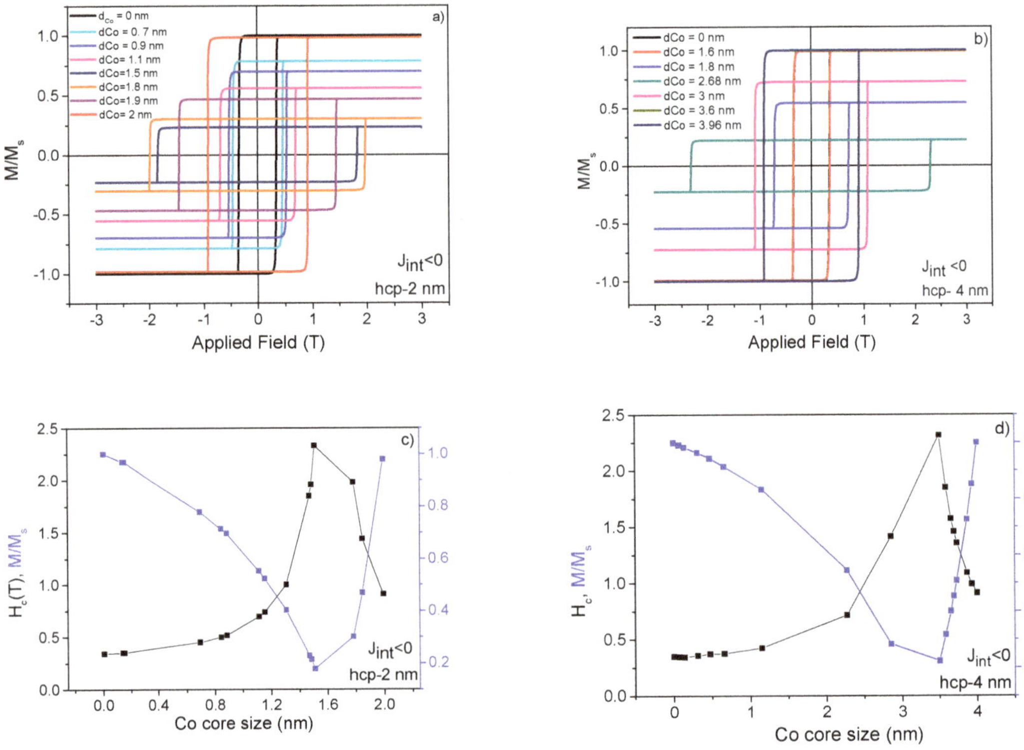

Figure 6. Normalized *M(H)* curves for different Co core size for a total core–shell size of 2 nm (**a**) and 4 nm (**b**) for negative interfacial exchange constant. Dependence of the Co core size of the *Hc* and M/M_s values for 2 nm (**c**) and 4 nm (**d**). In (**c**,**d**), full black and blue lines are guides for eyes.

4. Conclusions

In this work, we first performed atomistic spin model simulations (using free VAMPIRE software) of the temperature-dependent magnetization of Co and Gd single NPs with spherical, cubic, and cylindrical shapes. Regarding the magnetic properties of the FM Co, Gd, and Co@Gd core–shell systems, we have to emphasize that the finite-size effect was observed for all the geometries with exponential critical parameters related to ordinary ferromagnets. We also demonstrated that the T_C^* value of a core–shell bi-magnetic nanoframework can be tuned by changing the Co core size with spherical geometry. No significant influence in the T_C^* determination was detected when we changed the J_{int} between the Co and Gd interface, i.e., by adjusting the J_{int} value for the Co@Gd core–shell system with total diameters of 2 and 4 nm. Then, we can conclude that the compensation temperature is not exclusive for the J_{int} sign, since it is seen in all *M(T)* curves for all tested sizes. Indeed, it seems to depend on the fraction of FM structures and also the thermal effects on the individual magnetizations in Co@Gd core–shell structures. Further simulations with the presence of an external field will be carried out to better interpret this issue. On the other hand, it was shown that there is a strong dependence on the sign of J_{int} for the shape of the hysteresis curves and for the extracted magnetic parameters. We demonstrated that the sign of the magnetic interfacial exchange constant is an important parameter in bi-magnetic core–shell systems that may lead the nanohybrid materials to an exchange spring behavior or not ($J_{int} > 0$ yields an enhancement of the coercive field, for $J_{int} < 0$ the enhancement is more pronounced, and in addition the core–shell behaves as a magnetic spring). Therefore, on the one hand, it is shown that the interfacial exchange energy is the key to improving the magnetic response for a desired application where a critical temperature can be tuned, adjusting the sizes of core and shell contributions. On the other hand, all the theoretical results presented in this work have to be checked experimentally to fully demonstrate their impact in future biomedical applications.

Supplementary Materials: The following are available online at https://www.mdpi.com/article/s1, Figure S1: Normalized $M(T)/M_s$ curves for the core–shell bi-magnetic Co@Gd NP with total diameter of 2, 4, and 6 nm (a) for a positive interfacial exchange. Dependence of the T_C^* on the Co core size for a total particle size: 2 nm (b), 4 nm (c), and 6 nm (d). Figure S2. Normalized $M(T)/M_s$ curves for the core–shell bi-magnetic Co@Gd NP with total diameter of 2, 4, and 6 nm (a) for a zero interfacial exchange. Dependence of the T_C^* on the Co core size for a total particle size: 2 nm (b), 4 nm (c), and 6 nm (d). Figure S3: Normalized $M(H)/M_s$ curves under different Δt values at 0 K (a) and Δt dependence of the H_c and M/M_s parameters (b). Figure S4. Normalized $M(H)/M_s$ curves for different Co core size for a total core–shell size of 2 nm (a) and 4 nm (b) for positive interfacial exchange constant. Dependence of the Co core size of the H_c and M/M_s values for 2 nm (c) and 4 nm (d). Figure S5. Normalized $M(H)/M_s$ curves for different Co core size for a total core–shell size of 2 nm (a) and 4 nm (b) for zero interfacial exchange constant. Dependence of the Co core size of the H_c and M/M_s values for 2 nm (c) and 4 nm (d).

Author Contributions: Conceptualization, J.A.R.-G. and C.A.T.-V.; methodology, J.A.R.-G. and C.A.T.-V.; software, J.A.R.-G., C.A.T.-V., F.J.L. and E.C.P.; formal analysis, J.A.R.-G., C.A.T.-V. and E.C.P.; investigation, C.A.T.-V.; resources, J.A.R.-G.; data curation, J.A.R.-G. and C.A.T.-V.; writing—original draft preparation, J.A.R.-G.; writing—review and editing, J.A.R.-G., C.A.T.-V., F.J.L. and E.C.P.; visualization, J.A.R.-G. and C.A.T.-V.; supervision, J.A.R.-G. and E.C.P.; project administration, J.A.R.-G.; funding acquisition, J.A.R.-G. All authors have read and agreed to the published version of the manuscript.

Funding: This research was funded by Fondo Nacional de Desarrollo Científico, Tecnológico y de Innovación Tecnológica (FONDECYT-CONCYTEC), Project number 177-2020-FONDECYT.

Data Availability Statement: The simulated data of the present research can be provided upon reasonable request at the email juan.ramos5@unmsm.edu.pe.

Acknowledgments: The authors thank the Fondo Nacional de Desarrollo Científico, Tecnológico y de Innovación Tecnológica (FONDECYT-CONCYTEC), Project number 177-2020-FONDECYT, UFES and FAPES, CNPq for financial support.

Conflicts of Interest: The authors declare no conflict of interest.

References

1. Evans, R.F.L.; Atxitia, U.; Chantrell, R.W. Quantitative simulation of temperature-dependent magnetization dynamics and equilibrium properties of elemental ferromagnets. *Phys. Rev. B* **2015**, *91*, 144425. [CrossRef]
2. Evans, R.F.L.; Fan, W.J.; Chureemart, P.; Ostler, T.A.; Ellis, M.O.A.; Chantrell, R.W. Atomistic spin model simulations of magnetic nanomaterials. *J. Phys. Condens. Matter* **2014**, *26*, 103202. [CrossRef]
3. Penny, C.; Muxworthy, A.R.; Fabian, K. Mean-field modelling of magnetic nanoparticles: The effect of particle size and shape on the Curie temperature. *Phys. Rev. B* **2019**, *99*, 174414. [CrossRef]
4. Fei, C.L.; Dan, X.; Ming-Xing, G.; Park, H.S.; Fujita, T. Size and shape effects on Curie temperature of ferromagnetic nanoparticles. *Trans. Nonferrous Met. Soc. China* **2007**, *17*, 1451–1455.
5. He, L.; Zhang, H.; Liu, Y.; Sun, F.; Yu, X.; Li, X.; Zhang, L.; Wang, L.; Mao, K.; Wang, G.; et al. Maximizing Specific Loss Power for Magnetic Hyperthermia by Hard–Soft Mixed Ferrites. *Small* **2018**, *14*, 1800135. [CrossRef] [PubMed]
6. Herynek, V.; Turnovcová, K.; Veverka, P.; Dědourková, T.; Žvátora, P.; Jendelová, P.; Gálisová, A.; Kosinová, L.; Jiráková, K.; Syková, E. Using ferromagnetic nanoparticles with low Curie temperature for magnetic resonance imaging-guide thermoablation. *Int. J. Nanomed.* **2016**, *11*, 3801–3811. [CrossRef]
7. Obaidat, I.; Issa, B.; Mohite, V.; Haik, Y. Controlling the Curie-temperature of Magnetic Nanoparticles for Hyperthermia. *Dyn. Biochem. Process Biotechnol. Mol. Biol.* **2011**, *5*, 85–88.
8. Graham, C.D., Jr. Magnetic Behavior of Gadolinium Near the Curie Point. *J. Appl. Phys.* **1965**, *36*, 1135. [CrossRef]

9. Hankiewicz, J.H.; Celinski, Z.; Stupic, K.F.; Anderson, N.R.; Camley, R.E. Ferromagnetic particles as magnetic resonance imaging temperature sensors. *Nat. Commun.* **2016**, *7*, 12415. [CrossRef]
10. Budnyk, A.P.; Lastovina, T.A.; Bugaev, A.L.; Polyakov, V.A.; Vetlitsyna-Novikova, K.S.; Sirota, M.A.; Abdulvakhidov, K.G.; Fedorenko, A.G.; Podlesnaya, E.O.; Soldatov, A.V. Gd^{3+}-Doped Magnetic Nanoparticles for Biomedical Applications. *J. Spectrosc.* **2018**, *1–9*, 1412563. [CrossRef]
11. Anderson, N.R.; Camley, R.E. Temperature-dependent magnetization in bimagnetic nanoparticles with antiferromagnetic interfacial exchange. *Phys. Rev. B* **2016**, *94*, 134432. [CrossRef]
12. Velasco, S.; Román, F.L. Determining the Curie temperature of iron and nickel. *Phys. Teach.* **2007**, *45*, 387–389. [CrossRef]
13. Serantes, D.; Baldomir, D. Superparamagnetism and Monte Carlo Simulations. *Open Surf. Sci. J.* **2012**, *4*, 71–84. [CrossRef]
14. Arantes, F.R.; Cornejo, D.R. Monte Carlo study of the magnetic properties of frozen and non-interacting nanoparticles. *J. Nanopart. Res.* **2013**, *15*, 1859. [CrossRef]
15. Coehoorn, R.; de Mooij, D.B.; De Waard, C. Meltspun permanent magnet materials containing Fe_3B as the main phase. *J. Magn. Magn. Mater.* **1989**, *80*, 101. [CrossRef]
16. Nehme, Z.; Labaye, Y.; Hassan, R.S.; Yaacoub, N.; Greneche, J.M. Modeling of hysteresis loops by Monte Carlo simulation. *AIP Adv.* **2015**, *5*, 127124. [CrossRef]
17. Hovorka, O.; Devos, S.; Coopman, Q.; Fan, W.J.; Aas, C.J.; Evans, R.F.L.; Chen, X.; Ju, G.; Chantrell, R.W. The Curie temperature distribution of FePt granular magnetic recording media. *Appl. Phys. Lett.* **2012**, *101*, 052406. [CrossRef]
18. Huang, F.; Mankey, G.J.; Kief, M.T.; Willis, R.F. Finite-size scaling behavior of ferromagnetic thin films. *J. Appl. Phys.* **1993**, *73*, 6760–6762. [CrossRef]
19. Fisher, M.E.; Barber, M.N. Scaling Theory for finite-size effects in the critical region. *Phys. Rev. Lett.* **1972**, *28*, 1516–1519. [CrossRef]
20. Benford, S.M.; Brown, G.V. T-S diagram of Gadolinium near the Curie temperature. *J. Appl. Phys.* **1981**, *52*, 2110–2112. [CrossRef]
21. Percharsky, V.K.; Gschneidner, K.A. Tunable magnetic regenerator alloys with a giant magnetocaloric effect for magnetic refrigeration from ~20 to 290 K. *Appl. Phys. Lett.* **1997**, *70*, 3299. [CrossRef]
22. Bertaut, F.; Pauthenet, R. Temperature dependencies of the magnetic moments of the rare earth garnets. *Proc. Inst. Electr. Eng.* **1957**, *104* (Suppl. 5), 261.
23. Bsoul, I.; Hawamdeh, K.; Mahmood, S.H. Effect of Al-substitution on Structural and Magnetic Properties of $Er_3Fe_{5-x}Al_xO_{12}$ Garnets. In *Magnetic Oxides and Composites II*; Materials Research Forum LLC: Millersville, PA, USA, 2019; Chapter 3. [CrossRef]
24. Chamati, H.; Shopova, D.V. Ferrimagnetism in a Two-sublattice Bilinearly Coupled Heisenberg model. *AIP Conf. Proc.* **2019**, *2075*. [CrossRef]
25. Binder, M.; Weber, A.; Mosendz, O.; Woltersdorf, G.; Izquierdo, M.; Neudecker, I.; Dahn, J.R.; Hatchard, T.D.; Thiele, J.-U.; Back, C.H.; et al. Magnetization dynamics of the ferrimagnet CoGd near the compensation of magnetization and angular momentum. *Phys. Rev. B* **2006**, *74*, 134404. [CrossRef]
26. Aurélio, D.; Vejpravova, J. Understanding Magnetization Dynamics of a Magnetic Nanoparticle with a d Disordered Shell using Micromagnetic Simulations. *Nanomaterials* **2020**, *10*, 1149. [CrossRef] [PubMed]
27. Kneller, E.F.; Luborsky, F.E. Particle Size Dependence of Coercivity and Remanence of Single-Domain Particles. *J. Appl. Phys.* **1963**, *34*, 656. [CrossRef]
28. Quesada, A.; Delgado, G.; Pascual, L.; Aragón, A.M.; Marín, P.; Granados-Miralles, C.; Foerster, M.; Aballe, L.; Prieto, J.E.; de la Figuera, J.; et al. Exchange-spring behavior below the exchange length in hard-soft bilayers in multidomain configurations. *Phys. Rev. B* **2018**, *98*, 214435. [CrossRef]
29. Soares, J.M.; Galdino, V.B.; Conceição, O.L.A.; Morales, M.A.; de Araújo, J.H.; Machado, F.L.A. Critical dimension for magnetic exchange-spring coupled core/shell $CoFe_2O_4/CoFe_2$ nanoparticles. *J. Magn. Magn. Mater.* **2013**, *326*, 81–84. [CrossRef]

nanomaterials

Article

Tunable Transmissive Terahertz Linear Polarizer for Arbitrary Linear Incidence Based on Low-Dimensional Metamaterials

Zhenyu Yang [1], Dahai Yu [2], Huiping Zhang [1], Anqi Yu [1,3,*], Xuguang Guo [1,3], Yuxiang Ren [1], Xiaofei Zang [1,3], Alexei V. Balakin [1,4,5] and Alexander P. Shkurinov [1,4,5]

1 Shanghai Key Lab of Modern Optical System, Terahertz Spectrum and Imaging Technology Cooperative Innovation Center, Terahertz Technology Innovation Research Institute, University of Shanghai for Science and Technology, 516 Jungong Road, Shanghai 200093, China; 192380304@st.usst.edu.cn (Z.Y.); hpzhang@usst.edu.cn (H.Z.); xgguo@usst.edu.cn (X.G.); 182390295@st.usst.edu.cn (Y.R.); xfzang@usst.edu.cn (X.Z.); a.v.balakin@physics.msu.ru (A.V.B.); ashkurinov@physics.msu.ru (A.P.S.)
2 Focused Photonics (Hangzhou) Inc., No: 760, Bin'an Road, Binjiang District, Hangzhou 310052, China; dahai_yu@fpi-inc.com
3 Shanghai Institute of Intelligent Science and Technology, Tongji University, Shanghai 200092, China
4 Faculty of Physics and International Laser Center, Lomonosov Moscow State University, Leninskie Gory 1-2, 19991 Moscow, Russia
5 ILIT RAS–Branch of the FSRC "Crystallography and Photonics" RAS, Svyatoozerskaya 1, 140700 Shatura, Russia
* Correspondence: anqiyu@usst.edu.cn

Abstract: In this work, we propose a structure consisting of three metamaterial layers and a metallic grating layer to rotate the polarization of arbitrary linearly polarized incidence to the y-direction with high transmissivity by electrically tuning these metamaterials. The transfer matrix method together with a harmonic oscillator model is adopted to theoretically study the proposed structure. Numerical simulation based on the finite difference time-domain method is performed assuming that the metamaterial layers are constituted by graphene ribbon arrays. The calculation and simulation results show that the Drude absorption is responsible for the polarization rotation. Fermi level and scattering rate of graphene are important for the transmissivity. For a polarization rotation of around 90°, the thickness of either the upper or lower dielectric separations influences the transmission window. For a polarization rotation of around 45° and 135°, the lower dielectric separations decide the frequency of the transmission window, while the upper dielectric separations just slightly influence the transmissivity.

Keywords: terahertz; graphene; plasmons; Drude absorption; polarization conversion

Citation: Yang, Z.; Yu, D.; Zhang, H.; Yu, A.; Guo, X.; Ren, Y.; Zang, X.; Balakin, A.V.; Shkurinov, A.P. Tunable Transmissive Terahertz Linear Polarizer for Arbitrary Linear Incidence Based on Low-Dimensional Metamaterials. *Nanomaterials* **2021**, *11*, 1851. https://doi.org/10.3390/nano11071851

Academic Editor: Vladimir S. Bystrov

Received: 8 July 2021
Accepted: 16 July 2021
Published: 18 July 2021

Publisher's Note: MDPI stays neutral with regard to jurisdictional claims in published maps and institutional affiliations.

1. Introduction

Terahertz waves, electromagnetic waves with frequencies ranging from 0.1 THz to 10 THz, have extraordinary characteristics such as being non-ionizing, having a large bandwidth, and molecular fingerprints, which offer terahertz waves potential applications in medical examination, high-speed communication, environmental monitoring, etc. These applications require terahertz sources, detectors, and other functional devices such as polarization rotators or wave plates with high performances. Conventionally, polarization rotators use birefringence [1,2], total internal reflection effects [3,4], and the Faraday effect [5,6] to change the polarization. However, birefringent rotators are bulky, total internal reflection rotators need complex designs and fabrication, and Faraday rotators require the application of an external magnetic field.

In recent years, metamaterial-based polarization rotators have attracted a lot of attention as metamaterials can realize birefringence with a total device thickness less than the incident light. For transmissive polarization rotators, polarization rotation is usually achieved by deliberate design of the metamaterial to tune the phase and the amplitude of the two orthogonal components [7–15]. However, since the phase difference and the amplitude of two orthogonal components change as the frequency of the incidence changes, the bandwidths of such transmissive polarization rotators are usually limited [7–14]. Metallic gratings can help to enhance the bandwidth and the transmissivity [16–22]. Electromagnetic waves transmitted through or reflected by the metamaterials with unwanted polarization are reflected by metallic gratings and interact with the metamaterials again, so the transmission window grows taller and wider.

As the material properties of dielectrics and metals can hardly be tuned by an applied electric field, polarization rotators based on these metamaterials lack post-fabrication tunability. This weakness can be compensated by low-dimensional materials, e.g., graphene. The tuning on the phase and amplitude caused by graphene plasmons or Drude absorption (which is considered as 0th order plasmon resonance) can be actively tuned by the graphene Fermi level, offering the opportunity to realize actively tunable polarization rotators. Recently, Zhang et al. combined two metallic gratings and two orthogonal graphene gratings to actively tune the polarization of the transmitted waves from 20° to 70° by changing the Fermi level on each graphene grating [23]. Qi et al. demonstrated a reflective quarter-wave plate and half-wave plate by changing the graphene Fermi level [24]. Peng et al. proposed a reflective half wave-plate whose functioning frequency can be tuned by changing the graphene Fermi level [25]. Although there has been a plethora of simulation works realizing polarization conversion, little has been done on theoretical models which can explain the influence of physical and structural parameters on the efficiency of the polarization conversion, yield good correspondence with simulated results, and offer quick optimization of the proposed structures, especially for transmissive polarization converters.

In this work, we propose an actively tunable polarization converter consisting of a metamaterial layer (MML) on the top, a metallic grating layer in the bottom, and two MMLs in between, to realize electrically tunable polarization conversion in linearly polarized transmission for arbitrary linearly polarized incidence with high transmission efficiency. A theoretical model with the abovementioned functionalities is constructed by the transfer matrix method (TMM). Then, we numerically simulate the structure by the finite difference time-domain (FDTD) method by assuming that the MMLs are constituted by graphene ribbon arrays. Fermi level and the scattering rate of graphene are important for the efficiency of the polarization rotation. In particular, the 1st MML functions well as a linear polarizer only when the high carrier density is high, which is vital for efficient rotation angles around 90°. This result may find applications in THz communication, imaging, manipulation, etc.

2. Model Descriptions

The proposed structure is schematically shown in Figure 1. The 1st MML is assumed to have resonances along the y-coordinate. The 2nd and 3rd MMLs in the middle are assumed to have resonances along the u-coordinate and v-coordinate, respectively, with $\vec{u} = \frac{1}{\sqrt{2}}(\vec{x} - \vec{y})$ and $\vec{v} = \frac{1}{\sqrt{2}}(\vec{x} + \vec{y})$. The metallic gratings in the bottom extend along the x-coordinate and function as a linear polarizer, allowing only the transmission of y-polarized electromagnetic waves and reflecting the x-polarized waves. The 2nd and 3rd MMLs are separated from the 1st MML and the metallic gratings by an upper dielectric layer with thickness h_1 and a lower dielectric layer with thickness h_2. The refractive indices of the two dielectric layers and the substrate are all n_1 for simplicity. The boundary matrix T_{top} at the surface of the proposed structure can be calculated as:

$$T_{top} = \begin{pmatrix} E_x^+ \\ E_x^- \\ E_y^+ \\ E_y^- \end{pmatrix} = M_{top} M_{upper} M_{uv \to xy} M_{res} M_{xy \to uv} M_{lower} T_{bottom}, \quad (1)$$

where

$$M_{res} = \frac{1}{2}\begin{pmatrix} 1+\frac{n_2}{n_1}+\frac{\sigma_u(\omega)}{\varepsilon_0 c} & 1-\frac{n_2}{n_1}+\frac{\sigma_u(\omega)}{\varepsilon_0 c} & 0 & 0 \\ 1-\frac{n_2}{n_1}-\frac{\sigma_u(\omega)}{\varepsilon_0 c} & 1+\frac{n_2}{n_1}-\frac{\sigma_u(\omega)}{\varepsilon_0 c} & 0 & 0 \\ 0 & 0 & 1+\frac{n_2}{n_1}+\frac{\sigma_v(\omega)}{\varepsilon_0 c} & 1-\frac{n_2}{n_1}+\frac{\sigma_v(\omega)}{\varepsilon_0 c} \\ 0 & 0 & 1+\frac{n_2}{n_1}-\frac{\sigma_v(\omega)}{\varepsilon_0 c} & 1+\frac{n_2}{n_1}-\frac{\sigma_v(\omega)}{\varepsilon_0 c} \end{pmatrix}, \quad (2)$$

is the transfer matrix describing the 2nd and 3rd MMLs by considering them as a whole [22]. $\sigma_u(\omega)$ and $\sigma_v(\omega)$ are the effective conductivities in the x- and y-coordinates, respectively, ε_0 is the dielectric constant in vacuum, and c is the speed of light in vacuum. M_{upper} and M_{lower} are the transfer matrices of the upper and lower dielectric separations, $M_{uv\rightarrow xy}$ and $M_{xy\rightarrow uv}$ are the transfer matrices describing the coordinate transformation, M_{top} is the transfer matrix describing plasmon resonances of the 1st MML at the surface of the proposed structure, and T_{bottom} describes the electric field at the upper surface of the metallic gratings. The details of these transfer matrices and boundary matrix have been given in our previous work [22]. Here, the spacing between the 2nd and 3rd MMLs (~100 nm) is ignored because it is far less than the relevant wavelength (~ 300 μm).

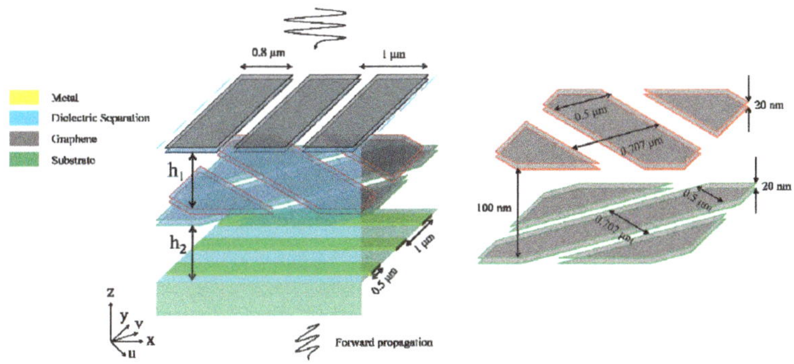

Figure 1. Schematics of the proposed structure.

In our previous work [22], it was pointed out that plasmon resonances or Drude absorption should not be excited simultaneously with the same frequency and strength in the u- and v-coordinates. Therefore, the 2nd and 3rd MMLs cannot be in the "ON" state (the MML is electrically biased and has plenty of charge carriers) simultaneously. When the 1st MML is in the "ON" state, it is supposed to function as a linear polarizer, allowing only the transmission of x-polarized electromagnetic waves and reflecting the y-polarized waves. Then, the proposed structure functions differently as the state of the 1st MML changes. If the 1st and the 2nd/3rd MML are in the "ON" state, the x-component of the incidence will pass through the 1st MML. Then, the change in the amplitude and phase induced by the plasmon resonance or Drude absorption of the 2nd/3rd MML will change the polarization of the electromagnetic waves, generating the y-component by transforming the x-polarized waves into elliptically polarized waves. When the elliptically polarized waves in the lower dielectric separation arrives at the metallic gratings, the y-component will pass through while the x-component will be reflected. The reflected x-polarized waves will again be transformed into elliptically polarized waves, whose y-component will be reflected back into the cavity by the 1st MML. The total transmission will be enhanced if constructive interaction takes place in the cavity formed between the 1st and the 2nd/3rd MML and the cavity formed between the 2nd/3rd MML and the metallic gratings. In this case, the proposed structure is similar to the one studied in our previous work [22], and polarization rotation around 90° is anticipated. However, there remains the problem of how well the 1st MML works as a linear polarizer, which is important for the efficiency of the cross-polarization conversion. Due to the fact that the model is too complicated and a clear conclusion can hardly be drawn, in this section, we mainly focus on the theoretical study of the second case, that is, when the 1st and the 3rd MMLs are in the "OFF" state (the MML is not electrically biased and has few charge carriers), while the 2nd MML is in the "ON" state. In the second case, the v-component can pass through the 2nd MML with little loss, while the u-component can hardly pass through. The v-component can be divided into x- and y-components with equal amplitude, with the former reflected back by, and the latter passing through, the metallic gratings. Then, the reflected x-component is divided into u- and v-components, with the former strongly reflected by the 2nd MML. The total transmission will be enhanced if constructive interaction takes place in the cavity formed between the 2nd MML and the metallic gratings. Then, it is anticipated that the polarization conversion is most efficient for incidence polarized along the v-coordinate because there is no reflection when the incidence arrives at the 2nd MML for the first time.

Normalizing the transmitted energy as 1, the amplitude of the incidence is calculated as:

$$E_v^+ = \frac{e^{-ik_0 n_1 (h_1+h_2)}}{2\sqrt{2}} \left[\frac{1+n_1+\frac{\sigma_u(\omega)}{\varepsilon_0 c} e^{ik_0 n_1 h_1} \mathrm{B}}{1-i\frac{\sigma_u(\omega)}{\varepsilon_0 c} \sin(k_0 n_1 h_2) \frac{\mathrm{B}}{\mathrm{A}}} + 1 + n \right] \tag{3}$$

$$\mathrm{A} = n_1 \cos[k_0 n_1 (h_1+h_2)] - i \sin[k_0 n_1 (h_1+h_2)] \tag{4}$$

$$\mathrm{B} = n_1 \cos(k_0 n_1 h_1) - i \sin(k_0 n_1 h_1) \tag{5}$$

where k_0 is the wavevector of the incidence in the vacuum. Mathematically, $|E_v{}^+|$ should be small enough to achieve high transmissivity, so the numerator in the square bracket in Equation (3) should be small while the denominator should be large. Note that $|\sigma_u(\omega)/\varepsilon_0 c|$ >> 1 if the plasmon resonance or the Drude absorption in the u-coordinate is strong, then the denominator is possibly dominated by the 2nd term. Since there is "$\sin(k_0 n_1 h_2)$", a large denominator requires that $\sin(k_0 n_1 h_2)$ cannot be 0. Assuming that $k_0 n_1 h_2$ is an odd multiple of $\pi/2$, $|\mathrm{B}/\mathrm{A}|$ achieves the maximum value of n_1 when $k_0 n_1 h_1$ is an even multiple of $\pi/2$. Although the 3rd term in the numerator also achieves the maximum value if $k_0 n_1 h_1$ is an even multiple of $\pi/2$, the denominator benefits more than the numerator, so the transmissivity is enhanced under this condition. Undoubtedly, h_2 plays a much more important role that h_1 does. Other important factors affecting the transmissivity, according to our previous work [22], may include E_F and the scattering rate.

3. Simulation Results and Discussion

To verify the theoretical analyses, we perform FDTD simulations with Lumerical FDTD Solutions. The metal is modeled as perfect electric conductor (whose dielectric function is treated as 1 + 10^6i in the software with i the imaginary unit) with 100 nm thickness. MMLs are assumed to be paired graphene ribbon arrays (GRAs) such that they can gate themselves. Graphene is characterized by the Kubo formula [23] and the scattering rate is initially set as 10^{12} Hz. The thickness of graphene is set as 1 nm, and the minimum meshes at the graphene boundaries are set as 0.1 nm to ensure the accuracy. The widths of the 1st GRA pair, the 2nd/3rd GRA pairs, and the bottom metallic gratings are 0.8 μm, 0.5 μm, and 0.5 μm, respectively, and the periods are 1 μm, 0.707 μm, and 1 μm, respectively. n_1 is set as 1.4, h_1 and h_2 are initially set as 56 μm. For the given parameters, $k_0 n_1 h_2 = \mathrm{m}\pi/2$ with m being an integer yields $f_0 \approx$ m*0.957 THz with f_0 being the frequency of the incidence. In the theoretical calculation, for the 1st, the 2nd, and the 3rd GRA pairs, Drude absorption is included only in $\sigma_y(\omega)$, $\sigma_u(\omega)$, and $\sigma_v(\omega)$, respectively, while high-order plasmon resonances are excluded.

Firstly, we keep the 1st and 3rd GRA pairs in the "OFF" state (E_F = 0 eV) and the 2nd pair in the "ON" state (E_F = 0.9 eV). Figure 2a,b show the calculated and simulated transmission, reflection, and absorption spectra for incidence polarized along the u-coordinate and the v-coordinate, respectively, in the absence of the metallic gratings. As shown in Figure 2a, the reflection/transmission is very high/low at the low frequencies, and gradually falls/rises as the frequency increases. The peaks and dips result from the resonances in the Fabry–Pérot-like cavity formed by the surface of the proposed structure and the 2nd GRA pair. Obviously, u-polarized incidence with a frequency around 1 THz can hardly pass through the 2nd GRA pair. For v-polarized incidence, the reflection/transmission is nearly 0%/100%, as shown in Figure 2b, which means that v-polarized incidence can pass through the 2nd GRA pair with little loss. The slight fluctuation in the simulated spectra, which is not observed in the calculated spectra, is probably caused by the plasmon resonances in the v-coordinate above 10 THz. The excellent correspondence between the calculated and simulated spectra below 4 THz shows that the treatment on $\sigma_u(\omega)$ and $\sigma_v(\omega)$ works well in the concerned frequency range. Since u-polarized incidence is blocked while v-polarized incidence passes through, the polarization conversion is first studied for v-polarized incidence (that is, in the presence of the metallic gratings). As shown in Figure 2c, both the calculated and the simulated transmission spectra show high transmission at odd multiples of 0.93 THz and low transmission in close vicinity to even multiples of 0.93 THz, corresponding to the calculated 0.957 THz. The correspondence between the theoretical analysis and the transmission spectra demonstrates the effectiveness of our model. In our previous work [22], it was shown that either Drude absorption or plasmon resonances can tune the phase and amplitude of the incidence with polarization in the same coordinate. As there is only Drude absorption in the u-coordinate and no plasmon resonance in either the u- or v-coordinate, the u-directional Drude absorption is the cause of the polarization conversion. Since both the real part and the imaginary part of the Drude absorption become smaller as the frequency increases, the ability of the polarization conversion gets weaker, and then the transmission windows become weak as m increases. The polarization of the incidence is then rotated by θ from 0° to 180°, with θ representing the counterclockwise angle from the y-axis to the polarization direction. As shown in Figure 2d, the transmission around 0.93 THz exceeds 70% for θ = 115° ~ 160°.

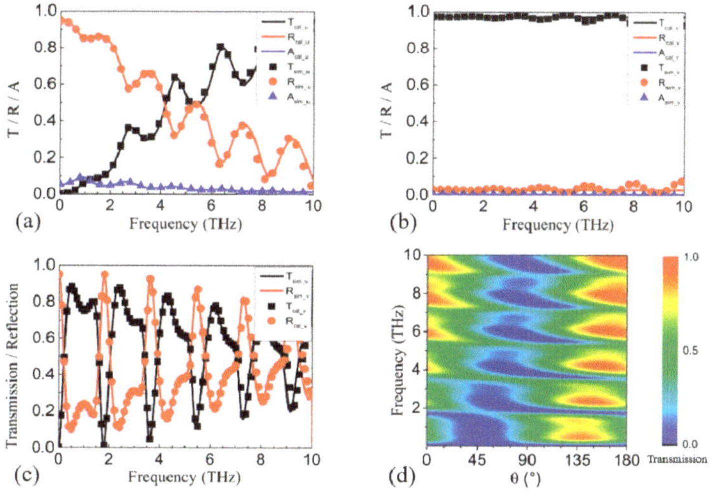

Figure 2. The calculated and simulated transmission, reflection, and absorption spectra for incidence polarized along (**a**) the u-coordinate and (**b**) the v-coordinate in the absence of the 1st GRA pair, the 3rd GRA pair, and the metallic gratings. (**c**) The calculated and simulated transmission spectra for incidence polarized along the v-coordinate with the 1st and 3rd GRA pairs in the "OFF" state. (**d**) The contour by rotating θ from 0° to 180° with the 1st and 3rd GRA pairs in the "OFF" state.

If the 1st and 2nd GRA pairs are in the "OFF" state and the 3rd pair is in the "ON" state, it can be predicted from the symmetry that the polarization rotation is most efficient for incidence with polarization in the u-coordinate. As shown in Figure 3a, the transmission around 0.93 THz exceeds 70% for θ = 20° ~ 65°. Note that the transmission is only ~ 40% for θ ~ 90°, as shown in Figures 2d and 3a. The reason is that the 1st GRA pair is in the "OFF" state and it cannot block the reflected waves. When the 1st GRA pair is in the "ON" state, the Drude absorption causes strong reflection within 2 THz for *y*-polarized electromagnetic waves, as shown by the black curve in Figure 4e, and then the 1st GRA can work as a linear polarizer in the frequency range. Consequently, the proposed structure becomes similar to the one in our previous work [22]. Then, if either the 2nd or the 3rd GRA pair is in the "ON" state, the transmission for θ = 60° ~ 120° is larger than 70%, as shown in Figure 3b,c. Note that the 2nd and the 3rd GRA pair cannot be in the "ON" state simultaneously, or tuning on the phase and the amplitude of the *u*-directional and the *v*-directional electric field components will be the same, which will lead to no polarization conversion. When all the GRAs are in the "OFF" state, the structure contains nothing but *x*-directional gratings, and θ = 0° ~ 30° and 150° ~ 180° can pass through with high transmission for all frequencies, as shown in Figure 2d. Therefore, for arbitrary linear incidence, we can selectively turn on some of the GRA pairs to rotate the incidence to *y*-polarized electromagnetic waves with transmission higher than 70%.

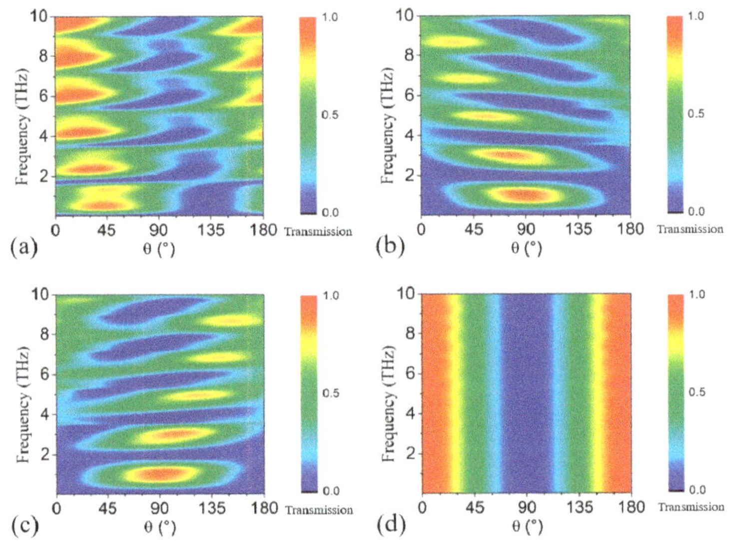

Figure 3. (**a**) The contour by rotating θ from 0° to 180° with (**a**) the 1st and 2nd GRA pairs in the "OFF" state, (**b**) the 2nd GRA pair in the "OFF" state, (**c**) 3rd GRA pair in the "OFF" state, (**d**) all the GRA pairs in the "OFF" state.

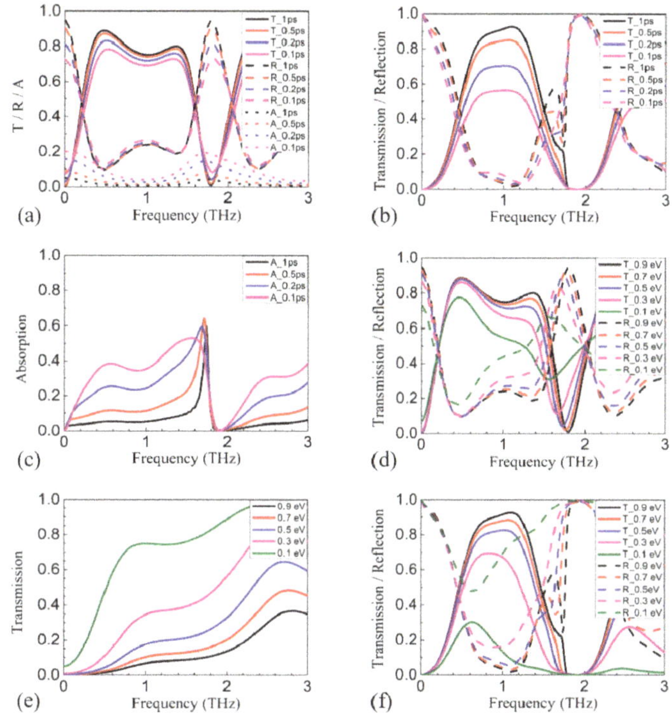

Figure 4. (**a**) The simulated transmission, reflection, and absorption spectra of the v-polarized incidence with only the 2nd GRA pair in the "ON" state and γ increased from 10^{12} Hz to 10^{13} Hz. (**b**) The simulated transmission, reflection, and (**c**) absorption spectra of the x-polarized incidence with the 1st and the 2nd GRA pairs in the "ON" state and γ increased from 10^{12} Hz to 10^{13} Hz. (**d**) The simulated transmission spectra of the v-polarized incidence with only the 2nd GRA pair in the "ON" state and E_F is reduced from 0.9 eV to 0.1 eV. (**e**) The simulated transmission spectra of the *u*-polarized incidence with E_F reduced from 0.9 eV to 0.1 eV in the absence of the 1st GRA pair, the 3rd GRA pair, and the metallic gratings. (**f**) The simulated transmission spectra of the x-polarized incidence with the 1st and the 2nd GRA pair in the "ON" state and E_F is reduced from 0.9 eV to 0.1 eV.

Since the polarization rotation depends on the tuning on the phase and amplitude by the Drude absorption, the excitation efficiency of the Drude absorption will influence the efficiency of the polarization rotation. Then, it can be inferred that E_F and γ will influence the transmission. Figure 4a shows the transmission, reflection, and absorption spectra of the v-polarized incidence with only the 2nd GRA pair in the "ON" state and γ increased from 10^{12} Hz to 10^{13} Hz. The increase in γ not only weakens the tuning on the phase and the amplitude [22], but also causes more dissipative damping. As a result, both reflection and absorption are increased, and the transmission is reduced by ~ 5%. Figure 4b shows the transmission and reflection spectra of the *x*-polarized incidence with the 1st and 2nd GRA pairs in the "ON" state and γ increased from 10^{12} Hz to 10^{13} Hz. Obviously, the transmission decreases by about 30% ~ 40%, while the reflection remains nearly constant. Figure 4c shows that the increase in absorption is the cause of the decrease in transmission. The reason is that the 90° polarization rotation requires more reflections inside the proposed structure, and the amount of lossy material nearly doubles compared to the former case. As a result, the absorption is greatly enhanced. Comparatively, E_F influences the transmission more than γ, which is similar to our previous results [22]. The transmission of the v-polarized incidence with only the 2nd GRA pair in the "ON" state is reduced by ~ 16% at 0.7 THz and ~ 32% at 1.3 THz when E_F is reduced from 0.9 eV to 0.1 eV, as shown in Figure 4d. Comparatively, the reflection increases by nearly the same amount. The coupling efficiency between the incidence and the free charge carriers in graphene will be reduced as E_F is reduced. Accordingly, the Drude absorption will be weakened, and its broadening will be reduced, as shown in Figure 4e. Consequently, both the tuning in phase and amplitude will be weakened, especially for higher frequencies. For y-polarized incidence with the 1st and 2nd GRA pairs in the "ON" state, the influence of E_F is more severe. Apart from the weakened Drude absorption in the 2nd GRA pair, the Drude absorption of the 1st GRA pair is also weakened, which will reduce the reflection on the y-polarized waves of the 1st GRA pair within the concerned frequency range, as shown in Figure 4e. Then, the proposed structure gradually becomes dissimilar to the one in our previous work [22] because the blocking of the 1st GRA pair on the y-polarized reflected waves from the 2nd GRA pair is greatly reduced, breaking the cavity formed between the 1st and the 2nd GRA pairs. As a result, the reflection increases a lot while the transmission is reduced by ~ 54% at 0.7 THz and ~ 75% at 1.3 THz, as shown in Figure 4f.

Finally, we consider the influence of h_1 and h_2 on the transmission. For x-polarized incidence with both the 1st and the 2nd GRA pair in the "ON" state, the proposed structure is similar to that in our previous work [22], so that each transmission dip will be divided into two if $h_1 \neq h_2$, as shown in Figure 5a,b. For v-polarized incidence with only the 2nd GRA pair in the "ON" state, the case is somewhat different. As the x-polarized electromagnetic waves will be completely reflected by the metallic gratings, constructive/destructive interaction will take place when $k_0 n_1 h_2$ is close to an odd/even multiple of $\pi/2$, which will enhance/suppress the Drude absorption and thus enhance/suppress the transmission. The corresponding k_0 for constructive/destructive interaction decreases as h_2 increases, inducing the red-shift of the transmission window, as shown in Figure 5c. Comparatively, at the surface of the upper dielectric layer, the reflection for the reflected waves is not so high, so the change in h_1 just slightly influences the transmission. In addition, the phase change is 0 instead of π (at the metallic gratings), so that constructive interaction takes place when $k_0 n_1 h_1$ is close to even multiple of $\pi/2$. Therefore, the transmission at 0.93 THz keeps growing as h_1 increases from 56 μm to 112 μm, as shown in Figure 5d. As the reflection at the surface of the proposed structure is low, the increase at 0.93 THz is only about 20%. In order for the proposed structure to work well for all the linear incidences, h_1 is set equal to h_2.

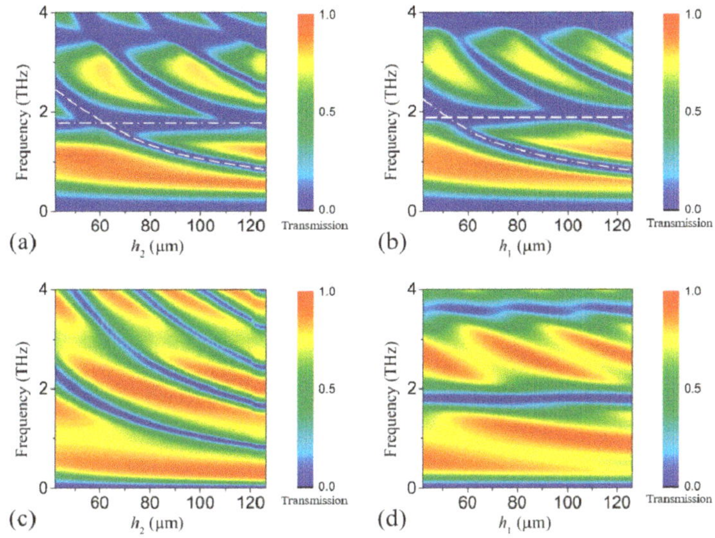

Figure 5. The transmission spectra for x-polarized incidence with both the 1st and the 2nd GRA pair in the "ON" state as (**a**) h_2 and (**b**) h_1 increase from 42 μm to 112 μm. The white dashed curve and the white dash-dotted curve indicate the transmission dip caused by h_2 and h_1, respectively. The transmission spectra for v-polarized incidence with only the 2nd GRA pair in the "ON" state as (**c**) *h2* and (**d**) h_1 increase from 42 μm to 112 μm.

Finally, we would like to stress that, although GRA pairs are employed as the metamaterial in the simulation, the MMLs can be constituted by other materials. Considering that the E_F influences more than γ, few-layer graphene (gated by ion gel) may be better than GRA pairs. Note that the theoretical model just takes the optical conductivity into account, and the model is expected to work for semiconductor spoof surface plasmons [26] or even nanowire arrays [27], which also show a Drude-like dielectric function.

4. Conclusions

In this work, we proposed a sandwich structure consisting of an MML on the top, two MMLs extending along the u-coordinate and v-coordinate in the middle, and metallic gratings extending along the x-coordinate at the bottom. Assuming that GRA pairs constitute the MMLs, for arbitrary linearly polarized incidence, broadband transmission greater than 70% can be obtained by selectively tuning the GRA pairs on or off by electrical bias. The transfer matrix method is applied to theoretically study the polarization rotation. The theoretical analysis and FDTD simulation show that the polarization rotation results from the tuning on the amplitude and phase by the Drude absorption and the cavity effect formed by the upper and lower dielectric separation. A large scattering rate will lead to high absorption, thus reducing the transmission. A low Fermi level will weaken the Drude absorption and its broadening, severely reducing the transmission. In order to enhance the efficiency of the polarization conversion, graphene with high mobility is needed, and the Fermi level or carrier density should also be high. For polarization around 90°, the carrier density of the first MML must be high such that it functions well as a linear polarizer. The proposed structure may find applications in systems which need devices with post-fabrication tunability, such as THz communication, imaging, manipulation, etc.

Author Contributions: Conceptualization, A.Y., Z.Y.; methodology, A.Y.; software, Z.Y., D.Y., Y.R., H.Z.; formal analysis, A.Y., X.G., X.Z.; investigation, A.Y., Z.Y., D.Y., H.Z.; resources, A.Y., Z.Y.; data curation, Z.Y.; writing—original draft preparation, A.Y., Z.Y.; visualization, Z.Y.; supervision, A.V.B., A.P.S.; project administration, A.Y.; funding acquisition, X.G., A.V.B. All authors have read and agreed to the published version of the manuscript.

Funding: This work was supported in part by the National Natural Science Foundation of China (61731020, 61988102, 61722111), in part by the 111 Project (D18014), in part by the International Joint Lab Program supported by Science and Technology Commission Shanghai Municipality (17590750300), in part by the Ministry of Science and Higher Education (075-15-2019-1950), and in part by the State assignment FSRC "Crystallography and Photonics" RAS.

Institutional Review Board Statement: Not Applicable.

Informed Consent Statement: Not Applicable.

Data Availability Statement: The data presented in this study are available on request from the corresponding author.

Conflicts of Interest: The authors declare no conflict of interest.

References

1. Wu, H.-Y.; Hsieh, C.-F.; Tang, T.-T.; Pan, R.-P.; Pan, C.-L. Electrically tunable room-temperature 2π liquid crystal terahertz phase shifter. *IEEE Photonics Technol. Lett.* **2006**, *18*, 1488–1490.
2. Scherger, B.; Scheller, M.; Vieweg, N.; Cundiff, S.T.; Koch, M. Paper terahertz wave plates. *Opt. Express* **2011**, *19*, 24884–24889. [CrossRef]
3. Petrov, N.I. Achromatic polarization rotator. *Appl. Opt.* **2007**, *46*, 6340–6343. [CrossRef]
4. Messaadi, A.; Sánchez-López, M.M.; Vargas, A.; García-Martínez, P.; Moreno, I. Achromatic linear retarder with tunable retardance. *Opt. Lett.* **2018**, *43*, 3277–3280. [CrossRef] [PubMed]
5. Chin, J.Y.; Steinle, T.; Wehlus, T.; Dregely, D.; Weiss, T.; Belotelov, V.I.; Stritzker, B.; Giessen, H. Nonreciprocal plasmonics enables giant enhancement of thin-film Faraday rotation. *Nat. Commun.* **2013**, *4*, 1599. [CrossRef] [PubMed]
6. Fan, F.; Xu, S.-T.; Wang, X.-H.; Chang, S.-J. Terahertz polarization converter and one-way transmission based on dou-ble-layer magneto-plasmonics of magnetized InSb. *Opt. Express* **2016**, *24*, 26431–26443. [CrossRef] [PubMed]
7. Dai, Y.; Cai, H.; Ding, H.; Ning, Z.; Pan, N.; Zhu, H.; Shi, Q.; Wang, X. Near-infrared quarter-waveplate with near-unity polarization conversion efficiency based on silicon nanowire array. *Opt. Express* **2015**, *23*, 8929–8938. [CrossRef]
8. Wang, D.; Gu, Y.; Gong, Y.; Qiu, C.-W.; Hong, M. An ultrathin terahertz quarter-wave plate using planar babinet-inverted metasurface. *Opt. Express* **2015**, *23*, 11114–11122. [CrossRef] [PubMed]
9. Cao, W.; Yang, X.; Gao, J. Broadband polarization conversion with anisotropic plasmonic metasurfaces. *Sci. Rep.* **2017**, *7*, 1–9. [CrossRef]
10. Wang, D.-C.; Sun, S.; Feng, Z.; Tan, W.; Qiu, C.-W. Multipolar-interference-assisted terahertz waveplates via all-dielectric metamaterials. *Appl. Phys. Lett.* **2018**, *113*, 201103. [CrossRef]
11. Zhao, X.; Schalch, J.; Zhang, J.; Seren, H.; Duan, G.; Averitt, R.D.; Zhang, X. Electromechanically tunable metasurface transmission waveplate at terahertz frequencies. *Optica* **2018**, *5*, 303–310. [CrossRef]
12. Li, T.; Hu, X.; Chen, H.; Zhao, C.; Xu, Y.; Wei, X.; Song, G. Metallic metasurfaces for high efficient polarization conversion control in transmission mode. *Opt. Express* **2017**, *25*, 23597–23604. [CrossRef] [PubMed]
13. Li, S.; Wei, M.; Feng, X.; Wang, Q.; Xu, Q.; Xu, Y.; Liu, L.; Ouyang, C.; Zhang, W.; Hu, C.; et al. Polarization-insensitive tunable terahert polarization rotator. *Opt. Express* **2019**, *27*, 16966–16974. [CrossRef]
14. Wang, J.; Tian, H.; Li, S.; Li, L.; Wang, G.; Gao, J.; Guo, W.; Zhou, Z. Efficient terahertz polarization conversion with hybrid coupling of chiral metamaterial. *Opt. Lett.* **2020**, *45*, 1276–1279. [CrossRef] [PubMed]
15. Hu, S.; Yang, S.; Liu, Z.; Li, J.; Gu, C. Broadband cross-polarization conversion by symmetry-breaking ultrathin metasur-faces. *Appl. Phys. Lett.* **2017**, *111*, 241108. [CrossRef]
16. Grady, N.K.; Heyes, J.E.; Chowdhury, D.R.; Zeng, Y.; Reiten, M.T.; Azad, A.; Taylor, A.J.; Dalvit, D.A.R.; Chen, H.-T. Terahertz Metamaterials for Linear Polarization Conversion and Anomalous Refraction. *Science* **2013**, *340*, 1304–1307. [CrossRef]
17. Xu, S.-T.; Hu, F.-T.; Chen, M.; Fan, F.; Chang, S.-J. Broadband Terahertz Polarization Converter and Asymmetric Transmission Based on Coupled Dielectric-Metal Grating. *Ann. Phys.* **2017**, *529*, 1700151. [CrossRef]
18. Gao, X.; Singh, L.; Yang, W.; Zheng, J.; Li, H.; Zhang, W. Bandwidth broadening of a linear polarization converter by near-field metasurface coupling. *Sci. Rep.* **2017**, *7*, 6817. [CrossRef]
19. Sun, H.S.H.; Wang, L.W.L.; Zhang, Y.Z.Y.; Liang, S.L.S.; Han, J.H.J.; Lan, F.L.F.; Zhou, X.Z.X.; Yang, Z.Y.Z. Arbitrary linear THz wave polarization converter with cracked cross meta-surface. *Chin. Opt. Lett.* **2019**, *17*, 041602. [CrossRef]
20. Xu, K.-K.; Xiao, Z.-Y.; Tang, J.-Y. Linearly Polarized Converters for Reflected and Transmitted Waves Based on Double-Split Ring Resonator. *Plasmononics* **2017**, *12*, 1869–1874. [CrossRef]

21. Zhao, J.; Song, J.; Xu, T.; Yang, T.; Zhou, J. Controllable linear asymmetric transmission and perfect polarization conversion in a terahertz hybrid metal-graphene metasurface. *Opt. Express* **2019**, *27*, 9773–9781. [CrossRef]
22. Yu, A.; Yu, D.; Yang, Z.; Guo, X.; Ren, Y.; Zhang, X.; Balakin, A.V.; Shkurinov, A.P.; Zhu, Y. Terahertz Broadband Polarization Conversion for Transitted Waves Based on Graphene Plasmon Resonances. *Nanomaterials* **2021**, *11*, 56. [CrossRef] [PubMed]
23. Zhang, Y.; Feng, Y.; Jiang, T.; Cao, J.; Zhao, J.; Zhu, B. Tunable broadband polarization rotator in terahert frequency based on graphene metamaterial. *Carbon* **2018**, *133*, 170–175. [CrossRef]
24. Qi, X.; Zou, J.; Li, C.; Zhang, J.; Guo, C.; Zhu, Z. Graphene-based electrically controlled terahertz polarization switching between a quarter-wave plate and half-wave plate. *Opt. Express* **2020**, *28*, 39430–39442. [CrossRef]
25. Peng, J.; Zhu, Z.; Zhang, J.; Yuan, X.; Qin, S. Tunable terahertz half-wave plate based on hybridization effect in coupled graphene nanodisks. *Appl. Phys. Express* **2016**, *9*, 055102. [CrossRef]
26. Gric, T.; Wartak, M.; Cada, M.; Wood, J.; Hess, O.; Pištora, J. Spoof plasmons in corrugated semiconductors. *J. Electromagn. Waves Appl.* **2015**, *29*, 1899–1907. [CrossRef]
27. Gric, T.; Hess, O. Surface plasmon polaritons at the interface of two nanowire metamaterials. *J. Opt.* **2017**, *19*, 085101. [CrossRef]

nanomaterials

MDPI

Article

Investigation on Plastic Flow Behaviors of FCC Polycrystalline Aluminum under Pre-Cyclic Tension-Compression Loading: Experiments and Crystal Plasticity Modeling

Damin Lu [1], Keshi Zhang [1] and Guijuan Hu [2,*]

[1] Key Laboratory of Disaster Prevention and Structural Safety/Guangxi Key Lab Disaster Prevention and Engineering Safety, College of Civil and Architectural Engineering, Guangxi University, Nanning 530004, China; ludamin@st.gxu.edu.cn (D.L.); zhangks@gxu.edu.cn (K.Z.)

[2] School of Landscape Architecture, Zhejiang A & F University, Hangzhou 311300, China

* Correspondence: guijuan_hu@zafu.edu.cn

Abstract: The plastic flow behaviors of FCC polycrystalline aluminum after pre-cyclic tension-compression deformation are mainly investigated in tension–torsion stress space by the physically based crystal plasticity model introducing a back-stress. A global finite element model (GFEM) constructed of sufficient grains was established to simulate the same-size thin-walled tube specimen constrained and loaded as the experiments of yield surfaces. The computational results showed that the shape of subsequent yield surfaces and the plastic flow directions directly depended on the given offset strain levels and the applied re-loading paths under different pre-cyclic deformations. The angle deviation between the plastic flow direction and the theoretical orthogonal direction further indicated that there was a large difference between them in the inverse pre-straining direction, but the difference was negligible in the pre-straining direction. From the influence of the anisotropic evolution of the subsequent yield surfaces on plastic flow, we found that the plastic normality rule followed the smooth yield locus; conversely, the significant non-associated flow was attributed to the distorted yield locus. Furthermore, it was also demonstrated that the anisotropic evolution and the plastic flow trend of the subsequent yield surfaces obtained by experiments can be better reproduced by the crystal plasticity model.

Keywords: yield surface; plastic flow; crystal plasticity; polycrystalline aluminum

Citation: Lu, D.; Zhang, K.; Hu, G. Investigation on Plastic Flow Behaviors of FCC Polycrystalline Aluminum under Pre-Cyclic Tension-Compression Loading: Experiments and Crystal Plasticity Modeling. *Nanomaterials* **2021**, *11*, 2397. https://doi.org/10.3390/nano11092397

Academic Editor: Olivier Thomas

Received: 21 July 2021
Accepted: 3 September 2021
Published: 14 September 2021

1. Introduction

During a sheet forming operation, sheet metals generally exhibit significant plastic anisotropy due to rolling and annealing operations [1,2]; moreover, the plastic anisotropy causes directional dependency of the yield stress, and its degree of anisotropy can be geometrically characterized by the distortion and the non-associated flow rule of the yield surfaces. The plastic anisotropy of materials has tremendous influence on material flow, wrinkling behavior, forming limit, springback, and failure properties of the sheet metal [3,4]; further, such material properties require accurate estimation, which depends on a precise knowledge of the yield surface and plastic flow rule [4]. In general, the material models are dominated by three relations: a yield criterion characterizing a yield surface that defines the critical states between the elastic and elastoplastic ranges, flow rule presenting the direction of the plastic strain rate, and a strain-hardening rule describing the yield surface evolution [5,6]. For the repeated loading and unloading case, the direction change of plastic strain accumulation results in the "sharp corner" direction shift of yield surfaces in the loading direction [7]. Therefore, it is imperatively necessary to develop constitutive models capable of describing the yield surface and the direction of the plastic strain rate under complex cyclic loading history. Namely, more accurate yield criterion and plastic flow theories play a crucial role in exploring the influence of physically based anisotropy on the yield surface [2,8].

Many experimental investigations were conducted to indicate the geometrical shapes of yield surface distortion with the remarkable corner effects in the pre-strain direction [9,10] and non-normality of plastic flow on the smooth yield surface [11,12]. At the same time, two main material constitutive models were employed to predict the anisotropic yielding and plastic flow behaviors of polycrystalline materials: the phenomenological approach introducing the anisotropic yield functions with associated or non-associated flow rule [13–15] and the micromechanical approach based on the polycrystalline plasticity models [16–18].

For the phenomenological approach, various anisotropic yield functions based on the associated flow rule (AFR) or non-associated flow rule (non-AFR) to account for the anisotropy of yield surface evolution, Bauschinger effect, and strain-hardening behavior have been proposed [2,16,19,20]. The earliest and most widely used yield criteria under the associated flow rule with a single function in a quadratic or a non-quadratic form to describe both yield behavior and the direction of plastic strain were initiated by Hill [21] and Hosford [22], and the yield locus based on these yield criteria was only restricted to the plane stress states excluding shear stress terms and orthotropic symmetry assuming principal stress axes superimposed onto orthotropic axes [3,4,19]. Hill's yield criterion was further extended by Lin and Ding [23]. To overcome the limitation stated above, the tricomponent and a six-component yield criterion by coupling shear and normal stress components was established to describe the yield surfaces under a full plane stress condition by Barlat et al. [19]. The evolution of the yield surface depends significantly on the deformation history; however, the endochronic theory is inherently suited to describe the responses in general loading histories [24]. The original endochronic theory proposed by Valanis [24–27] did not use the concept of a yield surface, which deals with the elastic–plastic response of materials in terms of the memory kernel functions that lead to an implicit derived result of a yield surface. The deformation history is described with respect to intrinsic time defined as the path length in the strain space and the rate equations through hereditary function [28]. Further, an endochronic plasticity theory was developed by Yeh et al. [29] to describe the distortion of yield surface by the combination of two ellipses with different aspect ratios governed by the aspect ratio function. Moreover, the non-AFR models with the non-normality of plastic flow as experimental evidence to describe the anisotropic plastic strain rates and yield behavior with two different functions were proposed by Stoughton and Yoon [30]. According to the classical theories of plasticity, the direction of plastic strain is normality or non-normality due to the yield surface evolution exhibiting an ellipse or a rough blunt-nosed corner under proportional loading paths [8], namely, the yield functions and plastic flow rules are either inadequate or inaccurate for describing the non-normality of plastic flow with the complicated sharper curvature of the yield surface obtained with the polycrystalline plasticity theory as experimental observations. However, the yield criterion and flow rule lack a physical basis and the microscopic mechanism to account for the microstructural effects, leading to the difficulties and limitations in the application of the phenomenological models in complex stress conditions.

For the micromechanical approach, since the essential physical mechanisms of anisotropic properties are mainly caused by the crystallography of slip, the texture, or the preferred grain orientation, the yield surfaces are calculated with polycrystalline plasticity models on the basis of the plastic deformation mechanisms using some averaging schemes [18]. Barlat et al. [4] analyzed the effect of crystallographic texture on planar anisotropic yield surfaces using the Taylor/Bishop and Hill model based on the pencil glide system of polycrystalline plasticity. Schurig et al. [31] used the classical Taylor–Lin polycrystal model and a viscous grain model to simulate the development of a yield vertex for different strain processes. Beradai et al. [8] used the self-consistent approach to effectively simulate a very complex evolution of subsequent yield surfaces for real FCC metals under complex loading path. The Taylor model is based on the assumption of the same macroscopic strain state for individual grain in the aggregate. Indeed, in addition to the simplicity, Taylor-based and self-consistent models cannot account for the effect of the grain shape and their interactions. For the solving of these simplified problems, the crystal plasticity finite element method

(CPFEM) is well suited to consider the microstructure effects and their interactions of each grain under the displacement continuity assumption of the finite elements at the grain boundaries [32]. Canova et al. [33] pointed out that crystallographic texture gives rise to the anisotropic behavior of yield surfaces for textured materials by polycrystal plasticity simulation. Cazacu et al. [34,35] introduced the texture and grain shape in the modelling of the yield surface using crystal plasticity associated with the representative volume element (RVE). Primary engineering objectives of a crystal plasticity finite element method (CPFEM) applications in macroscopic forming simulations are the prediction of the precise material shape after forming optimization of material flow, elastic spring-back, forming limits, and texture evolution of the formed part [36–39]. Therefore, it was demonstrated that micromechanical approach based on the crystallography of slip, the texture, or the preferred grain orientation can be used to predict the complicated anisotropic evolution of the yield surfaces as experimental results more accurately than the phenomenological approach under complex loading paths [17,18].

In this study, new results concerning the yield surfaces and the plastic flow direction under cyclic loading paths are presented and are investigated by the crystal plasticity model. A crystal plasticity finite element (CPFE) scheme is established with a new GFEM method, illustrated in Section 4.1 in detail. The dependence of the yield surfaces and the plastic flow directions on unloading point position, pre-strain directions, and the levels of yield definitions are investigated in detail by computational simulation. Furthermore, the capabilities of the models are verified by comparing the plastic flow direction of the yield surfaces simulated by the crystal plasticity model with experimentally observed results using the statistical methods.

2. Material and Experimental Procedure

2.1. Experimrntal Material and Specimen

The experimental material was commercially pure aluminum with the purity of 99.89% weight. Its chemical composition and basic mechanical properties are given in Table 1. The cross-sectional EBSD map, the grain size distribution, and pole figure observed by electron backscatter diffraction (EBSD) technique are presented in Figure 1. The initial microstructure of the material contains a large number of grains, and the average grain size was about 65 μm, as shown in Figure 1a,b. It can be further seen that the pure aluminum had a weak texture and the grain orientations were almost random from the EBSD orientation map as per the grain orientation represented by the different colors in Figure 1a and pole figure showing its maximum intensity in 3.93 times random in Figure 1c. The tests were carried out by the MTS809 electro-hydraulic servo testing machine (MTS Systems Corporation, Eden Prairie, MN, USA), applying the tension and torsion load at the two gripping ends of the specimen, and an extensometer with a 25 mm gauge length was mounted on the outer surface of the gauge section of specimens to measure the axial and torsion strain. The geometry size of thin-walled tube specimens of aluminum materials were machined for the experiments shown in Figure 2, and the $\mathbf{n}_1$, $\mathbf{n}_2$, and $\mathbf{n}_3$ axes were set along the axial, thickness, and circumferential directions, respectively.

Table 1. Chemical compositions and mechanical properties of polycrystalline aluminum.

Chemical Composition						Mechanical Properties				
Al/%	**Cu/%**	**Mg/%**	**Si/%**	**Mn/%**	**Zn/%**	**E/GPa**	**G/GPa**	**$\sigma_{0.2}$/MPa**	**σ_u/MPa**	**ε_f**
99.89	0.02	0.03	0.03	0.02	0.01	57	25	20	81	24%

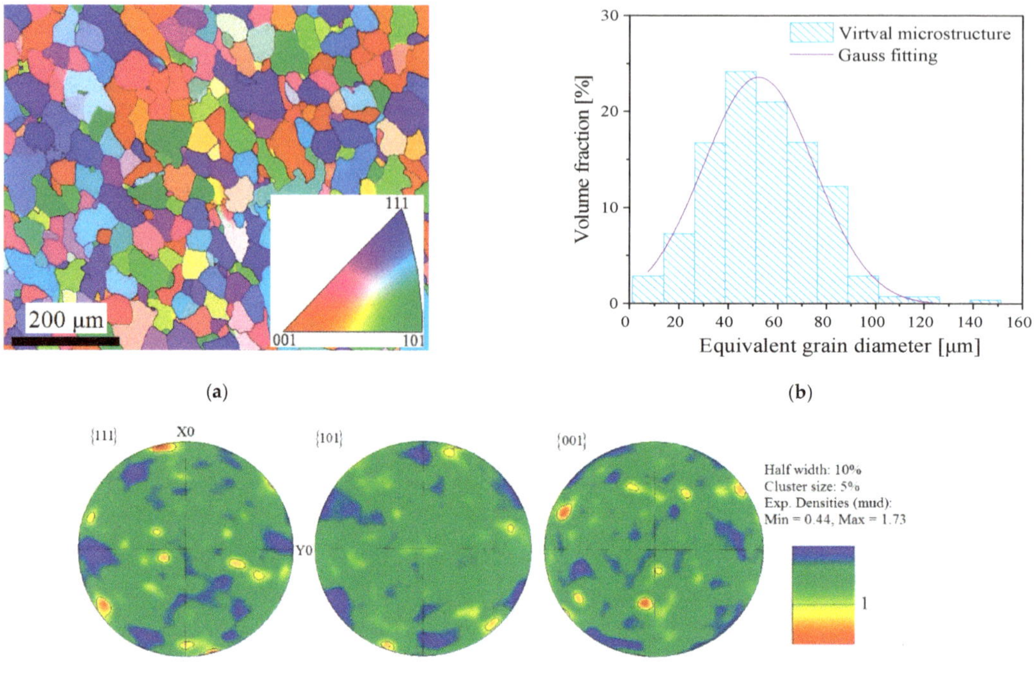

Figure 1. Cross-sectional microstructure of the specimen for pure aluminium observed by EBSD: (**a**) grain orientation map of pure aluminium; (**b**) the statistical distribution of the grain size; (**c**) the pole figure.

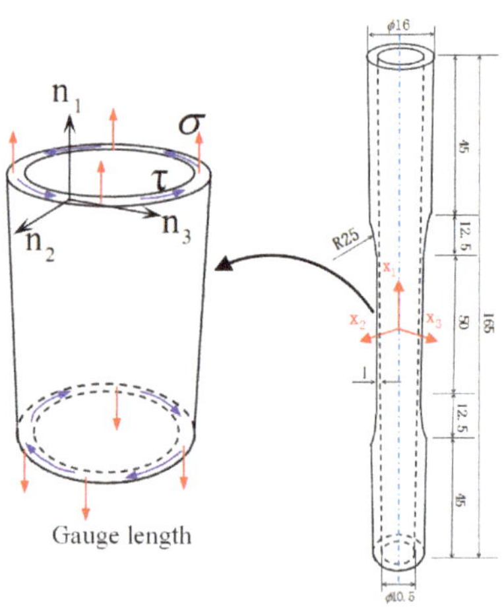

Figure 2. The geometrical figure for thin-walled tube specimens of pure aluminum (unit: mm).

2.2. Experiment Produce

The entire loading process of determining subsequent yield is described according to three steps in detail in this section. First, the specimen underwent the pre-cyclic deformation of 30 cycles to achieve the saturated stress–strain state under symmetrical tension-compression loading at the strain amplitude of 0.3% and remained the constant strain for 2 min corresponding to the respective unloading starting points (A (0.3%), B (0.12%), C (0%), or D (−0.3%)) on the hysteresis loop to eliminate the effect of strain rate or viscoplasticity caused by pre-loading [8]. Then, the specimen was subsequently unloaded to the unloading endpoint (O_{A1}, O_{B1}, O_{C1}, or O_{D1}) within the elastic domain or the reverse yield endpoint (O_{A2}) from unloading the starting point on the saturated stress–strain hysteresis curve by the stress-controlled mode, namely, the five different unloading paths were A (0.3%) → O_{A1}, A (0.3%) → O_{A2}, B (0.12%) → O_B, C (0%) → O_C or D (−0.3%) → O_D, as shown in Figure 3a. Finally, the specimen was reloaded to probe the yield point corresponding to different target offset strain along a prescribed tension–torsion proportional reloading path, as shown in Figure 3b.

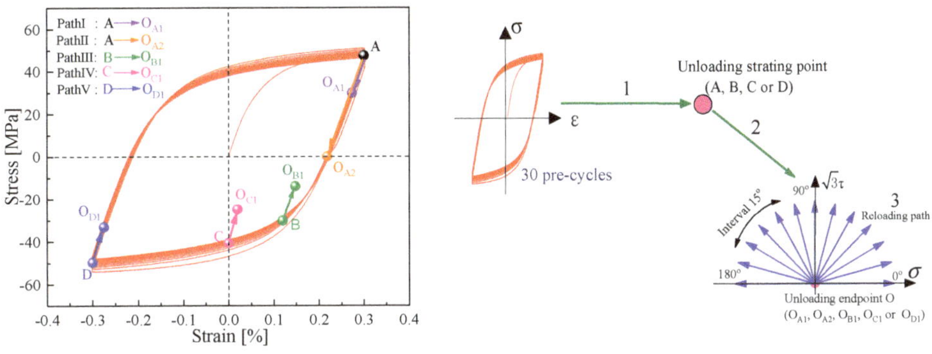

Figure 3. The probing paths of the subsequent yield surface including pre-cycle, unloading, and reloading cases: (**a**) different unloading starting points and end positions after tension-compression pre-cycle; (**b**) schematic diagram of pre-cycle, unloading, and reloading paths of yield surface.

Here, the target offset strain $\Delta \mathbf{E}^{p}_{offset}$ is prescribed as 20 μ, 50 μ, 100 μ, 200 μ, 600 μ, and 1000 μ under each reloading path as shown in Figure 3b. In this study, the multiple-specimen strategy was chosen, for which the yield pointed along a specified tension–torsion proportional loading path by different offset strain definition, determined using a thin-walled tubular specimen; therefore, a complete yield surface could be obtained using several same specimens under the reloading paths at intervals of 15° ranging from 0° to 180°, as shown in Figure 3b.

In this study, the unloading stiffness method was adopted to determine the yield point; since the unloading stiffness $\mathbf{E}_{eff}$ decreased as plastic deformation proceeded during the reloading [9,10], the determination of yield points were affected by the unloading stiffness. To capture yield points more accurately at different offset strains along a reloading path, we fitted the unloading stiffness $\mathbf{E}_{eff}$ by the linear part of the equivalent stress–strain curves during the unloading process of gradual reloading and unloading, as shown in Figure 4a. During gradual reloading and unloading, as soon as an accumulated effective plastic strain increment $\Delta \mathbf{E}^{p}_{offset}$ approached the prescribed target offset strain value ($\Delta \mathbf{E}^{p}_{offset}$ = 20 μ, 50 μ, 100 μ, 200 μ, 600 μ, or 1000 μ, μ = 10^{-6}), the intersection point of the straight line $\mathbf{Y} = \mathbf{E}_{eff}(\mathbf{X} - \Delta \mathbf{E}^{p}_{offset})$ parallel to the unloading stiffness $\mathbf{E}_{eff}$ and the equivalent stress–strain curve was defined as the subsequent yield point represented by **Y** in Figure 4b. By repeating this method, we were able to determine the subsequent yield points corresponding to each prescribed target strain along a reloading path.

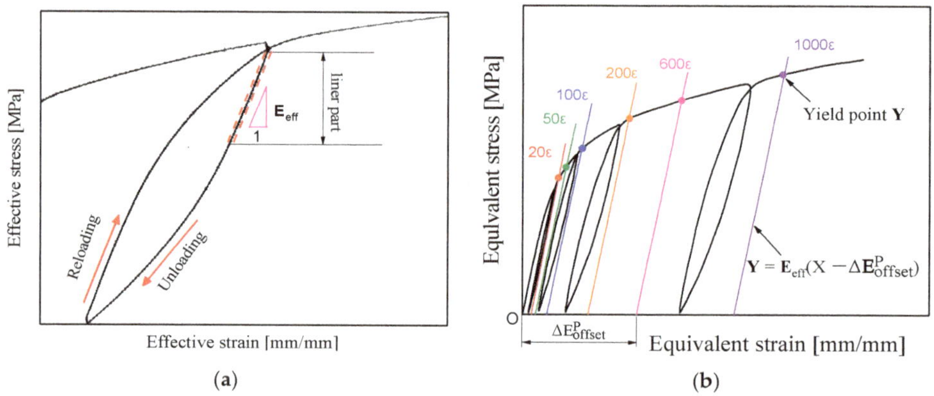

Figure 4. The yield point definition in terms of unloading modulus: (**a**) the determination method of unloading stiffness; (**b**) the variation of unloading stiffness $\mathbf{E}_{\text{eff}}$ with the increasing offset strain during reloading and unloading processes.

2.3. Determining the Orthogonal Direction and Plastic Strain Incremental Direction of Yield Point

For the investigated plastic flow behavior for polycrystalline aluminum, we assumed that the stress space was consistent with the plastic strain space. Figure 2 shows the geometric dimensions of the test specimen. The nominal tension stress σ and shear stress τ on the cross-section of the gauge length range under the combined tension–torsion loading were set along the $\mathbf{n}_1$ axis and $\mathbf{n}_3$ axis, respectively, and the direction of the nominal strain was assumed to be coincided with that of the stress. The schematic illustration of plastic strain incremental direction and its normal direction to the yield surface is shown in Figure 5 [12]. The normal direction angle θ_2 to yield surface was approximately obtained by geometrical calculation. Taking a yield point B located on the yield locus as an example, first, the yield point B is connected with its adjacent yield points of A and C on the same yield surface; then, the intersection D is obtained by making the respective midperpendicular of the straight lines AB and BC, and finally, the vector $\vec{\mathbf{n}}$ that passes through the intersection D and the yield point B is referred to as the normal direction at the yield point B, and the vector $\vec{\mathbf{n}}$ representing plastic flow direction is orthotropic to the yield locus. The plastic flow direction angle θ_1 to the yield surface is calculated by the following Equations (1)–(6). The vector of plastic strain increment in the $\sigma - \sqrt{3}\tau$ stress space can be expressed as follows:

$$\Delta\vec{\varepsilon}^{\mathrm{p}} = \Delta\varepsilon_1^{\mathrm{p}}\vec{\mathbf{n}}_1 + \Delta\varepsilon_3^{\mathrm{p}}\vec{\mathbf{n}}_3 \tag{1}$$

$$\Delta\varepsilon_1^{\mathrm{p}} = \Delta\varepsilon^{\mathrm{p}} = \Delta\varepsilon - \frac{\Delta\sigma}{\mathrm{E}} \tag{2}$$

$$\Delta\varepsilon = \varepsilon - \varepsilon_0; \Delta\sigma = \sigma - \sigma_0 \tag{3}$$

$$\Delta\varepsilon_3^{\mathrm{p}} = \frac{\Delta\gamma^{\mathrm{p}}}{\sqrt{3}} = \frac{\Delta\gamma - \frac{\Delta\tau}{\mathrm{G}}}{\sqrt{3}} \tag{4}$$

$$\Delta\gamma = \gamma - \gamma_0;\ \Delta\tau = \tau - \tau_0 \tag{5}$$

where $\Delta\sigma$ and $\Delta\tau$ are axial stress increment and shear stress increment, respectively; $\Delta\varepsilon^{\mathrm{p}}$ and $\Delta\gamma^{\mathrm{p}}$ are axial plastic strain increment and shear plastic strain increment, respectively; ε_0 and γ_0 are axial residual strain and shear residual strain at the unloading point, respectively; σ_0 and τ_0 correspond to the axial stress and the shear stress at the unloading point, respectively; and E and G are elastic modulus and shear modulus, respectively.

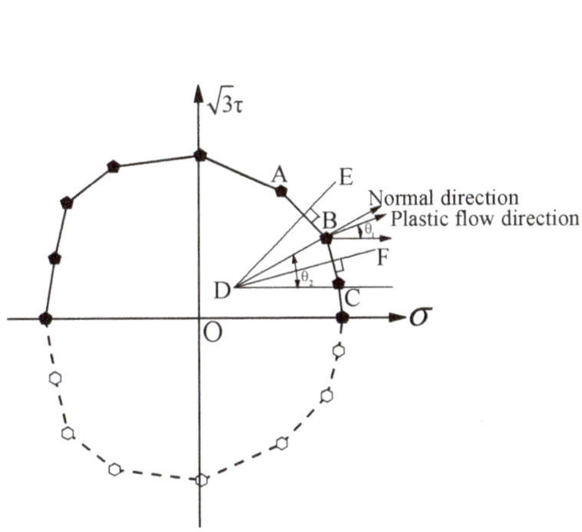

Figure 5. The schematic illustration about the determination of normal direction of yield point on yield surface and plastic strain incremental direction.

The plastic flow direction in the $\sigma - \sqrt{3}\tau$ stress space is defined as

$$\theta = \arctan\left(\frac{\Delta\varepsilon_2^p}{\Delta\varepsilon_1^p}\right) \tag{6}$$

3. Constitutive Relationships and Coupling Methodology

3.1. Flow Rule and Strain Hardening Modeling

According to the classical multiplicative decomposition of the deformation gradient, the total deformation gradient $\boldsymbol{F}$ within the crystal plasticity framework can be decomposed into elastic and plastic components [40,41]:

$$\boldsymbol{F} = \boldsymbol{F}^* \cdot \boldsymbol{F}^p \tag{7}$$

where the elastic component $\boldsymbol{F}^* = \boldsymbol{R}^L \cdot \boldsymbol{U}^e$ is caused by the elastic stretching $\boldsymbol{U}^e$ and rigid body rotation $\boldsymbol{R}^L$ of a crystal lattice, and the plastic component $\boldsymbol{F}^p$ describes the dislocation glide along the specific slip planes.

The velocity gradient $\boldsymbol{L}$ can be decomposed into the elastic gradient tensor $\boldsymbol{L}^*$ and the plastic gradient tensor $\boldsymbol{L}^p$ [40–42]:

$$\boldsymbol{L} = \dot{\boldsymbol{F}} \cdot \boldsymbol{F}^{-1} = \boldsymbol{L}^* + \boldsymbol{L}^p = \dot{\boldsymbol{F}}^* \cdot \boldsymbol{F}^{*^{-1}} + \boldsymbol{F}^* \cdot \dot{\boldsymbol{F}}^p \cdot \boldsymbol{F}^{p^{-1}} \cdot \boldsymbol{F}^{*^{-1}} \tag{8}$$

where $\boldsymbol{L}^* = \dot{\boldsymbol{F}}^* \cdot \boldsymbol{F}^{*^{-1}}$ represents elastic deformation gradient and $\boldsymbol{L}^p = \dot{\boldsymbol{F}}^p \cdot \boldsymbol{F}^{p^{-1}}$ stands for plastic deformation gradient.

Further, because only the activated slip systems contribute to the plastic deformation in each grain, the plastic velocity gradient $\boldsymbol{L}^p$ is expressed as the sum of shear strain rates $\dot{\gamma}^{(\alpha)}$ on all active systems [43]:

$$\boldsymbol{L}^p = \dot{\boldsymbol{F}}^p \cdot \boldsymbol{F}^{p^{-1}} = \sum_{\alpha=1}^{n} (\boldsymbol{m}^{(\alpha)} \otimes \boldsymbol{n}^{(\alpha)}) \dot{\gamma}^{(\alpha)} = \sum_{\alpha=1}^{n} s^{(\alpha)} \dot{\gamma}^{(\alpha)} \tag{9}$$

where $\dot{\gamma}^{(\alpha)}$ is the plastic shear rate along the crystallographic slip directions, and $\boldsymbol{n}$ is the total number of slip systems considering the typical 12 slip systems $\{111\} < 110 >$ for a single crystal of the face centered cubic (FCC) structure materials. The unit vectors $\boldsymbol{m}^{(\alpha)}$ and $\boldsymbol{n}^{(\alpha)}$ stand for the slip direction and the normal direction with respect to the slip plane in the intermediate configuration, respectively; their dyadic product ($s^{(\alpha)} = \boldsymbol{m}^{(\alpha)} \otimes \boldsymbol{n}^{(\alpha)}$) defines the Schmid tensor.

The crystalline lattice rotation will lead to the transformation of these vectors $\boldsymbol{m}^{(\alpha)}$ and $\boldsymbol{n}^{(\alpha)}$ in the reference configuration; further, the corresponding vectors $\boldsymbol{m}^{(\alpha)*}$ and $\boldsymbol{n}^{(\alpha)*}$ in the current configuration can be obtained by the following equations:

$$\boldsymbol{m}^{(\alpha)*} = \boldsymbol{F}^{*} \cdot \boldsymbol{m}^{(\alpha)} \tag{10}$$

$$\boldsymbol{n}^{(\alpha)*} = \boldsymbol{n}^{(\alpha)} \cdot \boldsymbol{F}^{*^{-1}} \tag{11}$$

where the vectors $\boldsymbol{m}^{(\alpha)}$ and $\boldsymbol{n}^{(\alpha)}$ are assumed to be orthogonal to each other.

$$\boldsymbol{m}^{(\alpha)*} \cdot \boldsymbol{n}^{(\alpha)*} = \boldsymbol{m}^{(\alpha)} \cdot \boldsymbol{n}^{(\alpha)} = 0 \tag{12}$$

3.2. Hardening Rules

The mesoscopic deformation mechanism of polycrystalline materials follows the deformation law of each individual crystal. Because the subsequent yield surfaces are studied under pre-cyclicloading, in order to address the influence of Bauschinger effect, a back-stress $x^{(\alpha)}$ is here introduced by Zhang [7] into the viscoplastic constitutive model. On the basis of the viscous regularization proposed by Hutchinson [44], the slip rate for a given slip system is written as

$$\dot{\gamma}_{(\alpha)} = \dot{\gamma}_0 \mathrm{sgn}\left(\tau^{(\alpha)} - x^{(\alpha)}\right)\left|\frac{\tau^{(\alpha)} - x^{(\alpha)}}{g^{(\alpha)}}\right|^{k} \tag{13}$$

where $\dot{\gamma}_0$ is a reference shear strain rate on the α-th slip system, k is the rate sensitivity exponent, $\tau^{(\alpha)}$ is the resolved shear stress on the α-th slip system, $g^{(\alpha)}$ is the critical shear stress on the activated slip system α to govern the isotropic hardening of the crystal, and the back-stress $x^{(\alpha)}$ is especially introduced here to characterize the nonlinear directional hardening of the crystal on the α-th slip system.

The critical shear stress $g^{(\alpha)}$ evolves with plastic strain on a given activated slip system according to the following form proposed by Hill [45]:

$$\dot{g}^{(\alpha)}(\gamma) = \sum_{\beta=1}^{n} h_{\alpha\beta}(\gamma)\left|\dot{\gamma}^{(\beta)}\right|;\ \gamma = \int \sum_{\beta}^{n}\left|\mathrm{d}\gamma^{(\beta)}\right| \tag{14}$$

where $\dot{\gamma}^{(\beta)}$ is the slip rate on the system β, n is the total number of crystal slip systems, and γ is the accumulated shear strains on all active slip systems. $h_{\alpha\beta}(\gamma)$ is the slip hardening modulus matrix governing the interaction between various slip systems, which is suggested by Hutchinson [44] as follows:

$$h_{\alpha\beta}(\gamma) = h(\gamma)[q + (1-q)\delta_{\alpha\beta}] \tag{15}$$

where $\delta_{\alpha\beta}$ is the Kronecker delta, and q refers to the latent-hardening ratio that has values in the range of 0 to 1. A function describing the self-hardening ratio $h(\gamma)$ is defined by Chang and Asaro [46] as follows:

$$h(\gamma) = h_0 \sec h^2\left(\frac{h_0\gamma}{\tau_s - \tau_0}\right) \tag{16}$$

where h_0 is the initial hardening rate, τ_0 is the critical resolved shear stress, and τ_s is the saturation shear stress.

3.3. Back Stress Evolution

In order to model cyclic plastic behavior reasonably, on the basis of the models of Walker [47] and Chaboche [48], the evolution of back-stress $\dot{x}^{(\alpha)}$ is given by [49].

$$\dot{x}^{(\alpha)} = a\dot{\gamma}^{(\alpha)} - c\,[1 - e_1(1 - \exp(-e_2\gamma))]x^{(\alpha)}\left|\dot{\gamma}^{(\alpha)}\right| - dx^{(\alpha)} \tag{17}$$

where a is the linear hardening constant of slip systems, c and d denote the nonlinear hardening constants, and e_1 and e_2 are constant parameters representing the law of the cyclic hardening or softening. The first term of the equation denotes a linear strain hardening term. The second one denotes a dynamic recovery term. The third one denotes a static recovery term. Moreover, the back-stress term can adequately cover the cyclic hardening, the Bauschinger, and similar strain path change effects.

The detailed constitutive equations of the crystal plasticity model have been calculated and implemented into the user-supplied subroutine UMAT from the ABAQUS finite element code [49].

4. Modeling

4.1. Finite Element Models

The global finite element model (GFEM) simulating the same thin-walled tube specimen as the experiments of yield surfaces was built, consisting of the 64,000 elements and 3600 grains in the center gauge region of the GFEM, and the grain distribution in the cross-section of gauge length is also shown in Figure 6, with the grain number of the GFEM being sufficient to reflect the macroscopic convergent stress–strain response shown in the previous study [50]. Each grain is assigned to random shape, size, and random crystal orientation [50]. The center gauge section of the specimen defines two reference points A and B, simulating a 25 mm axial-torsional extensometer to measure the macroscopic axial strain and torsion strain under the combine tension–torsion loading. The loading and boundary conditions of the GFEM model are as follows: all element nodes at the loading end or the fixing end (the region marked in the red dotted line in Figure 6 are coupled with the corresponding reference point to form a point set, respectively, and loads and boundary conditions are applied to the reference points). All six degrees of freedom for the reference point in the fixed clamping end were set as zero. An axial load **F** and a torsion load **T** (combined tension–torsion loading) were applied on the reference point in the load clamping end along the first axial and the fifth rotational degrees of freedom as the testing, and the remaining four degrees of freedom for the reference point were set to zero $U_2 = U_3 = UR_2 = UR_3 = 0$, as shown in Figure 6.

Computational information was introduced in detail as follows: two server-cluster computers were used with 32 CPU cores in the parallel computing environment. The total calculation time of 30 pre-cycles was 6.3 h for each cycle including 200 increments. The loading process simulation corresponding to different re-loading path with the "restart write" function can be started simultaneously under different unloading cases after 30 pre-cycles. Each re-loading path was calculated with 10 CPU cores, and the calculation time was about 45 min.

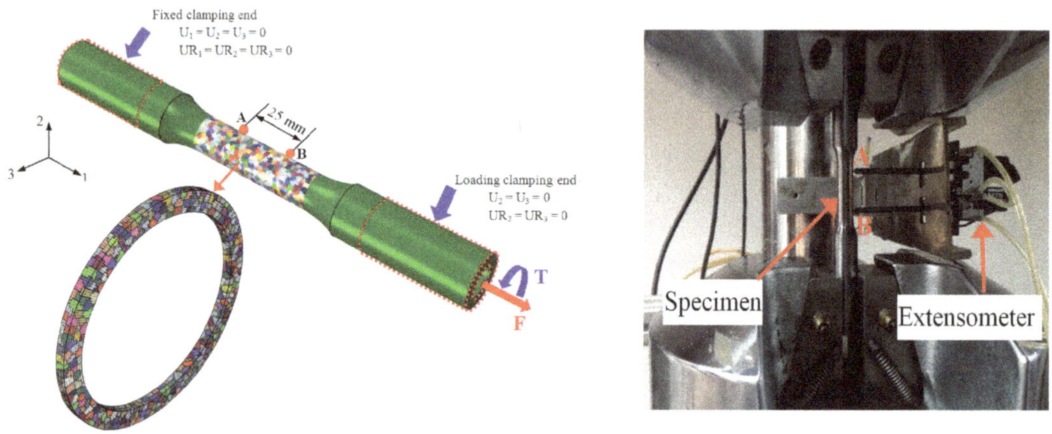

Figure 6. The loading process of the global finite element model (GFEM) containing 3600 grains under the combined tension–torsion loading.

4.2. Parameter Calibration of Crystalline Plasticity Models

The crystal lattice of a pure aluminum is a face-centered cubic. The crystal plastic model describing the elastic and plastic behavior of a single crystal contained 14 parameters (3 elastic constants and 11 viscoplastic parameters), and each parameter had physical meaning. The detailed determination process is as follows: three anisotropic elastic constants C_{11}, C_{12}, and C_{44} of single crystal aluminum are obtained by fitting elastic region of testing uniaxial tensile and torsion stress-strain curves [50]. τ_0 and τ_s are the critical resolved shear stress and the saturation value, respectively, which depend on the elastic range between the points of unloading and re-yielding on the stable stress–strain hysteresis loop. a and c describe the strain hardening of the elastoplastic region on the hysteresis loop, and e_1 and e_2 are used to describe cyclic hardening behavior during the initial 10 cycles. The reference strain rate was $\dot{\gamma}_0 = 0.001/s^{-1}$. The rate sensitivity exponent was defined as k = 200 to describe the insensitive strain rate of the pure aluminum. By taking into account the fact that creep is ignored under cyclic loading, the nonlinear hardening constants was set to $d = 0/s^{-1}$. The remaining material parameters τ_0, τ_s, h_0, a, c, e_1, and e_2 of the above crystal viscoplastic model are preliminarily estimated from the tested cyclic hysteresis curves, and more accurate material parameters can be fitted by trial-and-error method with test curve as the target curve. The summary of the final calibrated material parameters is listed in Table 2. Figure 7 shows the comparison of experimental curves (black lines) and simulated curves (red lines) of 30 pre-cycles, which confirms that the simulation results using the GFEM can match with the tested curves well. This good agreement suggests that the constitutive model can reasonably describe the cyclic plastic behavior of metal materials.

Table 2. The material parameters of single crystal aluminum.

C_{11} /GPa	C_{12} /GPa	C_{44} /GPa	τ_0 /MPa	τ_s /MPa	h_0 /MPa	a /GPa	c /GPa	e_1	e_2	d /s^{-1}	$\dot{\gamma}_0$ /s^{-1}	k
75.64	36.86	25.4	8.42	9.2	65	22.5	2.5	0.1	1.0	0	0.001	200

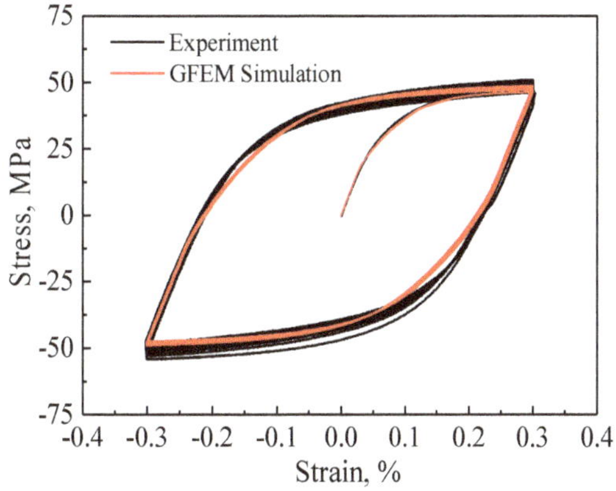

Figure 7. The comparison of the cyclic stress–strain curves under tension-compression loading by test and crystal plasticity simulation with the GFEM containing 3600 grains.

4.3. Effects of Loading Direction on Anisotropic Hardening Characteristics

Following the same experimental procedures for the case of unloading to the point O_D after pre-cyclic tension-compression loading, we conducted these predictions along the reloading directions at angles from 0° to 180° in steps of 30° in different proportions of tension (or compression) and torsion in the $(\sigma, \sqrt{3}\tau)$ plane by using the GFEM on the basis of the crystal plasticity theory introducing a back stress. Then, the simulated σ-ε curves were also compared with experimental observations, as shown in Figure 8. The results show that the anisotropic strain-hardening properties can be observed, and the trend towards strain hardening of the simulated curves was consistent with that of the experimental curves under all proportional tensile–torsion loading paths, which had slight difference despite the relative large error caused by the dispersion of the test results by multi-specimen method and the model error at 60° reloading paths. The above results indicate that the anisotropic hardening behavior of polycrystalline aluminum after deformation can be reasonably described by the crystal plasticity model introducing nonlinear strain hardening equation. Moreover, it can be clearly seen from Figure 8a,b that the plastic pre-deformation governed the strain hardening rate of the material, namely, an increase in the flow stress with increasing angle deviated from the pre-loading direction was correctly captured by the model as the experimental observations. In other words, the strain hardening rate was weak in the pre-loading direction compared with the other direction, as well as the higher hardening rate upon the reverse pre-loading direction. The directional hardening and distorted shape of the subsequent yield surfaces could be attributed to the developed anisotropic hardening of the materials. It should be pointed out that the difference between initial yield surface and subsequent yield surface was noticeable. After plastic pre-deformation, for example, axial tensile deformation, the conspicuous anisotropic hardening of material can be observed. Therefore, the evolution of the subsequent yield surfaces is focused on and discussed in the following section.

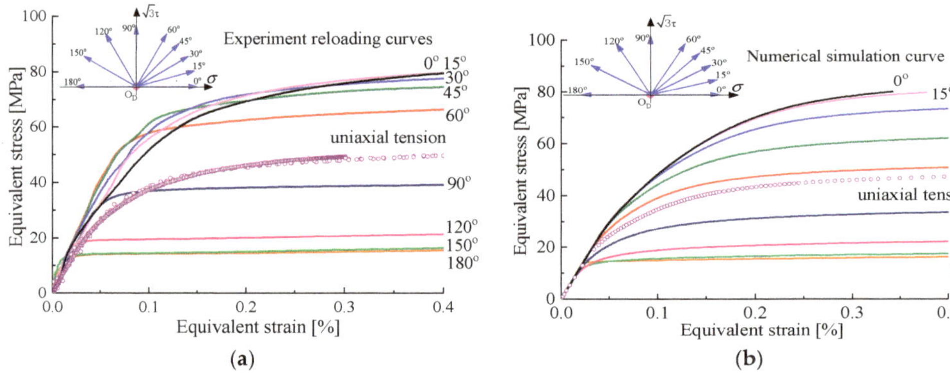

Figure 8. The strain-hardening curves under different tension–torsion reloading paths unloaded from the hysteresis loop D (−0.3%) to O_D after 30 pre-cycles by test and crystal plasticity simulation: (**a**) the experimental strain-hardening curves; (**b**) the simulated strain-hardening curves.

5. Results and Discussion

5.1. Initial Yield Surfaces and Plastic Flow Directions

The initial yield surfaces and its plastic flow directions were simulated under proportional tensile and torsional stress paths using the yield definitions with the offset strain of 20 μ, 200 μ, and 1000 μ by means of GFEM integrated into crystal plasticity theory, as illustrated in Figure 9. The initial yield surfaces of the ploycrystal aluminum exhibited approximately isotropic evolution and obeyed the Tresca yield criterion in the. $\sigma - \sqrt{3}\tau$. stress plane with arbitrarily specified offset strain in Figure 9a. The plastic flow directions of the initial yield surfaces were represented by the vector arrow denoting the plastic strain increment of 50 μ in the present study. The influence of different offset strain definitions on the plastic flow directions is evaluated in Figure 9b. When the offset strain was very small, the deviation angles between the plastic flow direction and normality direction regarding the initial yield surfaces had subtle differences, and with a larger offset strain definition, the directions of plastic strain increment were normal to the initial smooth yield surface, being not evidently affected by the offset strain definition, indicating that the plastic flow directions of the initial yield locus approximately followed the normality rule with any offset strain definition, as shown in Figure 9.

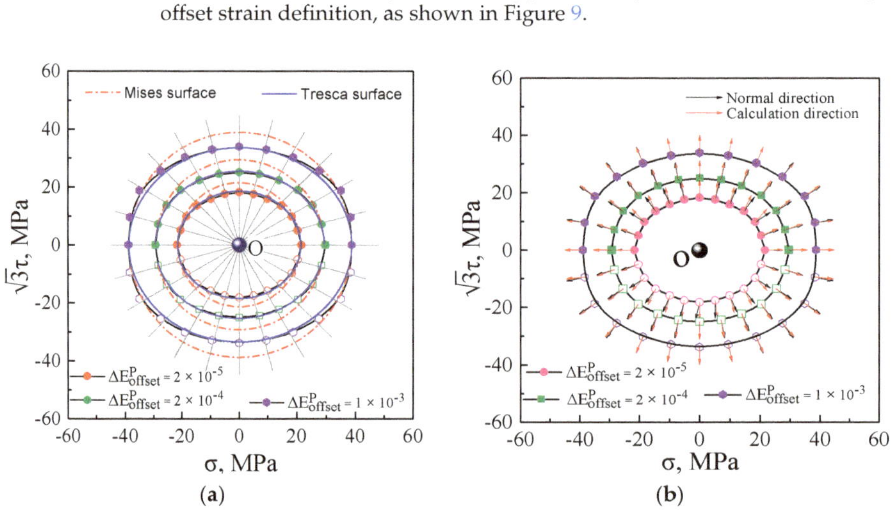

Figure 9. Initial yield surface by crystal plasticity simulation: (**a**) initial yield surfaces; (**b**) plastic flow directions.

5.2. Subsequent Yield Surfaces under Five Unloading Cases

As illustrated in Figures 3 and 4, the pre-loading, unloading, and reloading paths and the measured method of yield points for probing the subsequent yield surface are described in detail in Section 2.2. The radial light gray lines representing the different reloading paths are drawn in Figure 10. The subsequent yield surfaces were predicted by the crystal plasticity model associated with the GFEM containing 64,000 elements and 3600 crystalline grains under the five unloading paths with the offset strain of 20 μ, 50 μ, 100 μ, 200 μ, 600 μ, and 1000 μ, as shown in Figure 10. With reference to the hardening curves under different reloading paths in Figure 8, the curves appeared with the lower strain hardening rate in the pre-loading direction compared with the other direction and with higher hardening rate upon the reverse pre-loading direction. Such directional hardening behaviors related to loading paths resulted in the anisotropic expansion and distortion of subsequent yield surfaces. The yield surfaces with different offset strain definitions nearly overlapped in the pre-loading direction but they expanded greatly towards the reverse direction with the increase of the offset strains during the plastic deformation process. As seen in Figure 10, for the small offset strains between 20 and 200 μ, the subsequent yield surfaces were gradually altered from the extremely sharper corner to a rather blunt vertex in the pre-strained direction and remained flat in its reverse direction, whereas the subsequent yield surfaces evolved into a similar ellipse based on isotropic expansion with the larger offset strain of 600 and 1000 μ. As shown in Figure 10, following the variation of the pre-strain direction from tension to compression, the sharp vertex direction of the subsequent yield surfaces simultaneously changed to be consistent with it, and the subsequent yield surfaces were translated observably with it in stress space and were enlarged in contrast to the initial yield surfaces resulting from the previous cyclic hardening. Therefore, the size and shape of the subsequent yield surface depended on the offset strain level, direction of the accumulated plastic strain, and unloading point cases.

5.3. The Plastic Flow Direction of Subsequent Yield Surfaces

The plastic flow evolution of the subsequent yield surfaces was analyzed by the crystal plasticity simulation with the offset strain of 20 μ, 200 μ, and 1000 μ under five unloading cases, as shown in Figure 11. In addition, the comparison of between theoretical calculations and experimental results is shown in Figure 12. The vector arrows denoting the plastic flow direction were prescribed with the plastic strain increment of 50 μ. As shown in Figures 11 and 12, the plastic flow direction of the subsequent yield surfaces directly depended on the applied offset strain levels and the pre-strain paths. By yielding definition of the smaller offset strain, the plastic strain increment vectors (red arrows) corresponding to each yield point calculated by the Equations (1)–(6) according to the simulated and experimental data showed larger deviation from a vector normal to the yield surfaces (black arrow) and simultaneously rotated away from the pre-loading direction, whereas when a large offset strain definition was used to determine the yield surface, negligible deviation was observed between the computed plastic flow direction and orthogonal direction, namely, the vectors of plastic strain increment were nearly orthogonal to the yield surface, indicating that normality flow rule of a smooth yield surface was obeyed. The largest deviation of the plastic flow direction from orthogonal direction appeared in the inverse pre-strain direction, but the difference between them in the pre-strain direction was not significant.

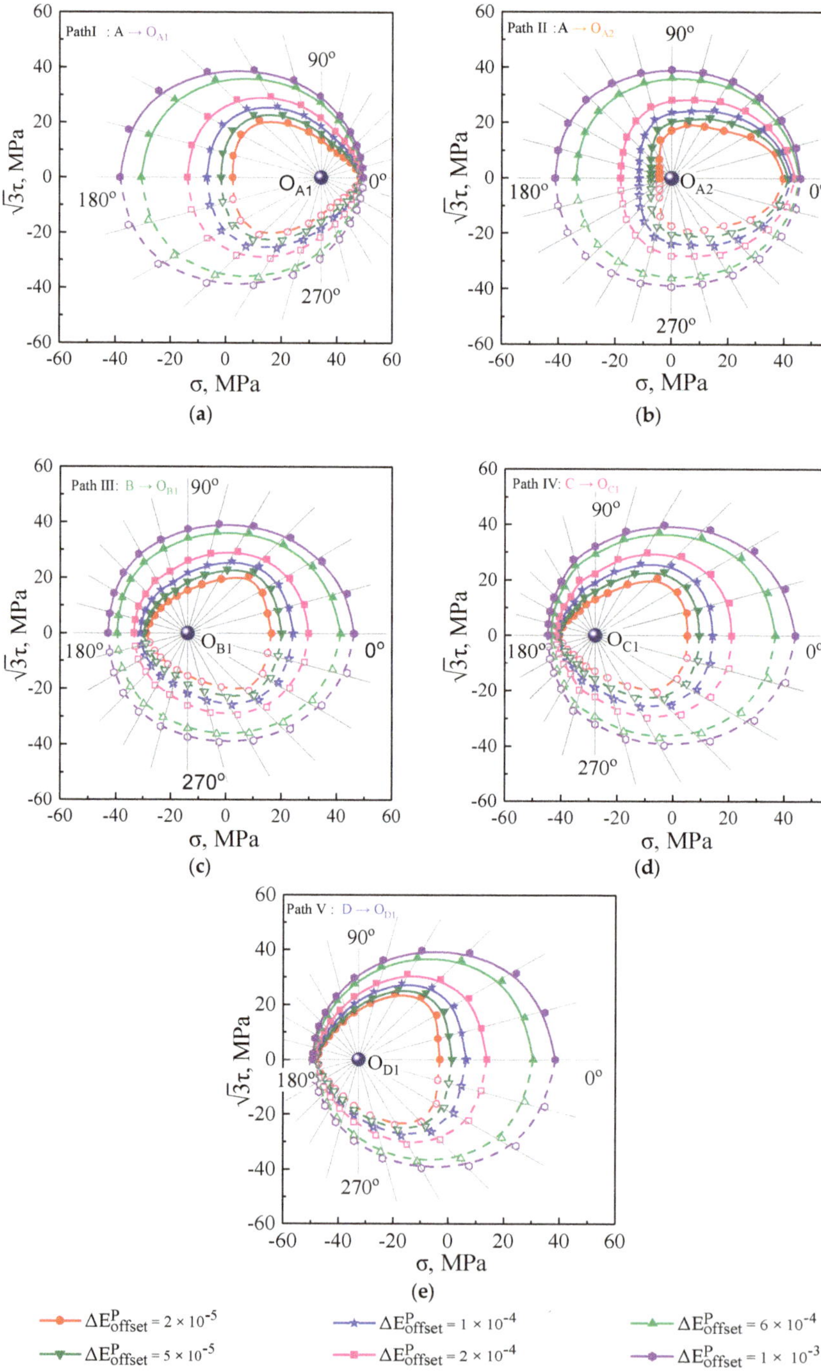

Figure 10. Subsequent yield surface at five different unloading paths by crystal plasticity simulation: (**a**) from A to O_{A1}; (**b**) from A to O_{A2}; (**c**) from B to O_B; (**d**) from C to O_C; (**e**) from D to O_D.

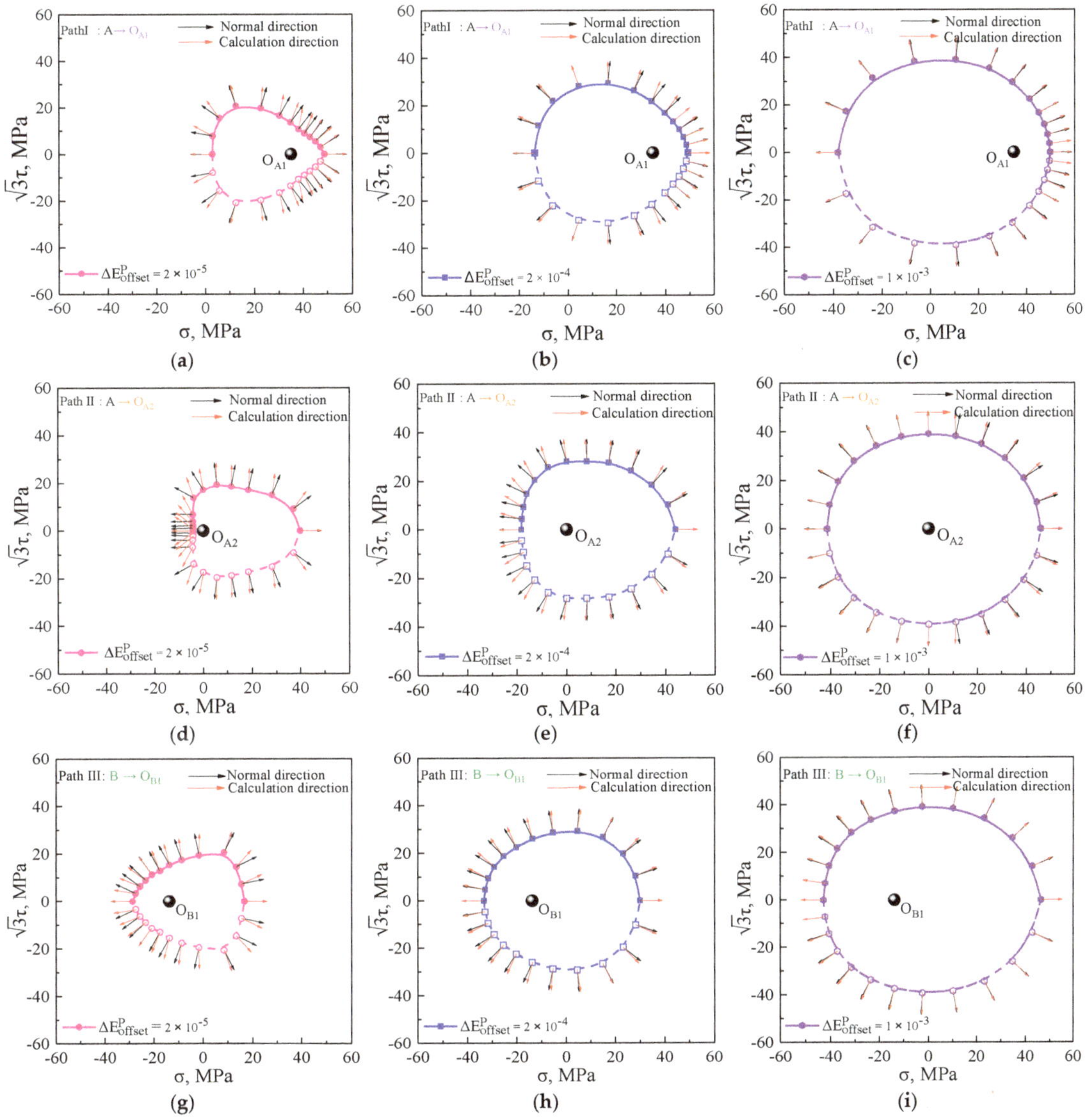

Figure 11. *Cont.*

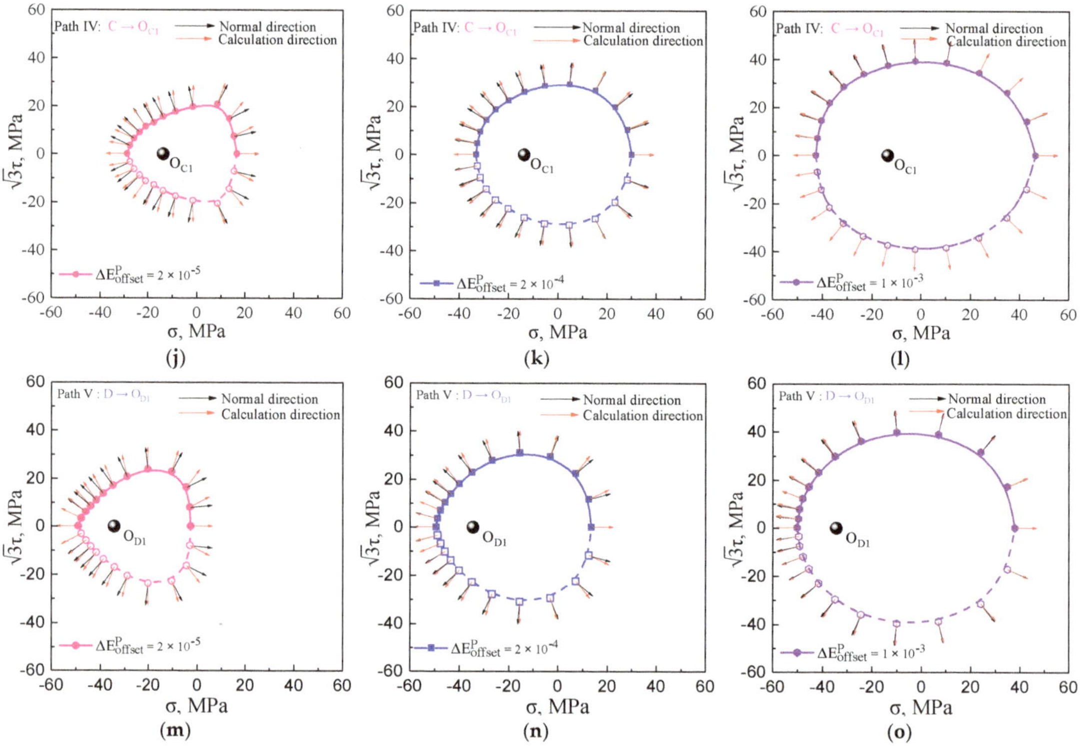

Figure 11. The plastic flow directions by theoretical calculations and simulated results with the offset strain of 20 μ, 200 μ, and 1000 μ under five unloading cases: (**a–c**) at unloading points O_{A1}; (**d–f**) at unloading points O_{A2}; (**g–i**) at unloading points O_{B1}; (**j–l**) at unloading points O_{C1}; (**m–o**) at unloading points O_{D1}.

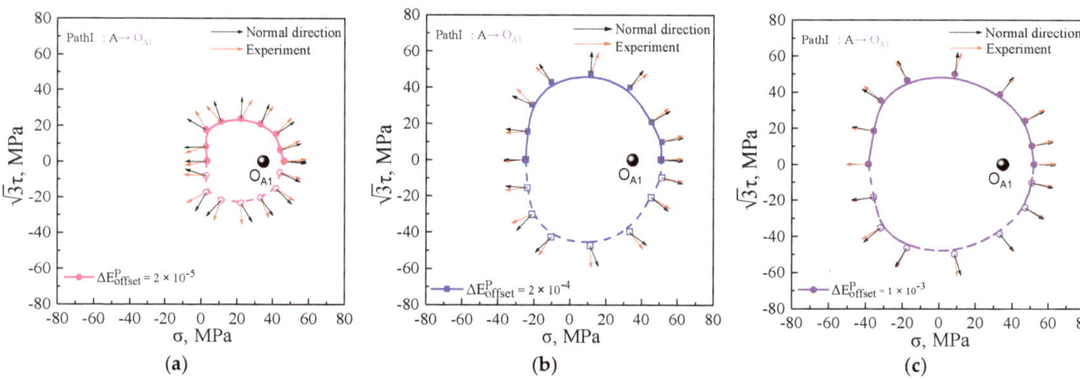

Figure 12. *Cont.*

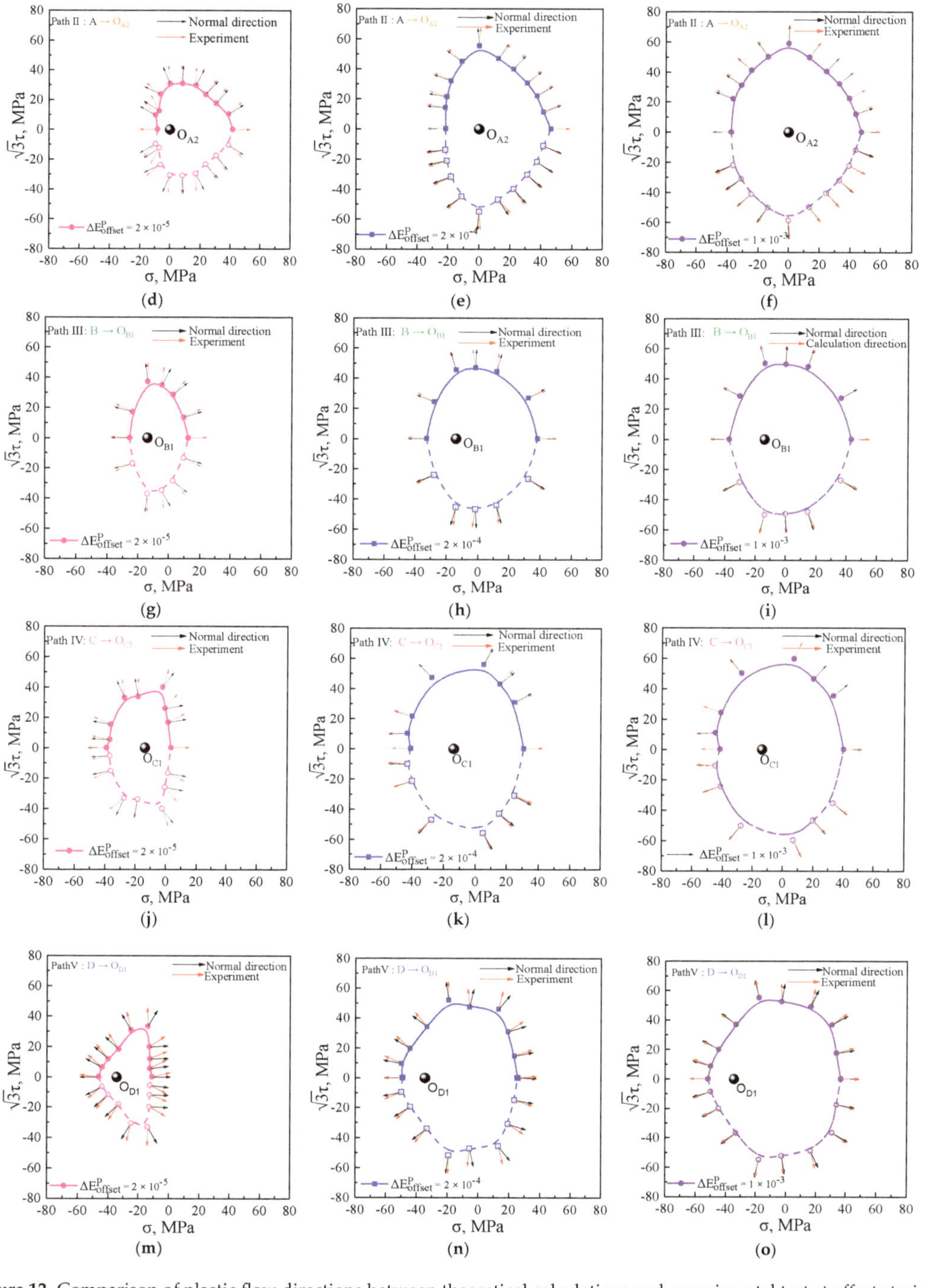

Figure 12. Comparison of plastic flow directions between theoretical calculations and experimental test at offset strain of 20 μ, 200 μ, and 1000 μ under five unloading cases: (**a–c**) at unloading points O_{A1}; (**d–f**) at unloading points O_{A2}; (**g–i**) at unloading points O_{B1}; (**j–l**) at unloading points O_{C1}; (**m–o**) at unloading points O_{D1}.

5.4. Statistical Analysis for Plastic Flow Direction of Subsequent Yield Surfaces

By statistical analysis, the deviation angles $\theta_2 - \theta_1$ between the plastic flow direction and the normal direction of the yield surfaces calculated in terms of the simulated and experimental data under the two unloading paths with the offset strain of 20 μ, 200 μ, and 1000 μ are shown in Figure 13. The results show that the plastic flow directions of subsequent yield surfaces directly depended on the applied offset strain levels. The deviation angles $\theta_2 - \theta_1$ of subsequent yield surfaces became small as the offset strain proceeded. The plastic flow direction was associated with the shape of the subsequent yield surfaces; furthermore, the more severe anisotropy of the yield surfaces with a sharp vertex resulted in the larger deviation from the normal direction with a small offset strain, whereas the normal flow rule was approximatively satisfied for the smooth yield surface with a larger offset strain, for which deviation angle $\theta_2 - \theta_1$ was less than 5°. In the case of the same path and yield definition, the slight difference of deviation angle $\theta_2 - \theta_1$ between the simulated results and experimental ones was mainly derived from the experimental yield surfaces determined by the fewer yield points; nevertheless, the trends of plastic flow direction by the crystal plasticity simulation were consistent with the experimental results.

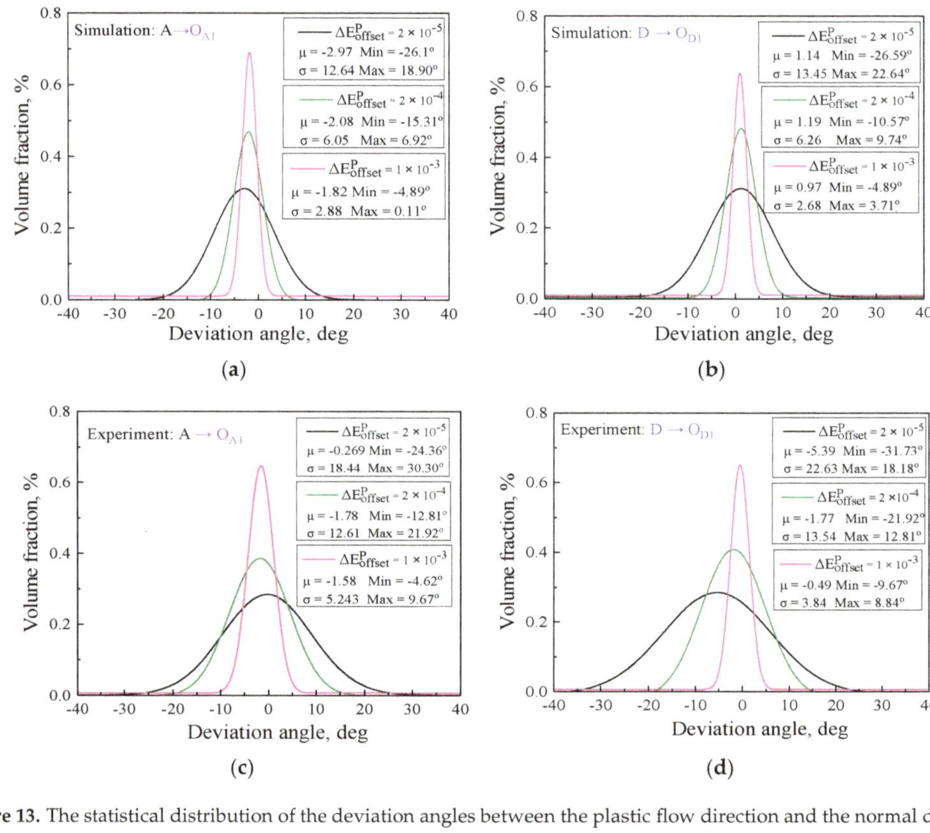

Figure 13. The statistical distribution of the deviation angles between the plastic flow direction and the normal direction of the yield surface: (**a**) simulated statistical distribution at unloading points O_{A1}; (**b**) simulated statistical distribution at unloading points O_{D1}; (**c**) experimental statistical distribution at unloading points O_{A1}; (**d**) experimental statistical distribution at unloading points O_{D1}.

6. Conclusions

In order to characterize the plastic anisotropy of metal deformation, we employed the crystal plasticity model introducing a back-stress to simulate the evolution of the subsequent yield surface of polycrystalline aluminum after pre-cyclic loading, with the simulated process being the same as the experimental approaches. At the same time, the plastic flow that reflects the direction of the plastic strain rate and a strain-hardening rule describing the yield surface evolution during plastic deformation of materials were also further studied, and the experimental observations and the simulation results were compared and analyzed. The main conclusions in this study are as follows:

- The crystal plasticity theory introducing a back-stress in conjunction with the GFEM provided the accurate simulated results with respect to the remarkable sharp corner and the non-associated flow direction of the subsequent yield surfaces as experimental observations under all explored cyclic pre-loading paths.
- For the arbitrarily specified offset strain definition, the initial yield surfaces of the ploycrystal aluminum exhibited approximate isotropic evolution and obeyed the Tresca yield criterion, noting that the performance was different from the Von Mises criterion predicted by the common software, in the $\sigma - \sqrt{3}\tau$ stress plane, and the plastic flow directions approximatively obeyed the normal flow rule.
- The evolution and plastic flow directions of the subsequent yield surfaces strongly depended on the offset strain levels. The subsequent yield surfaces defined with the smaller offset strain exhibited more severe distortion and larger deviation angles from the normal direction in comparison with those results using the larger offset strain. However, with the larger offset strain, the subsequent yield surfaces were similar to the convex ellipse, and the plastic flow directions approximatively obeyed the normal flow rule.
- The dependency of plastic flow on the anisotropic evolution of subsequent yield surfaces found that the normality rule of plasticity followed the smooth yield locus; conversely, the significant non-associated flow was attributed to the distorted yield locus. Furthermore, it is necessary to study the subsequent yield surfaces in depth and the plastic flow behaviors under multi-axial proportional and non-proportional cyclic loading conditions with different cycles in our future work.

Author Contributions: Conceptualization, D.L.; methodology, D.L. and G.H.; software, K.Z.; validation, D.L. and G.H.; formal analysis, D.L.; investigation, D.L.; resources, D.L.; data curation, K.Z. and G.H.; writing—original draft preparation, D.L.; writing—review and editing, D.L. and G.H.; visualization, D.L. and G.H.; supervision, K.Z.; project administration, D.L. and K.Z.; funding acquisition, K.Z. All authors have read and agreed to the published version of the manuscript.

Funding: This research was funded by The National Natural Scientific Foundation of China (fund no. 11632007). The financial support is gratefully acknowledged.

Data Availability Statement: The data presented in this research are available on request from the corresponding author.

Conflicts of Interest: The authors declare no conflict of interest.

References

1. Safaei, M.; Zang, S.-L.; Lee, M.-G.; Waele, W.D. Evaluation of anisotropic constitutive models: Mixed anisotropic hardening and non-associated flow rule approach. *Int. J. Mech. Sci.* **2013**, *73*, 53–68. [CrossRef]
2. Liao, K.C. Yield and Damage Criteria for Sheet Metal Forming Simulations. Ph.D. Thesis, The University of Michigan, Ann Arbor, MI, USA, 1997.
3. Barlat, F.; Richmond, O. Prediction of tricomponent plane stress yield surfaces and associated flow and failure behavior of strongly textured f.c.c. polycrystalline sheets. *Mater. Sci. Eng.* **1987**, *95*, 15–29. [CrossRef]
4. Barlat, F. Crystallographic texture, anisotropic yield surfaces and forming limits of sheet metals. *Mater. Sci. Eng.* **1987**, *91*, 55–72. [CrossRef]
5. Lademo, O.G.; Hopperstad, O.S.; Langseth, M. An evaluation of yield criteria and flow rules for aluminium alloys. *Int. J. Plast.* **1999**, *15*, 191–208. [CrossRef]

6. Estrada, N.; Taboada, A. Yield surfaces and plastic potentials of cemented granular materials from discrete element simulations. *Comput. Geotech.* **2013**, *49*, 62–69. [CrossRef]
7. Zhang, K.S.; Shi, Y.K.; Xu, L.B. Anisotropy of yielding/hardening and microinhomogeneity of deforming/rotating for a polycrystalline metal under cyclic tension-compression. *Acta Metall. Sin.* **2011**, *47*, 1292–1300.
8. Beradai, C.; Berveiller, M.; Lipinski, P. Plasticity of metallic polycrystals under complex loading paths. *Int. J. Plast.* **1987**, *3*, 143–162. [CrossRef]
9. Khan, A.S.; Kazmi, R.; Pandey, A.; Stoughton, T. Evolution of subsequent yield surfaces and elastic constants with finite plastic deformation. Part-I: A very low work hardening aluminum alloy (Al 6061-T6511). *Int. J. Plast.* **2009**, *25*, 1611–1625. [CrossRef]
10. Khan, A.S.; Pandey, A.; Stoughton, T. Evolution of subsequent yield surfaces and elastic constants with finite plastic deformation. Part III: Yield surface in tension–tension stress space (Al 6061–T 6511 and annealed 1100 Al). *Int. J. Plast.* **2010**, *26*, 1432–1441. [CrossRef]
11. Hu, G.J.; Zhang, K.S.; Huang, S.H.; Ju, J.-W. Yield surfaces and plastic flow of 45 steel under tension-torsion loading paths. *Acta Mech. Solida. Sin.* **2012**, *25*, 360. [CrossRef]
12. Chen, J.Y. Subsequent Yield Behavior of HRB400 Steel under Tension-Torsion Combined Loading and Anisotropic Yield Surface Model. Ph.D. Thesis, Guangxi University, Nanning, China, 2019.
13. Mollica, F.; Srinivasa, A.R. A general framework for generating convex yield surfaces for anisotropic metals. *Acta Mech.* **2002**, *154*, 61–84. [CrossRef]
14. Lee, E.H.; Stoughton, T.B.; Yoon, J.W. A yield criterion through coupling of quadratic and non-quadratic functions for anisotropic hardening with non-associated flow rule. *Int. J. Plast.* **2017**, *99*, 120–143. [CrossRef]
15. Hou, Y.; Min, J.Y.; Stoughton, T.B.; Lin, J.P.; Carsley, J.E.; Carlson, B.E. A non-quadratic pressure-sensitive constitutive model under non-associated flow rule with anisotropic hardening: Modeling and validation. *Int. J. Plast.* **2020**, *135*, 102808. [CrossRef]
16. Adzima, F.; Balan, T.; Manach, P.Y.; Bonnet, N.; Tabourot, L. Crystal plasticity and phenomenological approaches for the simulation of deformation behavior in thin copper alloy sheets. *Int. J. Plast.* **2017**, *94*, 171–191. [CrossRef]
17. Liu, W.; Chen, B.K.; Pang, Y.; Najafzadeh, A. A 3D phenomenological yield function with both in and out-of-plane mechanical anisotropy using full-field crystal plasticity spectral method for modelling sheet metal forming of strong textured aluminum alloy. *Int. J. Solids Struct.* **2020**, *193*, 117–133. [CrossRef]
18. Han, F.; Diehl, M.; Roters, F.; Raabe, D. Using spectral-based representative volume element crystal plasticity simulations to predict yield surface evolution during large scale forming simulations. *J. Mater. Process. Technol.* **2020**, *277*, 116449. [CrossRef]
19. Barlat, F.; Lian, K. Plastic behavior and stretchability of sheet metals. Part I: A yield function for orthotropic sheets under plane stress conditions. *Int. J. Plast.* **1989**, *5*, 51–66. [CrossRef]
20. Lee, E.-H.; Choi, H.; Stoughton, T.B.; Yoon, J.W. Combined anisotropic and distortion hardening to describe directional response with Bauschinger effect. *Int. J. Plast.* **2019**, *122*, 73–78. [CrossRef]
21. Hill, R. A theory of the yielding and plastic flow of anisotropic metals. *Proc. Roy. Soc. Lon.* **1948**, *193*, 281–297.
22. Hosford, W.F. A Generalized Isotropic Yield Criterion. *J. Appl. Mech.* **1972**, *39*, 607–609. [CrossRef]
23. Lin, S.B.; Ding, J.L. A modified form of Hill's orientation dash dependent yield criterion for orthotropic sheet metals. *J. Mech. Phys. Solids* **1996**, *44*, 1739–1764. [CrossRef]
24. Valanis, K.C.; Peters, J.F. An endochronic plasticity theory with shear-volumetric coupling. *Int. J. Numer. Anal. Met.* **1991**, *15*, 77–102. [CrossRef]
25. Valanis, K.C. Fundamental consequences of a new intrinsic time measure plasticity as a limit of the endochronic theory. *Arch. Mech.* **1980**, *32*, 171.
26. Valanis, K.C. A theory of viscoplasticity without yield surface, Part II. Application to mechanical behavior of metals. *Arch. Mech.* **1971**, *23*, 535–551.
27. Valanis, K.C.; Fan, J. A numerical algorithm for endochronic plasticity and comparison with experiment. *Comput. Struct.* **1984**, *19*, 717–724. [CrossRef]
28. Khoei, A.R.; Bakhshiani, A. A hypoelasto-plastic finite strain simulation of powder compaction processes with density-dependent endochronic model. *Int. J. Solids Struct.* **2004**, *41*, 6081–6110. [CrossRef]
29. Yeh, W.C.; Lin, H.Y. An endochronic model of yield surface accounting for deformation induced anisotropy. *Int. J. Plast.* **2006**, *22*, 16–38. [CrossRef]
30. Stoughton, T.B.; Yoon, J.W. Anisotropic hardening and non-associated flow in proportional loading of sheet metals. *Int. J. Plast.* **2009**, *25*, 1777–1817. [CrossRef]
31. Schurig, M.; Bertram, A.; Petryk, H. Micromechanical analysis of the development of a yield vertex in polycrystal plasticity. *Acta Mech.* **2007**, *194*, 141–158. [CrossRef]
32. Saai, A.; Dumoulin, S.; Hopperstad, O.S.; Lademo, O.-G. Simulation of yield surfaces for aluminium sheets with rolling and recrystallization textures. *Comput. Mater. Sci.* **2013**, *67*, 424–433. [CrossRef]
33. Canova, G.R.; Kocks, U.F.; Tomé, C.N.; Jonas, J.J. The yield surface of textured polycrystals. *J. Mech. Phys. Solids* **1985**, *33*, 371–397. [CrossRef]
34. Inal, K.; Mishra, R.K.; Cazacu, O. Forming simulation of aluminum sheets using an anisotropic yield function coupled with crystal plasticity theory. *Int. J. Solids Struct.* **2010**, *47*, 2223–2233. [CrossRef]

35. Watanabe, I.; Terada, K. A method of predicting macroscopic yield strength of polycrystalline metals subjected to plastic forming by micro-macro de-coupling scheme. *Int. J. Mech. Sci.* **2010**, *52*, 343–355. [CrossRef]
36. Roters, F.; Eisenlohr, P.; Hantcherli, L.; Tjahjanto, D.D.; Bieler, T.R.; Raabe, D. Overview of constitutive laws, kinematics, homogenization and multiscale methods in crystal plasticity finite-element modeling: Theory, experiments, applications. *Acta Mater.* **2010**, *58*, 1152–1211. [CrossRef]
37. Nakamachi, E.; Xie, C.L.; Harimoto, M. Drawability assessment of BCC steel sheet by using elastic/crystalline viscoplastic finite element analyses. *Int. J. Mech. Sci.* **2001**, *43*, 631–652. [CrossRef]
38. Zhao, Z.; Roters, F.; Mao, W.; Raabe, D. Introduction of a texture component crystal plasticity finite element method for anisotropy simulations. *Adv. Eng. Mater.* **2001**, *3*, 984–990. [CrossRef]
39. Xie, C.L.; Nakamachi, E. Investigations of the formability of BCC steel sheets by using crystalline plasticity finite element analysis. *Mater. Des.* **2002**, *23*, 59–68. [CrossRef]
40. Hill, R.; Rice, J.R. Constitutive analysis of elastic-plastic crystals at arbitrary strain. *J. Mech. Phys. Solids* **1972**, *20*, 401–413. [CrossRef]
41. Asaro, R.J.; Rice, J.R. Strain localization in ductile single crystals. *J. Mech. Phys. Solids* **1977**, *25*, 309–338. [CrossRef]
42. Needleman, A.; Asaro, R.J.; Lemonds, J.; Peirce, D. Finite element analysis of crystalline solids. *Comput. Methods Appl. Mech. Eng.* **1985**, *52*, 689–708. [CrossRef]
43. Rice, J.R. Inelastic constitutive relations for solids: An internal-variable theory and its application to metal plasticity. *J. Mech. Phys. Solids* **1971**, *19*, 433–455. [CrossRef]
44. Hutchinson, J.W. Bounds and self-consistent estimates for creep of polycrystalline materials. *Proc. R. Soc. Lond. A* **1976**, *348*, 101–127.
45. Hill, R. Generalized constitutive relations for incremental deformation of metal crystals by multislip. *J. Mech. Phys. Solids* **1966**, *14*, 95–102. [CrossRef]
46. Chang, Y.W.; Asaro, R.J. An experimental study of shear localization in aluminum-copper single crystals. *Acta Metall.* **1981**, *29*, 241–257. [CrossRef]
47. Walker, K.P. *Research and Development Program for Non-linear Structural Modeling with Advanced Time-temperature Dependent Constitutive Relationships*; Report PWA-5700-50; NASA CR-1655331981. Available online: https://ntrs.nasa.gov/archive/nasa/casi.ntrs.nasa.gov/19820008207.pdf (accessed on 1 July 2021).
48. Chaboche, J.L. On some modifications of kinematic hardening to improve the description of ratcheting effects. *Int. J. Plast.* **1991**, *7*, 661–678. [CrossRef]
49. Zhang, K.-S.; Shi, Y.-K.; Ju, J.W. Grain-level statistical plasticity analysis on strain cycle fatigue of a FCC metal. *Mech. Mater.* **2013**, *64*, 76–90. [CrossRef]
50. Lu, D.M.; Zhang, K.S.; Hu, G.J.; Lan, Y.T.; Chang, Y.J. Investigation of Yield Surfaces Evolution for Polycrystalline Aluminum After Pre-Cyclic Loading by Experiment and Crystal Plasticity Simulation. *Materials* **2020**, *13*, 3069. [CrossRef] [PubMed]

nanomaterials

MDPI

Article

Modeling of Self-Assembled Peptide Nanotubes and Determination of Their Chirality Sign Based on Dipole Moment Calculations

Vladimir Bystrov [1,*], Alla Sidorova [2], Aleksey Lutsenko [2], Denis Shpigun [2], Ekaterina Malyshko [2], Alla Nuraeva [3], Pavel Zelenovskiy [3,4], Svitlana Kopyl [4] and Andrei Kholkin [3,4,5]

1 Institute of Mathematical Problems of Biology, The Branch of Keldysh Institute of Applied Mathematics, RAS, 142290 Pushchino, Russia
2 Faculty of Physics, Lomonosov Moscow State University, 119991 Moscow, Russia; sky314bone@mail.ru (A.S.); aleksluchrus@yandex.ru (A.L.); denish.den@mail.ru (D.S.); katyamalyshko@mail.ru (E.M.)
3 School of Natural Sciences and Mathematics, Ural Federal University, 620000 Ekaterinburg, Russia; allanuraeva@gmail.com (A.N.); zelenovskiy@urfu.ru (P.Z.); kholkin@ua.pt (A.K.)
4 Department of Physics & CICECO-Aveiro Institute of Materials, University of Aveiro, 3810-193 Aveiro, Portugal; svitlanakopyl@ua.pt
5 Physical Materials Science and Composite Materials Centre, Research School of Chemistry & Applied Biomedical Sciences, National Research Tomsk Polytechnic University, 634050 Tomsk, Russia
* Correspondence: vsbys@mail.ru or bystrov@impb.ru

Citation: Bystrov, V.; Sidorova, A.; Lutsenko, A.; Shpigun, D.; Malyshko, E.; Nuraeva, A.; Zelenovskiy, P.; Kopyl, S.; Kholkin, A. Modeling of Self-Assembled Peptide Nanotubes and Determination of Their Chirality Sign Based on Dipole Moment Calculations. *Nanomaterials* **2021**, *11*, 2415. https://doi.org/10.3390/nano11092415

Academic Editor: Angelina Angelova

Received: 31 July 2021
Accepted: 9 September 2021
Published: 16 September 2021

Abstract: The chirality quantification is of great importance in structural biology, where the differences in proteins twisting can provide essentially different physiological effects. However, this aspect of the chirality is still poorly studied for helix-like supramolecular structures. In this work, a method for chirality quantification based on the calculation of scalar triple products of dipole moments is suggested. As a model structure, self-assembled nanotubes of diphenylalanine (FF) made of L- and D-enantiomers were considered. The dipole moments of FF molecules were calculated using semi-empirical quantum-chemical method PM3 and the Amber force field method. The obtained results do not depend on the used simulation and calculation method, and show that the D-FF nanotubes are twisted tighter than L-FF. Moreover, the type of chirality of the helix-like nanotube is opposite to that of the initial individual molecule that is in line with the chirality alternation rule general for different levels of hierarchical organization of molecular systems. The proposed method can be applied to study other helix-like supramolecular structures.

Keywords: dipeptides; diphenylalanine; helical structures; peptide nanotubes; self-assembly; molecular modeling; dipole moments; polarization; chirality

1. Introduction

Self-assembly of biomolecules such as amino acids, nucleotide bases, phospholipids, and oligo- and polypeptides is the basis for the formation of DNA, molecular motors, viruses, and many other biological systems [1–4]. Biomimetic self-assembly is also a promising bottom-up approach for nanomaterials design in nanobiotechnology [5,6]. Helical self-organizing structures of different levels of hierarchical organization is an often result of such self-assembly [1,4,5,7,8]. Among them, α-helices, a common type of regular secondary structure of many proteins, are the simplest and most energetically favorable structures [1,6]. For natural proteins consisting of L-amino acids, a chirality sign alternation from the "left-handed" type to the "right-handed" is observed at different levels of hierarchical organization [9–12]. In the case of DNA, the sequence of the chirality sign is "right-handed deoxyribose carbohydrate"–"left-handed nucleotides"–"right-handed DNA double helix"–"left-handed superhelix" with the increasing complexity of their level of organization. This feature of chirality is one of the key points in the hierarchy and self-organization of any biological system [9–12].

Despite the concept of chirality, in its qualitative sense, being widespread in natural sciences, its quantitative aspects (the magnitude and sign of chirality) are still poorly studied [13–24]. In structural biology, it is of great importance to obtain the quantitative estimates of the magnitude of chirality and the chirality sign to compare both molecular constructs with the same symmetry type and those with different types of the symmetry. The problem still lacks a solution, although many studies have been performed in the field. The detailed analysis of these studies can be found in [23].

Recently, a new method for the chirality quantification based on the scalar triple product of three consecutive vectors connecting Cα carbon atoms of neighboring amino acid residues in the polypeptide helical or superhelical structures was proposed [21–24]. Despite this approach being successfully applied to a variety of proteins [24] taken from the Protein Data Bank [25], until now it has not been used for helix-like supramolecular structures such as peptide nanotubes.

Peptide nanotubes (PNTs) is an important example of helix-like self-organizing supramolecular systems [5,26,27]. Peptides and their derivatives were recognized recently as biological and bio-inspired building blocks for the construction of various advanced functional materials for nanotechnology and biomedicine [28,29]. Short, linear peptides containing aromatic amino acid residues such as phenylalanine (F, H-Phe-OH) attract special attention due to their ability to mimic the self-organizing behavior of more complex proteins [26].

The simplest aromatic dipeptide capable to form helix-like nanotubes is diphenylalanine (FF, H-Phe-Phe-OH) [30–40]. Each turn of such helix PNT consists of six FF molecules (Figure 1). Fast self-assembly of such PNTs occurs in aqueous media, and the variation of external conditions allows tuning the PNT's growth rate, length, and their physical properties [33,38,39,41–45]. FF PNTs possess a wide range of useful functional properties [41,46–48] that make them promising material for various applications in nanotechnology [33,41,46–50], nanoelectronics [28,35,48,51,52], and biomedicine [34–36,47,50,53–55].

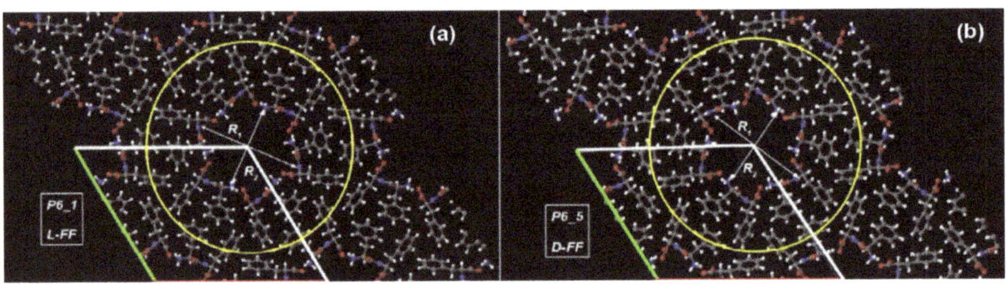

Figure 1. Images of molecular crystals composed of: (**a**) L-FF PNTs (space group $P6_1$), and (**b**) D-FF PNTs (space group $P6_5$). Hexagonal unit cells are marked with green, red, and white lines. The individual PNTs in crystal are highlighted by yellow circles. Atom colors: oxygen—red, nitrogen—blue, carbon—grey, and hydrogen—white.

The effect of chirality of FF molecules on the self-assembly and properties of PNTs has recently been studied in detail both experimentally and by computer simulation [37,38,42,56–59]. Lattice cell parameters for PNT made of "right-handed" FF molecules (H-D-Phe-D-Phe-OH, abbreviated hereafter D-FF) are close to those for PNTs made of "left-handed" FF molecules (H-L-Phe-L-Phe-OH, abbreviated hereafter L-FF), but their space groups are different [38]. Due to the difference of FF monomers chirality, the L-FF PNTs belong to the $P6_1$ space group, whereas D-FF PNTs belong to the $P6_5$. This $P6_1$–$P6_5$ pair is one of 11 pairs of enantiomorphic space groups [60] that are distinguished by the twisting direction of their 6-fold screw axis [61]. It was also shown that "left-handed" L-FF molecules form "right-handed" helix-like PNTs, whereas "right-handed" D-FF molecules form "left-handed" PNTs distinguished with their intermolecular interaction energies, self-assembly kinetics, and characteristic lengths [30,38].

In this work, we demonstrated that the method for chirality quantification proposed recently for protein helical structures [23,24] can be adopted for analysis of helix-like self-assembled PNTs. For the calculation of magnitude and sign of the chirality of L-FF and D-FF PNTs a set of sequential vectors of individual dipole moments of FF molecules comprising the turn of each helix of PNTs was used. The dipole moments were calculated using the HyperChem software [62].

2. Models Details and Computational Methods

2.1. Main Models and Used Software

Recently, we studied the structure and properties of empty L-FF and D-FF PNTs, as well as those with inner cavity filled with water molecules [56,57]. As water molecules do not affect the chirality of the PNTs, in this work, we considered empty (anhydrous) PNTs to simplify the calculations. The initial models of the PNTs were constructed using the same approach as in [56,57] based on X-ray crystallographic data for L-FF PNT (CCDC 16337, work [31]) and for D-FF PNT (CCDC 1853771, work [38]) taken from Cambridge Crystallographic Data Center (CCDC) [63]. The structural optimization and calculations were carried out using the density functional theory (DFT) methods (in Vienna Ab initio Simulation Package (VASP) program [64]), taking into account the Van der Waals interactions by "PBE + D3" method. The resulted molecular structures visualized by CCDC Mercury [65] are presented in Figure 1, whereas their main crystallographic parameters are summarized in Table 1.

Table 1. Experimental lattice cell parameters for L-FF [31] and D-FF [38] PNTs and inner cavity sizes R_1, R_2 [56].

	L-FF	D-FF
Space group	$P6_1$	$P6_5$
a, Å	24.0709	23.9468
b, Å	24.0709	23.9468
c, Å	5.4560	5.4411
V, Å^3	2737.7	2702.2
R_1, Å	15.3	15.2
R_2, Å	12.2	12.1

2.2. Models of FF Nanotubes

The molecular structures of both L-FF and D-FF PNTs were converted using OpenBabel software [66] from *.cif to *.hin format for their further analysis and calculations of their polar properties with various molecular mechanical and quantum-mechanical semi-empirical methods in HyperChem package [62] (Figure 2). These structures contain two coils of the helix arranged along with the c axis. Each coil consists of 6 FF molecules (258 atoms) and coils are separated with a lattice constant c around 5.45 Å [56]. The repetition of the coils along the c-axis leads to the formation of PNT with the corresponding chirality: right-handed helix for L-FF and left-handed helix for D-FF (Figure 3).

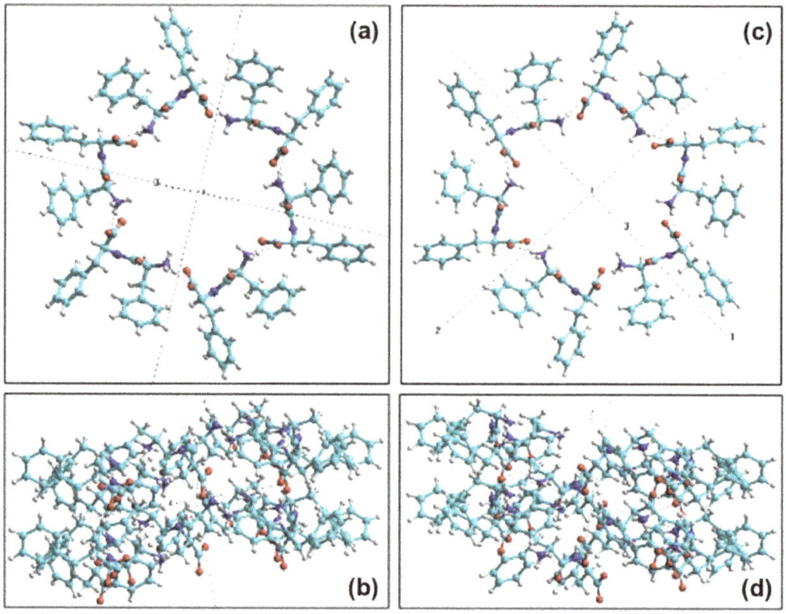

Figure 2. Two coils of FF PNT in HyperChem workspace based on experimental crystallographic data [31,38]: (**a**) L-FF in the Z-plane, (**b**) L-FF in the Y-plane, (**c**) D-FF in the Z-plane, and (**d**) D-FF in the Y-plane.

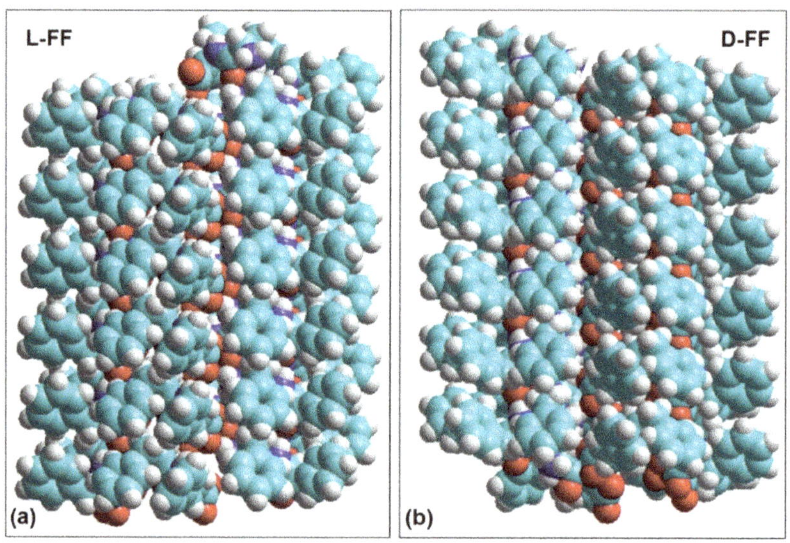

Figure 3. Images of helix-like FF PNTs based on experimental crystallographic data [31,38]: (**a**) L-FF and (**b**) D-FF. L-FF PNT shows right-handed twist, whereas D-FF shows left-handed.

It is known that water molecules in PNT nanochannels can affect both the structure and properties of the PNTs [41,44,56,57]. Therefore, in this work we considered in more detail the empty (anhydrous) nanotubes (Figure 1) to better understand how the dipole moments of FF molecules form the helix-like PNT with different chiralities and quantify the chirality.

The dipole moments, $\mathbf{D}_i$, of L-FF and D-FF molecules and corresponding helix-like structures were calculated using the semi-empirical quantum-mechanical method PM3 in the restricted Hartree-Fock approximation (RHF) and molecular mechanical force field method Amber from the HyperChem package [62]. Previous studies [37,38,42] have shown that for the dipole moments and energy calculated with other methods AM1 and BIOCHARM are similar to those obtained by PM3 and Amber. Therefore, in this work we used only PM3 and Amber methods. The calculated values of the dipole moments for individual L-FF and D-FF molecules are presented in Table 2. They are similar to the results obtained and analyzed earlier [30,37,38,42], and correspond well to the molecules' orientation in experimentally observed helix-like structures.

Table 2. The dipole moments and some other structural parameters of individual L-FF and D-FF molecules calculated by PM3 method.

Molecule	Dx, Debye	Dy, Debye	Dz, Debye	Dtot, Debye	Van der Waals Volume, Å^3	Polarization, C/m^2	Total Energy, a.u.	RMS Gradient, a.u./Å
L-FF	11.645	1.115	0.899	11.733	291.919	0.134	−133.959	~0.06
D-FF	−11.630	1.052	1.113	11.730	291.977	0.134	−133.959	~0.07

3. Results and Discussions

In contrast to α-helix proteins, supramolecular PNTs are comprised of individual FF molecules held by relatively weak hydrogen bonds [31–33]. Therefore, the chirality quantification method developed earlier for protein structures [21–24] cannot be directly applied for PNTs and requires some adaptations.

Briefly, the original method considered a helical polypeptide chain consisting of n amino acid residues, and a set of $(n-1)$ vectors $\mathbf{v}_i$ was built between each two adjacent Cα atoms in amino acid residues (Figure 4). For each three consecutive vectors, their scalar triple product was calculated:

$$([\mathbf{v}_1, \mathbf{v}_2], \mathbf{v}_3) = (y_1 z_2 - y_2 z_1)x_3 + (z_1 x_2 - z_2 x_1)y_3 + (x_1 y_2 - x_2 y_1)z_3, \quad (1)$$

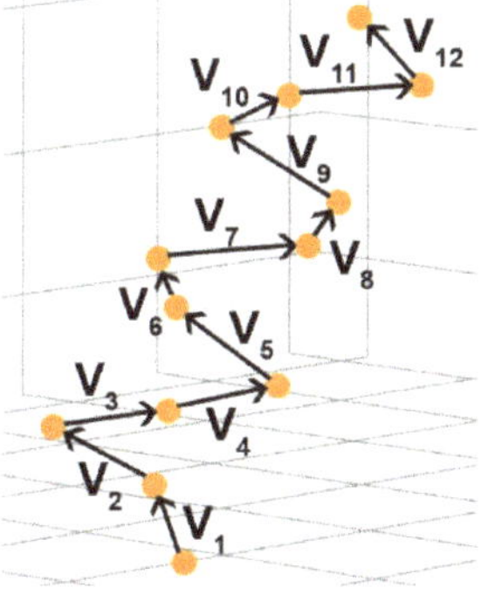

Figure 4. Graphical representation of vectors between neighboring Cα atoms in a helical protein used for calculating a scalar triple product.

The sum of all these scalar triple products (see Equation (2)) allowed us to estimate the chirality sign. If χ_{total} is positive, the structure is right-handed; for left-handed structures,

χ_{total} is negative [23,24]. This method has been validated for almost 1000 proteins [23,24] taken from the Protein Data Bank [25].

$$\chi_{total} = \sum_{i=1}^{n-3} ([\mathbf{v}_i, \mathbf{v}_{i+1}], \mathbf{v}_{i+2}) \quad (2)$$

In this work, the abovementioned approach was adopted for analysis of supramolecular FF PNTs. Instead of vectors between adjacent Cα atoms, a scalar triple product of dipole moments $\mathbf{D}_i$ of the successive individual FF molecules constituting a turn of the PNT helix-like nanotube was used. The origin of $\mathbf{D}_i$ vectors is taken relative to the center of mass of the corresponding molecules. The absolute value of each dipole moment $\mathbf{D}_i$ is

$$D_i = |\mathbf{D}_i| = \sqrt{D_{x,i}^2 + D_{y,i}^2 + D_{z,i}^2}, \quad (3)$$

where $D_{x,i}$, $D_{y,i}$, and $D_{z,i}$ are the components of the i-th vector $\mathbf{D}_i$ in the Cartesian coordinates. Similar to Equation (2), the sum of the scalar triple products of the dipole moments related to the PNT's chirality can be written as:

$$c_{total} = \sum_{i=1}^{n-2} ([\mathbf{D}_i, \mathbf{D}_{i+1}], \mathbf{D}_{i+2}). \quad (4)$$

It is necessary to note that the summation here is taken over i in the range from 1 to $(n-2)$, whereas in Equation (2), the i range is from 1 to $(n-3)$. This is because in supramolecular helixes i numerates the individual molecules instead of the Cα atoms in proteins. The c_{total} can be normalized over the average value of the total dipole momentum of the PNT's coil, $D_{av} = \frac{1}{6}\sum_{i=1}^{6} D_i$, to find a universal measure of the chirality:

$$c_{norm} = \frac{c_{total}}{D_{av}^3}. \quad (5)$$

Individual dipole moments of FF molecules in one coil of helix-like PNTs were obtained using semiempirical PM3 method in restricted Hartree–Fock (RHF) approximation and molecular mechanic Amber method (after PM3) implemented in HyperChem software [62]. The results are presented below in Tables 3 and 4 for L-FF and for D-FF, respectively. Schematic representation of the spatial arrangement of FF individual dipole moments $\mathbf{D}_i$ in two coils of PNTs are shown in Figure 5a,b for L-FF and in Figure 5c,d for D-FF PNTs.

Table 3. Values of dipole moments for a coil of the helix-like L-FF PNT computed using PM3 (RHF) and Amber (after PM3) methods. All values of dipole moments are given in Debye units.

i	PM3 RHF				Amber			
	Di	*Dx*	*Dy*	*Dz*	*Di*	*Dx*	*Dy*	*Dz*
1	24.022	14.576	−15.421	−11.261	23.458	14.901	−15.250	−9.781
2	22.549	−6.313	−18.923	−10.513	21.734	−6.280	−18.879	−8.748
3	22.389	−18.646	−3.636	−11.849	21.545	−18.698	−3.629	−10.070
4	22.381	−11.564	14.461	−12.573	21.530	−11.695	14.495	−10.801
5	22.441	7.555	17.308	−12.123	21.578	7.397	17.408	−10.384
6	22.587	18.767	2.568	−12.303	21.638	18.581	2.745	−10.742
Dcoil	70.851	4.376	−3.643	−70.622	60.752	4.206	−3.109	−60.526
Dav	22.728	0.729	−0.607	−11.770	21.914	0.701	−0.518	−10.088

Table 4. Values of dipole moments for a coil of the helix-like D-FF PNT computed using PM3 (RHF) and Amber (after PM3) methods. All values of dipole moments are given in Debye units.

i	PM3 RHF				Amber			
	Di	*Dx*	*Dy*	*Dz*	*Di*	*Dx*	*Dy*	*Dz*
1	22.523	−12.228	−15.267	−11.167	21.707	−12.299	−15.170	−9.475
2	22.340	7.302	−18.014	−11.072	21.527	7.210	−17.995	−9.360
3	22.372	19.234	−2.597	−11.125	21.520	19.168	−2.656	−9.416
4	22.475	11.905	15.905	−11.290	21.625	11.914	15.274	−9.612
5	22.629	−6.613	17.478	−12.761	21.703	−6.487	17.386	−11.256
6	23.855	−19.820	4.382	−12.531	23.271	−19.893	4.727	−11.112
Dcoil	69.971	−0.218	1.888	−69.945	60.253	−0.387	1.565	−60.231
Dav	22.704	−0.036	0.315	−11.658	21.892	−0.064	0.261	−10.038

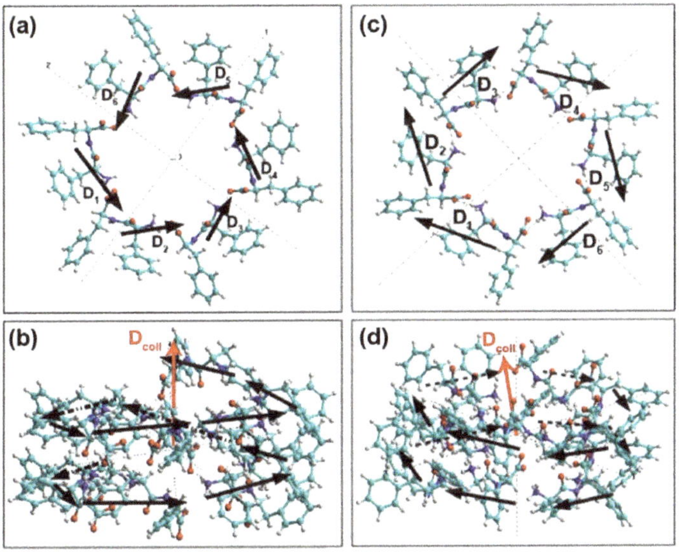

Figure 5. Schematic presentation of dipole moments D_i in two coils of (**a**,**b**) L-FF and (**c**,**d**) D-FF PNTs: (**a**,**c**) Z-plane projection, (**b**,**d**) Y-plane projection. For L-FF PNT dipole moments form a right-hand helix, whereas for D-FF PNT form a left-hand helix. Red arrows show the directions of the total dipole moments of the coil $\mathbf{D}_{coil}$.

It is important to note that, due to the helix-like structure of PNT, the dipole moment $\mathbf{D}_i$ of each next FF molecule in the coil is rotated by ~60° in the XOY plane. Therefore, at a full vector rotation at 360 degrees in a coil, the components $D_{i,x}$ and $D_{i,y}$ almost compensate for one another. Thus, the x and y components of the total dipole moment of the coil, D_{coil}, are much smaller than $D_{coil,z}$, which is always oriented along the OZ axis and increases D_{coil} (Tables 3 and 4). As a result, the total dipole moment of a coil D_{coil} is directed mainly along OZ axis with slight deviations (Figure 5), which corresponds to the previously obtained data [30,37,38,56,57].

It is worth noting that, in contrast to the original chirality quantification method developed for proteins [21–24], where the vectors were built between the carbon atoms Cα of each subsequent amino acid, in the current modification of the method, the vector of the dipole moment of each FF molecule in the PNT is taken relative to the center of mass of the corresponding molecule. The origin of $\mathbf{D}_i$ is a point in space defined by the vector $\mathbf{r}_{Di}$ with components $\{x_{Di}; y_{Di}; z_{Di}\}$ calculated as follows:

$$x_{Di} = \left\{\sum_{j=1}^{N} m_j \cdot x_j\right\} / \left\{\sum_{j=1}^{N} m_j\right\},$$

$$y_{Di} = \left\{\sum_{j=1}^{N} m_j \cdot y_j\right\} / \left\{\sum_{j=1}^{N} m_j\right\},$$

$$z_{Di} = \left\{\sum_{j=1}^{N} m_j \cdot z_j\right\} / \left\{\sum_{j=1}^{N} m_j\right\}.$$

Here, m_j, x_j, y_j, and z_j are the mass and coordinates, respectively, of the j-th atom in the i-th FF molecule in the PNT, and $N = 43$ is the number of atoms in one FF molecule. For example, the coordinates of the center of mass for the first FF molecule ($i = 1$) in a coil of L-FF PNT (Figure 6) are:

$$x_{D1} = 2.35 \text{ Å}; y_{D1} = -7.80 \text{ Å}; z_{D1} = 1.02 \text{ Å}.$$

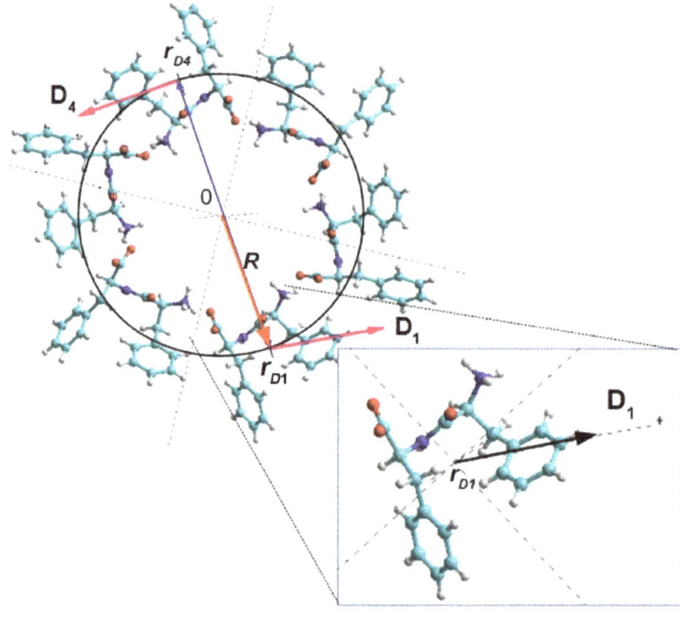

Figure 6. A coil of L-FF PNT with vectors $\mathbf{r}_{D1}$ and $\mathbf{r}_{D4}$ pointing the centers of mass of individual FF molecules with dipole moments $\mathbf{D}_1$ and $\mathbf{D}_4$, correspondingly.

This point is the origin for the $\mathbf{D}_1$ dipole moment vector. Similarly, the origins for other vectors $\mathbf{D}_i$ can be calculated. As a result, dipole moments form a helix with a pitch equal to $c = 5.456$ Å (for L-FF, $c = 5.441$ Å for D-FF PNTs), and the helix radius R is about 8.15 Å for L-FF PNT.

The obtained values of the dipole moments for L-FF and D-FF PNTs allow us to quantify their chiralities following the Equations (4) and (5). The calculated magnitudes of the PNTs chirality, c_{total}, and the normalized chirality, c_{norm}, are presented in Table 5. For each type of PNT, both PM3 and Amber methods c_{total} demonstrate close values with the divergence about 15%, whereas for c_{norm} the divergence is less than 5%. Therefore, c_{norm} can be considered as a universal value for chirality quantification that does not depend on the calculation method.

Table 5. Magnitudes and signs of the chirality obtained for L-FF and D-FF PNTs for various calculating methods.

Type of PNT	L-FF		D-FF	
Calculating Method	**PM3**	**Amber**	**PM3**	**Amber**
c_{total}, Debye^3	16,034.82	13,870.71	−14,497.03	−12,161.23
c_{norm}	1.37	1.32	−1.23	−1.16
Chirality sign	positive	positive	negative	negative
Chirality symbol	D	D	L	L

At the same time, the absolute value of c_{norm} for L-FF PNT is about 10% higher than that for D-FF PNT. This difference exceeds the calculation error and thus shows the difference in PNTs chiralities. Lower c_{norm} value observed for D-FF PNTs indicates that this PNT is twisted tighter than L-FF. This is also confirmed by the lower volume of the D-FF PNT unit cell (Table 1).

Following the original method [23,24], the sign of the c_{norm} corresponds to the PNT's chirality type. For L-FF PNT, c_{norm} is positive thus this PNT should be right-handed, whereas negative c_{norm} value for D-FF PNT indicates its left-handed twisting. This result is confirmed by the previous crystallographic studies [31,38] and the individual dipole moments completely follow this arrangement (Figure 5). It is worth noting that the chirality alternating observed earlier for natural proteins and DNA [9–12] preserves in the supramolecular PNTs as well. The type of chirality of the helix-like PNT is opposite to that of the individual dipeptide. This fact also can be a confirmation of the adequacy of the proposed method for supramolecular PNTs chirality quantification.

4. Conclusions

A method for quantification of the chirality of self-assembled helix-like FF nanotubes based on the scalar triple products of the individual FF molecules dipole moments is described. The dipole moments were calculated for nanotubes comprised of L-FF and D-FF molecules by quantum-chemical and molecular mechanics methods, and the independence of the magnitude and the sign of the chirality on the calculation method is demonstrated. The obtained magnitudes of the chirality for L-FF nanotubes are about 10% higher than those of L-FF, which indicates that D-FF nanotubes are twisted tighter than L-FF. The alternating of the chirality type observed earlier for natural proteins and DNA also preserves in the supramolecular PNTs. The type of chirality of the helix-like PNT is opposite to that of the individual dipeptide. This effect is in line with the chirality alternation rule, general for different levels of hierarchical organization of molecular systems, and additionally corroborates the validity of the proposed method.

The extension of the chirality quantification method to supramolecular helix-like nanostructures opens new facilities for comparing both molecular constructs of the same chirality type and those with different constructs. Moreover, it provides an opportunity to reveal the physical basis for the chirality sign formation, which is associated with the electrostatic dipole–dipole interaction of individual molecules. This approach can be applied to study other helical and helix-like supramolecular structures.

Author Contributions: V.B., A.S. and P.Z. wrote the manuscript. A.K. and E.M. supervised and supported this study. E.M. also developed the concept of chirality. A.L. and D.S. performed the calculations. S.K., A.N. and P.Z. participated in the experimental data processing and data analysis. All authors have read and agreed to the published version of the manuscript.

Funding: This research was funded by the Russian Foundation for Basic Research, grant n. 19-01-00519-a.

Institutional Review Board Statement: Not applicable.

Informed Consent Statement: Not applicable.

Data Availability Statement: The data presented in this study are available on request from the corresponding author.

Acknowledgments: The authors are grateful to the Russian Foundation for Basic Researches (RFBR): grants №№ 19–01–00519_A and 20-51-53014_GFEN_A. Part of this work (A.K.) was supported by the Ministry of Science and Higher Education of the Russian Federation (grant No. 075-15-2021-588). P.Z., S.K. and A.K. are grateful to the FCT project "BioPiezo"—PTDC/CTM−CTM/31679/2017 (CENTRO-01-0145-FEDER-031679) and to the project CICECO-Aveiro Institute of Materials, refs. UIDB/50011/2020 and UIDP/50011/2020, financed by national funds through the Portuguese Foundation for Science and Technology/MCTES. Part of this work was funded by national funds (OE), through FCT–Fundação para a Ciência e a Tecnologia, I.P., in the scope of the framework contract foreseen in the numbers 4, 5, and 6 of the article 23, of the Decree-Law 57/2016, of August 29, changed by Law 57/2017, of July 19. A.S. and E.M. are grateful to the the Interdisciplinary Scientific and Educational School of Moscow University "Fundamental and Applied Space Research".

Conflicts of Interest: The authors declare no conflict of interest.

References

1. Lehninger, A.L. *Biochemistry—The Molecular Basis of Cell Structure and Function*; Worth Publishers, Inc.: New York, NY, USA, 1972.
2. Stock, D.; Leslie, A.G.; Walker, J.E. Molecular architecture of the rotary motor in ATP synthase. *Science* **1999**, *286*, 1700–1705. [CrossRef]
3. Ban, N.; McPherson, A. The structure of satellite panicum mosaic virus at 1.9 Å resolution. *Nat. Struct. Biol.* **1995**, *2*, 882–890. [CrossRef]
4. Zhang, S. Emerging biological materials through molecular self-assembly. *Biotechn. Adv.* **2002**, *20*, 321–339. [CrossRef]
5. Levin, A.; Hakala, T.A.; Schnaider, L.; Bernardes, G.J.L.; Gazit, E.; Knowles, T.P.J. Biomimetic peptide self-assembly for functional materials. *Nat. Rev. Chem.* **2020**, *4*, 615–634. [CrossRef]
6. Mendes, A.C.; Baran, E.T.; Reis, R.L.; Azevedo, H.S. Self-assembly in nature: Using the principles of nature to create complex nanobiomaterials. *Wiley Interdiscip. Rev. Nanomed. Nanobiotechnol.* **2013**, *5*, 582–612. [CrossRef] [PubMed]
7. Whitesides, G.M.; Mathias, J.P.; Seto, C.T. Molecular self-assembly and nanochemistry: A chemical strategy for the synthesis of nanostructures. *Science* **1991**, *254*, 1312–1319. [CrossRef] [PubMed]
8. Mandal, D.; Shirazi, A.N.; Parang, K. Self-assembly of peptides to nanostructures. *Org. Biomol. Chem.* **2014**, *12*, 3544–3561. [CrossRef] [PubMed]
9. Tverdislov, V.A. Chirality as a primary switch of hierarchical levels in molecular biological systems. *Biophysics* **2013**, *58*, 128–132. [CrossRef]
10. Malyshko, E.V.; Tverdislov, V.A. Chirality as a physical aspect of structure formation in biological macromolecular systems. *IOP J. Phys. Conf. Ser.* **2016**, *741*, 012065. [CrossRef]
11. Tverdislov, V.A.; Malyshko, E.V. Chiral Dualism as an Instrument of Hierarchical Structure Formation in Molecular Biology. *Symmetry* **2020**, *12*, 587. [CrossRef]
12. Tverdislov, V.A.; Malyshko, E.V. On regularities in the spontaneous formation of structural hierarchies in chiral systems of nonliving and living matter. *Phys.-Uspekhi* **2019**, *62*, 354–363. [CrossRef]
13. Buda, A.B.; der Heyde, T.A.; Mislow, K. On Quantifying Chirality. *Angew. Chem. Int. Ed. Engl.* **1992**, *31*, 989–1007. [CrossRef]
14. Weinberg, N.; Mislow, K. On chirality measures and chirality properties. *Can. J. Chem.* **2000**, *78*, 41–45. [CrossRef]
15. Randic, M. Shape recognition and chirality measure: Reestablishing the link between similarity and dissimilarity in discrete space. *Chemom. Intell. Lab. Syst.* **1991**, *10*, 213–225. [CrossRef]
16. Zabrodsky, H.; Peleg, S.; Avnir, D. Continuous symmetry measures. *J. Am. Chem. Soc.* **1992**, *114*, 7843–7851. [CrossRef]
17. Mezey, P.G. Chirality Measures and Graph Representations. *Comput. Math. Appl.* **1997**, *34*, 105–112. [CrossRef]
18. Pinsky, M.; Dryzun, C.; Casanova, D.; Alemany, P.; Avnir, D. Analytical methods for calculating Continuous Symmetry Measures and the Chirality Measure. *J. Comput. Chem.* **2008**, *29*, 2712–2721. [CrossRef] [PubMed]
19. Petitjean, M. Chirality and Symmetry Measures: A Transdisciplinary Review. *Entropy* **2003**, *5*, 271–312. [CrossRef]
20. Zhao, T.; Zhang, Q.; Long, H.; Xu, L. Graph Theoretical Representation of Atomic Asymmetry and Molecular Chirality of Benzenoids in Two-Dimensional Space. *PLoS ONE* **2014**, *9*, e102043. [CrossRef]
21. Sidorova, A.E.; Malyshko, E.V.; Kotov, A.R.; Tverdislov, V.A.; Ustinin, M.N. Quantitative Criteria of Chirality in Hierarchical Protein Structures. *Biophysics* **2019**, *64*, 155–166. [CrossRef]
22. Sidorova, A.E.; Levashova, N.T.; Malyshko, E.V.; Tverdislov, V.A. Autowave Self-Organization in the Folding of Proteins. *Mosc. Univ. Phys. Bull.* **2019**, *74*, 213–226. [CrossRef]
23. Sidorova, A.E.; Lutsenko, A.O.; Shpigun, D.K.; Malyshko, E.V.; Tverdislov, V.A. Methods for Determining the Chirality sign of the Helical and Superhelical Protein Structures. *Biophysics* **2021**, *66*, 421. [CrossRef]
24. Sidorova, A.E.; Malyshko, E.V.; Lutsenko, A.O.; Shpigun, D.K.; Bagrova, O.E. Protein Helical Structures: Defining Handedness and Localization Features. *Symmetry* **2021**, *13*, 879. [CrossRef]
25. RCSB Protein Data Bank. Available online: http://www.rcsb.org (accessed on 30 August 2021).

26. Yuan, C.; Ji, W.; Xing, R.; Li, J.; Gazit, E.; Yan, X. Hierarchically oriented organization in supramolecular peptide crystals. *Nat. Rev. Chem.* **2019**, *3*, 567. [CrossRef]
27. Hamley, I.W. Peptide Nanotubes. *Angew. Chem. Int. Ed.* **2014**, *53*, 6866–6881. [CrossRef]
28. Tao, K.; Makam, P.; Aizen, R.; Gazit, E. Self-assembling peptide semiconductors. *Science* **2017**, *358*, eaam9756. [CrossRef] [PubMed]
29. Ghadiri, M.R.; Granja, J.R.; Milligan, R.A.; McRee, D.E.; Khazanovich, N. Self-assembling organic nanotubes based on cyclic peptide architecture. *Nature* **1993**, *366*, 324–327. [CrossRef]
30. Bystrov, V.S.; Zelenovskiy, P.S.; Nuraeva, A.S.; Kopyl, S.; Zhulyabina, O.A.; Tverdislov, V.A. Chiral peculiar properties of self-organization of diphenylalanine peptide nanotubes: Modeling of structure and properties. *Math. Biol. Bioinform.* **2019**, *14*, 94–124. [CrossRef]
31. Görbitz, C.H. Nanotube formation by hydrophobic dipeptides. *Chem. Eur. J.* **2001**, *7*, 5153–5159. [CrossRef]
32. Görbitz, C.H. The structure of nanotubes formed by diphenylalanine, the core recognition motif of Alzheimer's b-amyloid polypeptide. *Chem. Commun.* **2006**, *1*, 2332–2334. [CrossRef]
33. Kim, J.; Han, T.E.; Kim, Y.; Park, J.S.; Choi, J.; Churchill, D.G.; Kim, S.O.; Ihee, H. Role of water in directing diphenylalanine assembly into nanotubes and nanowires. *Adv. Mater.* **2010**, *22*, 583–587. [CrossRef]
34. Bystrov, V.S.; Bdikin, I.; Heredia, A.; Pullar, R.C.; Mishina, E.; Sigov, A.; Kholkin, A.L. Piezoelectricity and Ferroelectricity in biomaterials: From proteins to self-assembled peptide nanotubes. In *Piezoelectric Nanomaterials for Biomedical Applications*; Ciofani, G., Menciassi, A., Eds.; Springer: Berlin/Heidelberg, Germany, 2012; pp. 187–211.
35. Bystrov, V.S. *Computer Simulation Nanostructures: Bioferroelectric Peptide Nanotubes. Bioferroelectricity: Peptide Nanotubes*; LAP Lambert Academic Publishing: Saarbruecken, Germay, 2016.
36. Bystrov, V.S.; Paramonova, E.V.; Bdikin, I.K.; Kopyl, S.; Heredia, A.; Pullar, R.C.; Kholkin, A.L. Bioferroelectricity: Diphenylalanine peptide nanotubes computational modeling and ferroelectric properties at the nanoscale. *Ferroelectrics* **2012**, *440*, 3–24. [CrossRef]
37. Bystrov, V.S.; Zelenovskiy, P.S.; Nuraeva, A.S.; Kopyl, S.; Zhulyabina, O.A.; Tverdislov, V.A. Molecular modeling and computational study of the chiral-dependent structures and properties of the self-assembling diphenylalanine peptide nanotubes. *J. Mol. Modeling* **2019**, *25*, 199. [CrossRef]
38. Zelenovskiy, P.S.; Nuraeva, A.S.; Kopyl, S.; Arkhipov, S.G.; Vasilev, S.G.; Bystrov, V.S.; Gruzdev, D.A.; Waliszek, M.; Svitlyk, V.; Shur, V.Y.; et al. Chirality-Dependent Growth of Self-Assembled Diphenylalanine Microtubes. *Cryst. Growth Des.* **2019**, *19*, 6414–6421. [CrossRef]
39. Nuraeva, A.; Vasilev, S.; Vasileva, D.; Zelenovskiy, P.; Chezganov, D.; Esin, A.; Kopyl, S.; Romanyuk, K.; Shur, V.Y.; Kholkin, A.L. Evaporation-Driven Crystallization of Diphenylalanine Microtubes for Microelectronic Applications. *Cryst. Growth Des.* **2016**, *16*, 1472–1479. [CrossRef]
40. Yuan, T.; Xu, Y.; Fei, J.; Xue, H.; Li, X.; Wang, C.; Fytas, G.; Li, J. The Ultrafast Assembly of a Dipeptide Supramolecular Organogel and its Phase Transition from Gel to Crystal. *Angew. Chem. Int. Ed.* **2019**, *58*, 11072–11077. [CrossRef] [PubMed]
41. Andrade-Filho, T.; Martins, T.C.; Ferreira, F.F.; Alves, W.A.; Rocha, A.R. Water-driven stabilization of diphenylalanine nanotube structures. *Theor. Chem. Acc.* **2016**, *135*, 185. [CrossRef]
42. Bystrov, V.S.; Kopyl, S.A.; Zelenovskiy, P.; Zhulyabina, O.A.; Tverdislov, V.A.; Salehli, F.; Ghermani, N.E.; Shur, V.Y.; Kholkin, A.L. Investigation of physical properties of diphenylalanine peptide nanotubes having different chiralities and embedded water molecules. *Ferroelectrics* **2018**, *525*, 168–177. [CrossRef]
43. Dayarian, S.; Kopyl, S.; Bystrov, V.; Correia, M.R.; Ivanov, M.S.; Pelegova, E.; Kholkin, A. Effect of the Chloride Anions on the Formation of Self-Assembled Diphenylalanine Peptide Nanotubes. *IEEE Trans. Ultrason. Ferroelectr. Freq. Control* **2018**, *65*, 1563–1570. [CrossRef]
44. Zelenovskiy, P.; Yuzhakov, V.; Nuraeva, A.; Kornev, M.; Shur, V.Y.; Kopyl, S.; Kholkin, A.; Vasilev, S.; Tofail, S.A.M. The Effect of Water Molecules on Elastic and Piezoelectric Properties of Diphenylalanine Microtubes. *IEEE Trans. Dielectr. Electr. Insul.* **2020**, *27*, 1474–1477. [CrossRef]
45. Guo, C.; Luo, Y.; Zhou, R.; Wei, G. Probing the Self-Assembly Mechanism of Diphenylalanine-Based Peptide Nanovesicles and Nanotubes. *ACS Nano* **2012**, *6*, 3907–3918. [CrossRef]
46. Kholkin, A.; Amdursky, N.; Bdikin, I.; Gazit, E.; Rosenman, G. Strong piezoelectricity in bioinspired peptide nanotubes. *ACS Nano* **2010**, *4*, 610–614. [CrossRef] [PubMed]
47. Hereida, A.; Bdikin, I.; Kopyl, S.; Mishina, E.; Semin, S.; Sigov, A.; German, K.; Bystrov, V.; Gracio, J.; Kholkin, A.L. Temperature-driven phase transformation in self-assembled diphenylalanine peptide nanotubes. *J. Phys. D Appl. Phys.* **2010**, *43*, 462001. [CrossRef]
48. Bdikin, I.; Bystrov, V.S.; Delgadillo, I.; Gracio, J.; Kopyl, S.; Wojtas, M.; Mishina, E.; Sigov, A.; Kholkin, A.L. Polarization switching and patterning in self-assembled peptide tubular structures. *J. Appl. Phys.* **2012**, *111*, 074104. [CrossRef]
49. Reches, M.; Gazit, E. Controlled patterning of aligned self-assembled peptide nanotubes. *Nat. Nanotechnol.* **2006**, *1*, 195–200. [CrossRef] [PubMed]
50. Silva, R.F.; Araújo, D.R.; Silva, E.R.; Ando, R.A.; Alves, W.A. L-diphenylalanine microtubes as a potencial drug-delivery system: Characterization, release kinetics, and cytotoxicity. *Langmuir* **2013**, *29*, 10205–10212. [CrossRef]
51. Reches, M.; Gazit, E. Casting metal nanowires within discrete self-assembled peptide nanotubes. *Science* **2003**, *300*, 625–627. [CrossRef]

52. Nguyen, V.; Zhu, R.; Jenkins, K.; Yang, R. Self-assembly of diphenylalanine peptide with controlled polarization for power generation. *Nat. Commun.* **2016**, *7*, 13566. [CrossRef]
53. Emtiazi, G.; Zohrabi, T.; Lee, L.Y.; Habibi, N.; Zarrabi, A. Covalent diphenylalanine peptide nanotube conjugated to folic acid/magnetic nanoparticles for anti-cancer drug delivery. *J. Drug Deliv. Sci. Technol.* **2017**, *41*, 90–98. [CrossRef]
54. Gupta, S.; Singh, I.; Sharma, A.K.; Kumar, P. Ultrashort Peptide Self-Assembly: Front-Runners to Transport Drug and Gene Cargos. *Front. Bioeng. Biotechnol.* **2020**, *8*, 504. [CrossRef]
55. Marchesan, S.; Vargiu, A.V.; Styan, K.E. The Phe-Phe Motif for Peptide Self-Assembly in Nanomedicine. *Molecules* **2015**, *20*, 19775–19788. [CrossRef]
56. Bystrov, V.S.; Coutinho, J.; Zelenovskiy, P.; Nuraeva, A.; Kopyl, S.; Zhulyabina, O.; Tverdislov, V. Structures and properties of the self-assembling diphenylalanine peptide nanotubes containing water molecules: Modeling and data analysis. *Nanomaterials* **2020**, *10*, 1999. [CrossRef]
57. Bystrov, V.S.; Coutinho, J.; Zhulyabina, J.A.; Kopyl, S.A.; Zelenovskiy, P.S.; Nuraeva, A.S.; Tverdislov, V.A.; Filippov, S.V.; Kholkin, A.L.; Shur, V.Y. Modeling and physical properties of diphenylalanine peptide nanotubes containing water molecules. *Ferroelectrics* **2021**, *574*, 78–91. [CrossRef]
58. Tamamis, P.; Adler-Abramovich, L.; Reches, M.; Marshall, K.; Sikorski, P.; Serpell, L.; Gazit, E.; Archontis, G. Self-assembly of phenylalanine oligopeptides: Insights from experiments and simulations. *Biophys. J.* **2009**, *96*, 5020–5029. [CrossRef] [PubMed]
59. Zheng, Y.; Mao, K.; Chen, S.; Zhu, H. Chirality Effects in Peptide Assembly Structures. *Front. Bioeng. Biotechnol.* **2021**, *9*, 703004. [CrossRef] [PubMed]
60. Müller, U. *Symmetry Relationships between Crystal Structures. Applications of Crystallographic Group Theory in Crystal Chemistry*; University Press: Oxford, UK, 2013.
61. Hahn, T. (Ed.) *International Tables for Crystallography. Volume A: Space-Group Symmetry*; Springer: Berlin/Heidelberg, Germany, 2005.
62. HyperChem 8. *Tools for Molecular Modeling. Professional Edition for Windows AC Release 8.0 USB (on CD)*; Hypercube. Inc.: Gainesville, FL, USA, 2011.
63. The Cambridge Crystallographic Data Centre (CCDC). Available online: https://www.ccdc.cam.ac.uk/ (accessed on 30 August 2021).
64. VASP (Vienna Ab initio Simulation Package). Available online: https://www.vasp.at/ (accessed on 30 August 2021).
65. CCDC Mercury. Available online: https://www.ccdc.cam.ac.uk/solutions/csd-system/components/mercury/ (accessed on 30 August 2021).
66. Open Babel. Available online: https://openbabel.org/docs/dev/Installation/install.html (accessed on 30 August 2021).

nanomaterials

MDPI

Article

Low Resistance Asymmetric III-Nitride Tunnel Junctions Designed by Machine Learning

Rongyu Lin [1], Peng Han [2], Yue Wang [1], Ronghui Lin [1], Yi Lu [1], Zhiyuan Liu [1], Xiangliang Zhang [2,3,*] and Xiaohang Li [1,*]

[1] Advanced Semiconductor Laboratory, King Abdullah University of Science and Technology, Thuwal 23955, Saudi Arabia; rongyu.lin@kaust.edu.sa (R.L.); yue.wang@kaust.edu.sa (Y.W.); ronghui.lin@kaust.edu.sa (R.L.); yi.lu@kaust.edu.sa (Y.L.); zhiyuan.liu@kaust.edu.sa (Z.L.)

[2] Laboratory Machine, Intelligence and kNowledge Engineering (MINE), King Abdullah University of Science and Technology, Thuwal 23955, Saudi Arabia; peng.han@kaust.edu.sa

[3] Department of Computer Science and Engineering, University of Notre Dame, Notre Dame, IN 46556, USA

* Correspondence: xzhang33@nd.edu (X.Z.); xiaohang.li@kaust.edu.sa (X.L.)

Abstract: The tunnel junction (TJ) is a crucial structure for numerous III-nitride devices. A fundamental challenge for TJ design is to minimize the TJ resistance at high current densities. In this work, we propose the asymmetric p-AlGaN/i-InGaN/n-AlGaN TJ structure for the first time. P-AlGaN/i-InGaN/n-AlGaN TJs were simulated with different Al or In compositions and different InGaN layer thicknesses using TCAD (Technology Computer-Aided Design) software. Trained by these data, we constructed a highly efficient model for TJ resistance prediction using machine learning. The model constructs a tool for real-time prediction of the TJ resistance, and the resistances for 22,254 different TJ structures were predicted. Based on our TJ predictions, the asymmetric TJ structure (p-$Al_{0.7}Ga_{0.3}N$/i-$In_{0.2}Ga_{0.8}N$/n-$Al_{0.3}Ga_{0.7}N$) with higher Al composition in p-layer has seven times lower TJ resistance compared to the prevailing symmetric p-$Al_{0.3}Ga_{0.7}N$/i-$In_{0.2}Ga_{0.8}N$/n-$Al_{0.3}Ga_{0.7}N$ TJ. This study paves a new way in III-nitride TJ design for optical and electronic devices.

Keywords: tunnel junction; machine learning; III-nitride

Citation: Lin, R.; Han, P.; Wang, Y.; Lin, R.; Lu, Y.; Liu, Z.; Zhang, X.; Li, X. Low Resistance Asymmetric III-Nitride Tunnel Junctions Designed by Machine Learning. *Nanomaterials* **2021**, *11*, 2466. https://doi.org/10.3390/nano11102466

Academic Editors: Vladimir G. Dubrovskii and Vladimir S. Bystrov

Received: 7 August 2021
Accepted: 13 September 2021
Published: 22 September 2021

1. Introduction

The use of tunnel junction (TJ) is crucial for many advanced III-nitride electronic and optical devices, including tunnel field-effect transistors (TFETs), light-emitting diodes (LEDs), and solar cells [1–3]. For instance, TJs could replace the use of resistant and absorptive p-type layers and contacts in UV-LEDs. In addition, TJs enable cascading optoelectronic devices, which provide greater flexibility for functional designs [4]. A significant polarization effect provides more space to manipulate the performance of TJs [5]. Recently, GaN/AlGaN/GaN [6,7], metal/InGaN/GaN [8], and GaN/InGaN/GaN [9–11] TJs were investigated and proven to lead to significant improvements in devices.

The interplay of material compositions, polarization effects, and the thicknesses of each layer in the TJ structures provide enormous design parameter space for TJ designs, which increases the difficulties associated with TJ optimization. To this end, the Technology Computer-Aided Design (TCAD) software has been widely employed to validate TJ effectiveness [11–14]. However, massive calculations are required to investigate the enormous possibilities of TJ designs. In addition, the convergence issue caused by the self-consistent solution of classical and quantum equations has made a systematic investigation of TJ design even more difficult. Currently, there are no effective methods to circumvent these issues except the cumbersome trial-and-error method. Recently, the machine-learning (ML) technique has demonstrated its significant effectiveness to these TCAD convergence issues including for III-nitride LED and nanophotonics designs [15–18]. It is a global optimization tool that can search the whole parameter space and yield a viable design efficiently. We demonstrate, for the first time, machine learning can be applied to alleviate the above-mentioned issues in the designing of III-nitride TJs significantly. Although the computational resources are required at the development stage of the algorithm, it is a one-time cost and once the algorithm is developed, the algorithm runs on very few resources in real time without convergence issues.

The symmetric p-AlGaN/i-InGaN/n-AlGaN tunnel junctions have been widely applied in LEDs and HEMTs (high-electron-mobility transistors) for enhancing device performances [19–22]. The TJs are usually assumed with the same Al composition in both p-AlGaN and n-AlGaN layers without a detailed discussion of this symmetryA. Although the AlGaN grading designs with the change of the Al composition in p-layer or n-layer were introduced to enhance the device performance [2], it did not explain how the asymmetric Al composition in the p-AlGaN and n-AlGaN influence the p-AlGaN/i-InGaN/n-AlGaN TJ resistance.

In this work, we employed ML to explore the asymmetric p-AlGaN/i-InGaN/n-AlGaN TJ design by predicting TJ resistance. The XG-Boost strategy [23] was leveraged to directly predict the TJ resistance. We trained the ML model using cross-validated results by TCAD software with different TJ configurations. To exclude the outliers caused by the convergence issues, all the configurations were calculated twice with two different iterations and cross-validated by the results. The accuracy of the model is 90.5% in the test datasets, which suggested its great capacity to predict TJ resistance. We also used this model to generate 22,254 TJ configurations. Afterward, we investigated the configurations with relatively low resistance; and we discovered that the asymmetric TJ design such as p-$Al_{0.7}Ga_{0.3}N$/i-$In_{0.2}Ga_{0.8}N$/n-$Al_{0.3}Ga_{0.7}N$ with different Al compositions in the p-type and n-type layers, could lead to considerably lower (about seven times) TJ resistance compared with conventional symmetric p-$Al_{0.3}Ga_{0.7}N$/i-$In_{0.2}Ga_{0.8}N$/n-$Al_{0.3}Ga_{0.7}N$ design.

2. TCAD Calculations of p-AlGaN/i-InGaN/n-AlGaN TJ

We investigated the p-$Al_xGa_{1-x}N$/i-$In_yGa_{1-y}N$/n-$Al_zGa_{1-z}N$ TJs as shown in Figure 1a. Four features, including the Al compositions x and z of the p-$Al_xGa_{1-x}N$ and n-$Al_zGa_{1-z}N$ layers, the In composition y of the i-$In_yGa_{1-y}N$ layer, and the thickness t of the i-$In_yGa_{1-y}N$ layer, are selected to describe the TJ configurations. The doping concentrations of [Mg] and [Si] were set to be 5×10^{19} cm^{-3} and the In composition of the thin i-InGaN layer was in the range of 0–0.3, which was achievable in most growth systems [5,24]. The Al compositions were in the range of 0–0.8. We assumed a linear change in the acceptor activation energy from 140 meV in GaN to 630 meV in AlN, and a constant donor activation energy of 15 meV for AlGaN with an Al composition lower than 80% [25,26]. As shown in Figure 1b, because the band structures of p-AlGaN and n-AlGaN move into the flat band zone from the interface within 20 nm (usually less than 10 nm), we set the p-AlGaN and n-AlGaN layer to be 50 nm, a typical value which is thick enough for the band structure to form a flat band zone [11]. The thickness of the inserted InGaN layer is set within 2–7 nm [10,27]. This thin InGaN layer could induce a larger polarization charge at the interface, where the smaller bandgap of the InGaN increased the tunneling probability [9–11]. We carried out the TCAD calculations using Silvaco Atlas, a two-dimensional (2D) device simulator by consistently solving Schrodinger–Poisson equations, and used a non-local band-to-band tunneling model to calculate the TJ resistance [28]. The calculations were all conducted at a current density of 10 A/cm^2 for a fair comparison, which is a typical order in III-nitride tunnel junction research [2]. An example of a calculated current–voltage curve for the p-$Al_{0.3}Ga_{0.7}N$/i-$In_{0.3}Ga_{0.7}N$/n-$Al_{0.3}Ga_{0.7}N$ tunnel junction is shown in Figure 1c. The strain is considered in the polarization calculation, and the polarization parameters, electronic band structures, mobility parameters and other materials properties not mentioned above were using the default values recorded in ref. [28].

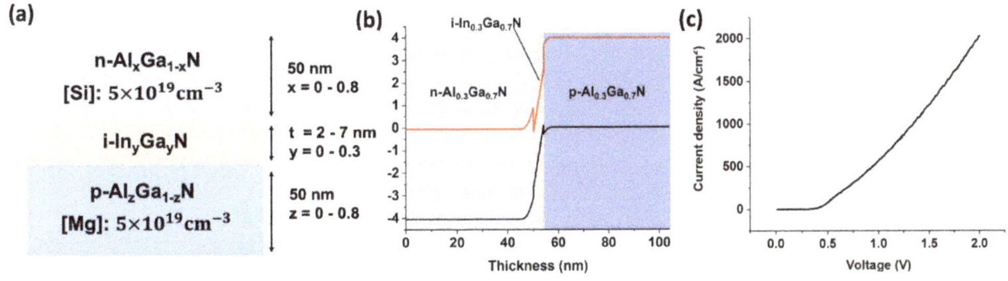

Figure 1. (**a**) Cross-sectional schematics of the proposed p-$Al_xGa_{1-x}N$/i-$In_yGa_{1-y}N$/n-$Al_zGa_{1-z}N$ TJ structures. (**b**) An example of the p-$Al_{0.3}Ga_{0.7}N$/i-$In_{0.3}Ga_{0.7}N$/n-$Al_{0.3}Ga_{0.7}N$ TJ band structure. (**c**) Calculated current–voltage curve for the p-$Al_{0.3}Ga_{0.7}N$/i-$In_{0.3}Ga_{0.7}N$/n-$Al_{0.3}Ga_{0.7}N$ tunnel junction.

We collected approximately 3500 results from the TCAD calculations. Because the convergence issue that occurs in the calculation may cause some data error, we calculated all the configurations twice with different iteration steps. These convergence issues usually lead to abrupt changes in the I–V curve [28]. This means that the corresponding resistances would suffer from convergence issues that usually exhibited a significant deviation from the other calculations according to the different iteration steps. Using this cross-validation method, we effectively ruled out the unreliable data that may have had a negative impact on the ML model and reduced the accuracy of the model prediction.

Specifically, we cross-validated the calculated data as credible when two calculation results had a data verification difference that was less than 5%, according to the following formula:

$$\frac{|\text{TJ resistance data 1} - \text{TJ resistance data 2}|}{\text{TJ resistance data 1}} = \begin{cases} (< 5\% \text{ (credible, no convergence issue)}) \\ (> 5\% \text{ (false, suffered from convergence issue)}) \end{cases} \quad (1)$$

After cross-validation, we had 2260 remaining data points, which were used to construct the TJ resistance database for the ML model. Considering the calculation resources, we set the iteration step as 0.01 Ω/cm^2 and set the terminated value of the maximum TJ resistance as 100 Ω/cm^2.

3. Machine Learning Model for TJ Resistance Prediction

Subsequently, the ML model was applied to the TJ resistance dataset. The schematic representation of the proposed ML model is shown in Figure 2a. By random partition, we used 80% of the dataset as the training data and considered the remaining to be the test data. We selected the XG-Boost classifier as the model [23], which is a boosting method that is widely used in many applications. Unlike many learning methods which have a fixed objective function and parameter format, the objective function and the format of the parameter in XG-Boost change iteratively in the process of optimization. For every iteration, the objective function of XG-Boost is to minimize the residuals of the last iteration. In this study, given the corresponding features $\{x_1, x_2, \ldots, x_n\}$ and the results $\{y_1, y_2, \ldots, y_n\}$, the objective function of the first iteration in XG-Boost is as follows:

$$Obj^{(1)} = \sum_{i=1}^{n} l(y_i, f_1(x_i)) + \Omega(f_1), \quad (2)$$

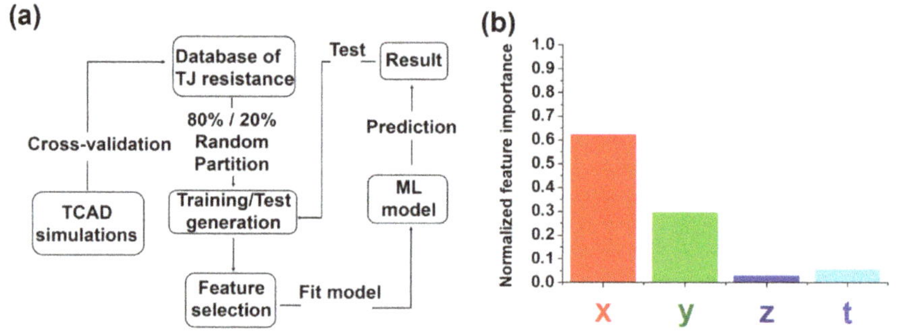

Figure 2. (**a**) Process flow of the ML model for TJ resistance prediction. (**b**) Feature importance retrieved from the XG-Boost model that was learned from the 2260 data samples.

$l(y_i, y_k)$ is the difference function between y_i and y_k; we set $l(y_i, y_k) = (y_i - y_k)^2$; and $\Omega(f_1)$ is the regularization term of function f_1, which is used to overcome the overfitting problem. Once we minimized $Obj^{(1)}$ to get the function f_1, we used $\hat{y}_i^{(1)} = f_1(x_i)$ to denote the predicted result of the i-th feature. The second iteration of XG-Boost is to minimize the residuals of the first iteration, as follows:

$$Obj^{(2)} = \sum_{i=1}^{n} l\left(y_i, \hat{y}_i^{(1)} + f_2(x_i)\right) + \Omega(f_2). \quad (3)$$

Generally, for the t + 1-th iteration, we used

$$Obj^{(n+1)} = \sum_{i=1}^{n} l\left(y_i, \hat{y}_i^{(t)} + f_{t+1}(x_i)\right) + \Omega(f_{t+1}), \quad (4)$$

where $\hat{y}_i^{(t)} = \sum_{j=1}^{t} f_j(x_i)$ and $\hat{y}_i^{(0)} = 0$. Once we finished the training process of XG-Boost with t iterations, we obtained the final predicted y_i^p of the i-th feature as the sum of all prediction functions as $y_i^p = \hat{y}_i^{(T)} = \sum_{j=1}^{T} f_j(x_i)$.

Under this framework, XG-Boost has better generalization and representation ability with a dense dataset, in which the number of samples is much larger than the dimension of features. By contrast, a sparse dataset easily leads to the overfitting problem [29]. To generalize our model extensible to the sparse dataset, a regularization term is added in the objective function to overcome the overfitting problem. Moreover, the computation of different functions in the inference process could be applied in parallel, which reduces the time complexity. The high average testing performance proves the feasibility of applying ML to TJ resistance prediction. Our predictions demonstrate that the XG-Boost model gives an average accuracy of 88.7%, with 90.5% accuracy being the best. We considered the prediction result to be true when it had an error of less than 10%.

The importance of the four features x, y, z, t could be extracted from the XG-Boost model. In Figure 2b, the Al composition of n-AlGaN and the In composition of the InGaN layer play the most important roles for TJ resistance optimization. To further verify our data, we compared our TJ resistance prediction results according to the proposed model. Results are shown in Figure 3.

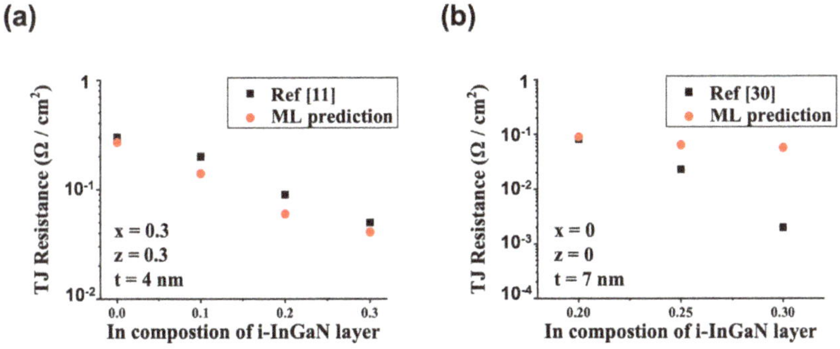

Figure 3. (**a**) p-$Al_{0.3}Ga_{0.7}N$/i-$In_yGa_{1-y}N$/n-$Al_{0.3}Ga_{0.7}N$ TJ structures with the 4 nm interlayer [11], and (**b**) p-GaN/i-$In_yGa_{1-y}N$/n-GaN TJ structures with the 7 nm interlayer [30].

As shown in Figure 3a, our ML prediction for p-$Al_{0.3}Ga_{0.7}N$/i-$In_yGa_{1-y}N$/n-$Al_{0.3}Ga_{0.7}N$ TJ structures with the 4 nm interlayer agreed well with the previous report [11]. Figure 3b shows the p-GaN/i-$In_yGa_{1-y}N$/n-GaN TJ structures with the 7 nm interlayer. Although the TJ resistances when In equals 0.2 and 0.25 predicted by ML agree with previous reports [30], it shows a relatively large deviation from earlier reference when In equals 0.3. The iteration step in our TCAD simulation was 0.01 Ω/cm^2, which meant that a TJ resistance less than 0.01 Ω/cm^2 could not be accurately recorded. This may have caused the inaccuracy in the exact TJ resistance value for the ultra-low TJ resistance configuration. The tendency of the TJ resistance variation, however, was still meaningful for TJ design instruction. In addition, we believed that the prediction could be improved significantly by reducing the iteration step and by using more powerful calculation resources.

4. TJ Resistance Prediction Results

Using the ML model, we predicted 22,254 combinations of TJ configurations with different x, y, z, t. The results of some of these configurations are shown in Figure 4. Higher Al composition in the n-AlGaN layer often leads to higher TJ resistance. Higher Al composition in the n-AlGaN leads to larger band discontinuity in the AlGaN/InGaN interface, and the tunneling probability decreases when the band discontinuity increases. By comparing these results with the different rows, i.e., different In compositions in Figure 4, we found that the higher In composition in the i-InGaN layer significantly reduced the TJ resistance, in particular for an Al composition in the n-AlGaN layer larger than 0.6. For higher-Al-composition p-AlGaN/i-InGaN/n-AlGaN TJs, the larger induced polarization charges in the AlGaN/InGaN surfaces reduce the depletion region, which assists the carriers in tunneling through the barriers. The thickness of the InGaN layer is critical in the p-AlGaN/i-InGaN/n-AlGaN TJs. If the InGaN layer is too thin, conduction and valence band extrema on either side do not align with enough region [31]. The extra-thick InGaN is introduced in a resistance series, which may have increased the resistance as well [10].

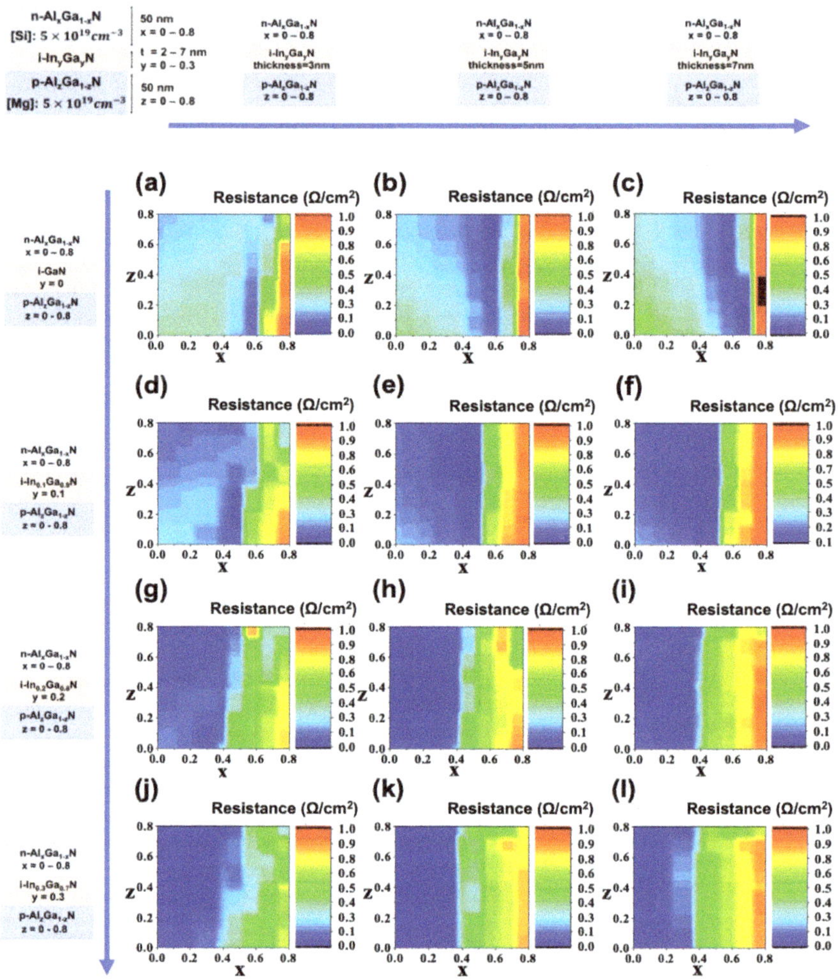

Figure 4. p-$Al_xGa_{1-x}N$/i-$In_yGa_{1-y}N$/n-$Al_zGa_{1-z}N$ TJ resistance prediction results with the values of the thickness of i-InGaN layer t and In composition of the i-InGaN layer y: (**a**) t = 3 nm, y = 0; (**b**) t = 5 nm, y = 0; (**c**) t = 7 nm, y = 0; (**d**) t = 3 nm, y = 0.1; (**e**) t = 5 nm, y = 0.1; (**f**) t = 7 nm, y = 0; (**g**) t = 3 nm, y = 0.2; (**h**) t = 5 nm, y = 0.2; (**i**) t = 7 nm, y = 0.2; (**j**) t = 3 nm, y = 0.3; (**k**) t = 5 nm, y = 0.3; (**l**) t = 7 nm, y = 0.3.

5. Asymmetric TJ with Different Al Compositions in p-AlGaN and n-AlGaN

Most previous TJ designs used a symmetrical structure, where the Al compositions of the p-AlGaN layer and n-AlGaN layer had the same values [32]. For a large part of the configurations, the effect of adjusting the Al composition of p-AlGaN was not significant. Thus, most studies have ignored the impact of this asymmetry. Our ML model, however, provides a novel asymmetric TJ design. We could identify many examples of the configurations, whose asymmetrical design could significantly reduce the TJ resistance and thus improve device performance. Figure 5a shows TJ resistance predictions for p-AlGaN/i-$In_{0.2}Ga_{0.8}N$/n-$Al_{0.3}Ga_{0.7}N$ and p-AlGaN/i-$In_{0.2}Ga_{0.8}N$/n-$Al_{0.5}Ga_{0.5}N$ with different Al compositions for the p-AlGaN layer. The thickness of the i-$In_{0.2}Ga_{0.8}N$ layer is 3 nm. We observed that the TJ resistances did not reach their minimum values when the Al composition in p-AlGaN was equal to that in n-AlGaN. For TJs with an n-$Al_{0.3}Ga_{0.7}N$ layer, in particular, the TJ resistance when the Al composition in p-AlGaN is equal to 0.3 is 0.14 Ω/cm^2, which is seven times larger than when that of p-AlGaN is equal to 0.7 (0.02 Ω/cm^2). Hence an asymmetric design is more superior. Moreover, for p-AlGaN/i-$In_{0.2}Ga_{0.8}N$/n-$Al_{0.5}Ga_{0.5}N$ TJ with high Al composition ($\geq$0.5) in the n-AlGaN layer, which is important in deep UV-LEDs to avoid light absorption from the active layer, p-$Al_{0.7}Ga_{0.3}N$/i-$In_{0.2}Ga_{0.8}N$/n-$Al_{0.5}Ga_{0.5}N$ has 25.2% lower TJ resistance compared to conventional p-$Al_{0.5}Ga_{0.5}N$/i-$In_{0.2}Ga_{0.8}N$/n-$Al_{0.5}Ga_{0.5}N$ TJ. Thus, applying the asymmetric tunnel junction into the UV-LED could lead to a reduction of the tunnel resistance, and the high Al composition p-layer could also adjust the electron blocking effect without additional light absorption due to its larger bandgap. As this might be the first work discussing III-nitride asymmetric tunnel junction, we believe that there is still huge potential to optimize the asymmetric III-nitride tunnel junction with different materials and grading designs.

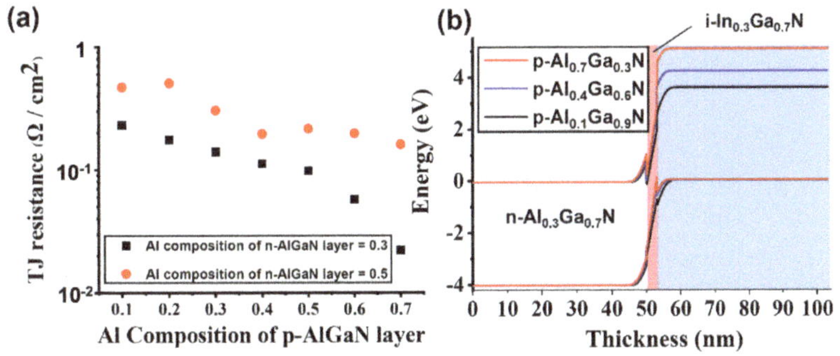

Figure 5. (**a**) TJ resistance prediction of p-AlGaN/i-$In_{0.2}Ga_{0.8}N$/n-$Al_{0.3}Ga_{0.7}N$ and p-AlGaN/i-$In_{0.2}Ga_{0.8}N$/n-$Al_{0.5}Ga_{0.5}N$ with different Al compositions of the p-AlGaN layer. (**b**) Band diagram of p-AlGaN/i-$In_{0.2}Ga_{0.8}N$/n-$Al_{0.3}Ga_{0.7}N$ when the Al composition of the p-AlGaN layer is equal to 0.1, 0.4, and 0.7. The thickness of this i-$In_{0.2}Ga_{0.8}N$ layer is 3 nm.

To better understand the asymmetric TJ designs, band diagrams of p-AlGaN/i-$In_{0.2}Ga_{0.8}N$/n-$Al_{0.3}Ga_{0.7}N$ are shown in Figure 5b. When the Al composition in p-AlGaN is equal to 0.7, the valence band across the zero-energy level in the tunneling region indicated a high tunneling possibility. The tunneling probability can be described by the WKB approximation [10,33].

$$T_n = \exp\left\{-2\int_0^{x_n}\left[\sqrt{\frac{m_e^* q^2 N_D t^2}{\hbar^2 \varepsilon}}\right]dt\right\},\ x_n = \sqrt{\frac{2\varepsilon\Delta E_C}{q^2 N_D}} \quad (5)$$

$$T_p = \exp\left\{-2\int_0^{x_p}\left[\sqrt{\frac{m_h^* q^2 N_D t^2}{\hbar^2 \varepsilon}}\right]dt\right\},\ x_p = \sqrt{\frac{2\varepsilon\Delta E_V}{q^2 N_D}} \quad (6)$$

$$T_{InGaN} = \exp\left\{-2\int_0^{\frac{E_g\varepsilon}{q\sigma}}\left[\sqrt{\frac{2m_{InGaN}^*\left(\left(\frac{E_{g,InGaN}}{2}\right)^2-\left(\frac{E_{g,InGaN}}{2}-\frac{E_{g,InGaN}}{\left(\frac{E_g\varepsilon}{q\sigma}\right)}\right)^2\right)}{\hbar E_{g,InGaN}}}\right]dt\right\} \quad (7)$$

$$T_{net} = T_n * T_p * T_{InGaN} \quad (8)$$

T_{InGaN}, T_n and T_p are the tunneling probabilities in i-InGaN, n-AlGaN and p-AlGaN layers. ΔE_C and ΔE_V represent the band discontinuities across the n-AlGaN/i-InGaN and i-InGaN/p-AlGaN interfaces. N_A and N_D are the acceptor and donor doping density,

m_e^* *and* m_h^* are the effective mass of electron and hole. ε is the permittivity and σ is the induced polarization charge density. ε is the permittivity and σ is the induced polarization charge density.

From the equations above, the tunneling probability across the devices is determined by both the band discontinuity and the induced polarization charge density. With the increasing Al composition of the p-AlGaN layer, the band discontinuity ΔE_V becomes larger, which decreases the tunneling probability. However, the induced polarization is increased according to the first-principal calculation [34], and the larger polarization charge σ assists the tunneling effect. The factors competing with each other; and in our case, the larger polarization-induced charge with an increase in Al composition in the p-AlGaN layer is more significant than the band offset increment, which leads to a reduction in TJ resistance.

6. Summary

In this work, we developed an efficient model for III-nitride TJ resistance prediction to instruct the TJ device design. We constructed the first ML model for p-AlGaN/i-InGaN/n-AlGaN TJ resistance prediction. Our ML model is based on the XG-Boost algorithm and trained with the data calculated by TCAD simulations, which rapidly predicted the p-AlGaN/i-InGaN/n-AlGaN TJ resistances with accuracy as high as 90.5% We compared our prediction with the previous report which shows good agreement. Furthermore, 22,254 combinations of p-AlGaN/i-InGaN/n-AlGaN TJ configurations with different Al or In compositions and InGaN layer thicknesses were investigated by our ML model, and the predicted TJ resistances are shown for future design instruction. Moreover, we demonstrated the effectiveness of the asymmetric tunnel junction design based on our TJ resistance prediction database. By increasing the Al composition in the p-AlGaN layer of the p-AlGaN/i-$In_{0.2}Ga_{0.8}N$/n-$Al_{0.3}Ga_{0.7}N$ TJ structure, the increased polarization charge enhances the tunneling effect, which leads to a significant reduction of the TJ resistance.

Author Contributions: Conceptualization, R.L. (Rongyu Lin), X.Z. and X.L.; methodology, R.L. (Rongyu Lin), P.H. and Y.W.; software, R.L. (Rongyu Lin), P.H. and Y.W.; validation, R.L. (Rongyu Lin), P.H. and Y.W.; formal analysis, R.L. (Rongyu Lin), P.H. and R.L. (Ronghui Lin); investigation, R.L. (Rongyu Lin); resources, X.Z. and X.L.; data curation, R.L. (Rongyu Lin), P.H. and Y.W.; writing—original draft preparation, R.L. (Rongyu Lin); writing—review and editing, R.L. (Rongyu Lin), P.H., Y.W., R.L. (Ronghui Lin), Y.L., Z.L., X.Z. and X.L.; visualization, R.L. (Rongyu Lin); supervision, X.Z. and X.L.; project administration, X.Z. and X.L.; funding acquisition, X.Z. and X.L. All authors have read and agreed to the published version of the manuscript.

Funding: This research was funded by KAUST Baseline Fund BAS/1/1664-01-01, GCC Research Council Grant REP/1/3189-01-01, Competitive Research Grants URF/1/3437-01-01 and URF/1/3771-01-01, and KAUST AI Initiative.

Data Availability Statement: The data presented in this study are available on request from the corresponding author.

Conflicts of Interest: The authors declare no conflict of interest.

References

1. Li, W.; Sharmin, S.; Ilatikhameneh, H.; Rahman, R.; Lu, Y.; Wang, J.; Yan, X.; Seabaugh, A.; Klimeck, G.; Jena, D. Polarization-Engineered III-Nitride Heterojunction Tunnel Field-Effect Transistors. *IEEE J. Explor. Solid-State Comput. Devices Circuits* **2015**, *1*, 28–34. [CrossRef]
2. Zhang, Y.; Deng, G.; Yu, Y.; Wang, Y.; Zhao, D.; Shi, Z.; Zhang, B.; Li, X. Demonstration of N-Polar III-Nitride Tunnel Junction LED. *ACS Photonics* **2020**, *7*, 1723–1728. [CrossRef]
3. Reichertz, L.A.; Gherasoiu, I.; Yu, K.M.; Kao, V.M.; Walukiewicz, W.; Ager, J.W., III. Demonstration of a III-Nitride/Silicon Tandem Solar Cell. *Appl. Phys. Express* **2009**, *2*, 122202. [CrossRef]
4. Akyol, F.; Krishnamoorthy, S.; Rajan, S. Tunneling-based Carrier Regeneration in Cascaded GaN Light Emitting Diodes to Overcome Efficiency Droop. *Appl. Phys. Lett.* **2013**, *103*, 081107. [CrossRef]
5. Zhang, Y.; Jamal-Eddine, X.; Rajan, S. Recent Progress of Tunnel Junction-based Ultra-violet Light Emitting Diodes. *Jpn. J. Appl. Phys.* **2019**, *58*, SC0805. [CrossRef]

6. Simon, J.; Zhang, Z.; Goodman, K.; Xing, H.; Kosel, T.; Fay, P.; Jena, D. Polarization-Induced Zener Tunnel Junctions in Wide-Band-Gap Heterostructures. *Phys. Rev. Lett.* **2009**, *103*, 026801. [CrossRef]
7. Schubert, M.F. Interband Tunnel Junctions for Wurtzite III-nitride Semiconductors Based on Heterointerface Polarization Charges. *Phys. Rev. B* **2010**, *81*, 035303. [CrossRef]
8. Zhang, Y.; Krishnamoorthy, S.; Akyol, F.; Johnson, J.M.; Allerman, A.A.; Moseley, M.W.; Armstrong, A.M.; Hwang, J.; Rajan, S. Reflective Metal/semiconductor Tunnel Junctions for Hole Injection in AlGaN UV LEDs. *Appl. Phys. Lett.* **2017**, *111*, 051104. [CrossRef]
9. Yan, X.; Li, W.; Islam, S.M.; Pourang, K.; Xing, H.; Fay, P.; Jena, D. Polarization-induced Zener Tunnel Diodes in GaN/InGaN/GaN Heterojunctions. *Appl. Phys. Lett.* **2015**, *107*, 163504. [CrossRef]
10. Krishnamoorthy, S.; Akyol, F.; Park, P.S.; Rajan, S. Low Resistance GaN/InGaN/GaN Tunnel Junctions. *Appl. Phys. Lett.* **2013**, *102*, 113503. [CrossRef]
11. Zhang, Y.; Krishnamoorthy, S.; Akyol, F.; Allerman, A.A.; Moseley, M.W.; Armstrong, A.M.; Rajan, S. Design and Demonstration of Ultra-wide Bandgap AlGaN Tunnel Junctions. *Appl. Phys. Lett.* **2016**, *109*, 121102. [CrossRef]
12. Gelinas, R. A Novel Approach to Modeling Tunnel Junction Diodes Using Silvaco Atlas Software. Master's Thesis, Naval Postgraduate School, Monterey, CA, USA, 2005.
13. Zhang, Z.; Tan, S.; Kyaw, Z.; Ji, Y.; Liu, W.; Ju, Z.; Hasanov, N.; Sun, X.; Demir, H. InGaN/GaN Light-Emitting Diode with a Polarization Tunnel Junction. *Appl. Phys. Lett.* **2013**, *102*, 193508. [CrossRef]
14. Sadaf, S.M.; Ra, Y.H.; Nguyen, H.P.T.; Djavid, M.; Mi, Z. Alternating-Current InGaN/GaN Tunnel Junction Nanowire White-Light Emitting Diodes. *Nano Lett.* **2015**, *15*, 6696–6701. [CrossRef] [PubMed]
15. Lin, R.; Alnakhli, Z.; Li, X. Engineering of Multiple Bound States in the Continuum by Latent Representation of Freeform Structures. *Photonics Res.* **2021**, *9*, B96–B103. [CrossRef]
16. Lin, R.; Zhai, Y.; Xiong, C.; Li, X. Inverse Design of Plasmonic Metasurfaces by Convolutional Neural Network. *Opt. Lett.* **2020**, *45*, 1362–1365. [CrossRef]
17. Rouet-Leduc, B.; Barros, K.; Lookman, T.; Humphreys, C.J. Optimisation of GaN LEDs and the Reduction of Efficiency Droop Using Active Machine Learning. *Sci. Rep.* **2016**, *6*, 24862. [CrossRef] [PubMed]
18. Rouet-Leduc, B.; Hulbert, C.; Barros, K.; Lookman, T.; Humphreys, C.J. Automatized Convergence of Optoelectronic Simulations Using Active Machine Learning. *Appl. Phys. Lett.* **2017**, *111*, 043506. [CrossRef]
19. Pandey, A.; Shin, W.J.; Gim, J.; Hovden, R.; Mi, Z. High-efficiency AlGaN/GaN/AlGaN Tunnel Junction Ultraviolet Light-Emitting Diodes. *Photon. Res.* **2020**, *8*, 331–337. [CrossRef]
20. Sadaf, S.M.; Zhao, S.; Wu, Y.; Ra, Y.H.; Liu, X.; Vanka, S.; Mi, Z. An AlGaN Core–Shell Tunnel Junction Nanowire Light-Emitting Diode Operating in the Ultraviolet-C Band. *Nano Lett.* **2017**, *17*, 1212–1218. [CrossRef] [PubMed]
21. Hwang, D.; Mughal, A.J.; Wong, M.S.; Alhassan, A.I.; Nakamura, S.; DenBaars, S.P. Micro-light-emitting Diodes with III-Nitride Tunnel Junction Contacts Grown by Metalorganic Chemical Vapor Deposition. *Appl. Phys. Express* **2017**, *11*, 012102. [CrossRef]
22. Takasuka, D.; Akatsuka, Y.; Ino, M.; Koide, N.; Takeuchi, T.; Iwaya, M.; Kamiyama, S.; Akasaki, I. GaInN-based Tunnel Junctions with Graded Layers. *Appl. Phys. Express* **2016**, *9*, 081005. [CrossRef]
23. Chen, T.; Guestrin, C. Xgboost: A Scalable Tree Boosting System. In Proceedings of the 22nd ACM SIGKDD International Conference on Knowledge Discovery and Data Mining, KDD '16, San Francisco, CA, USA, 13–17 August 2016; pp. 785–794.
24. Lund, C.; Romanczyk, B.; Catalano, M.; Wang, Q.; Li, W.; DiGiovanni, D.; Kim, M.J.; Fay, P.; Nakamura, S.; DenBaars, S.P.; et al. Metal-organic Chemical Vapor Deposition of High Quality, High Indium Composition N-polar InGaN Layers for Tunnel Devices. *J. Appl. Phys.* **2017**, *121*, 185707. [CrossRef]
25. Nam, K.B.; Nakarmi, M.L.; Li, J.; Lin, J.Y.; Jiang, H.X. Mg Acceptor Level in AlN Probed by Deep Ultraviolet Photoluminescence. *Appl. Phys. Lett.* **2003**, *83*, 878–880. [CrossRef]
26. Borisov, B.; Kuryatkov, V.; Kudryavtsev, Y.; Asomoza, R.; Nikishin, S.; Song, D.Y.; Holtz, M.; Temkin, H. Si-doped $Al_xGa_{1-x}N$ (0.56 $\leq x \leq$ 1) Layers Grown by Molecular Beam Epitaxy with Ammonia. *Appl. Phys. Lett.* **2005**, *87*, 132106. [CrossRef]
27. Akyol, F.; Zhang, Y.; Krishnamoorthy, S.; Rajan, S. Ultralow-Voltage-Drop GaM/InGaN/GaN Tunnel Junctions with 12% Indium Content. *Appl. Phys. Express* **2017**, *10*, 121003. [CrossRef]
28. Silvaco, I. ATLAS User's Manual. 2016. Available online: http://www.silvaco.com/ (accessed on 9 September 2021).
29. Zhang, D.; Qian, L.; Mao, B.; Huang, C.; Huang, B.; Si, Y. A Data-Driven Design for Fault Detection of Wind Turbines Using Random Forests and Xgboost. *IEEE Access* **2018**, *6*, 21020–21031. [CrossRef]
30. Tsai, M.C.; Leung, B.; Hsu, T.C.; Kuo, Y.K. Low Resistivity GaN-based Polarization-induced Tunnel Junction. *J. Lightwave Technol.* **2013**, *31*, 3575–3581. [CrossRef]
31. Krishnamoorthy, S.; Nath, D.N.; Akyol, F.; Park, P.S.; Esposto, M.; Rajan, S. Polarization-engineered GaN/InGaN/GaN Tunnel Diodes. *Appl. Phys. Lett.* **2010**, *97*, 203502. [CrossRef]
32. Jamal-Eddine, Z.; Zhang, Y.; Rajan, S. Recent Progress in III-nitride Tunnel Junction-based Optoelectronics. *J. High Speed Electron. Syst.* **2019**, *28*, 233–249. [CrossRef]
33. Lu, Y.; Wang, C.; Oliveira, V.P.D.; Liu, Z.; Li, X. UV Light-emitting Diode with Buried Polarization-induced N-AlGaN/InGaN/p-AlGaN Tunneling Junction. *IEEE Photon. Technol. Lett.* **2021**, *33*, 808–811. [CrossRef]
34. Liu, K.; Li, X. Polarization Properties of Wurtzite III-Nitride Indicate the Principle of Polarization Engineering. *arXiv* **2018**, arXiv:1808.07211.

nanomaterials

MDPI

Article

Dual-Spectral Plasmon-Induced Transparent Terahertz Metamaterial with Independently Tunable Amplitude and Frequency

Tong Wu [1], Guan Wang [1], Yang Jia [1], Yabin Shao [2], Chen Chen [2], Jing Han [1], Yang Gao [1,*] and Yachen Gao [1,*]

[1] Electronic Engineering College, Heilongjiang University, Harbin 150080, China; 124wutong@163.com (T.W.); wang2687220886@163.com (G.W.); jiayang_1990@163.com (Y.J.); hanjing1980@163.com (J.H.)

[2] Department of Computer & Electrical Engineering, East University of Heilongjiang, Harbin 150086, China; shao_yabin@163.com (Y.S.); rukawa1600@163.com (C.C.)

* Correspondence: gaoy_hit@163.com (Y.G.); gaoyachen@hlju.edu.cn (Y.G.)

Abstract: A bifunctional tunable metamaterial composed of pattern metal structure, graphene, and strontium titanate (STO) film is proposed and studied numerically and theoretically. The dual plasmon-induced transparency (PIT) window is obtained by coupling the bright state cut wire (CW) and two pairs of dark state dual symmetric semiring resonators (DSSRs) with different parameters. Correspondingly, slow light effect can also be realized. When shifting independently, the Fermi level of the graphene strips, the amplitudes of the two PIT transparency windows and slow light effect can be tuned, respectively. In addition, when independently tuning the temperature of the metamaterial, the frequency of the dual PIT windows and slow light effect can be tuned. The physical mechanism of the dual-PIT was analyzed theoretically by using a three-harmonic oscillator model. The results show that the regulation function of the PIT peak results from the change of the oscillation damping at the dark state DSSRs by tuning conductivity of graphene. Our design presents a new structure to realize the bifunctional optical switch and slow light.

Keywords: plasmon-induced transparency; terahertz; graphene; strontium titanate; slow light

Citation: Wu, T.; Wang, G.; Jia, Y.; Shao, Y.; Chen, C.; Han, J.; Gao, Y.; Gao, Y. Dual-Spectral Plasmon-Induced Transparent Terahertz Metamaterial with Independently Tunable Amplitude and Frequency. *Nanomaterials* **2021**, *11*, 2876. https://doi.org/10.3390/nano11112876

Academic Editor: Vladimir S. Bystrov

Received: 20 September 2021
Accepted: 25 October 2021
Published: 28 October 2021

1. Introduction

Electromagnetically induced transparency (EIT) is an effect resulting from quantum destructive interference. It can generate a narrow-band transparent window when light propagates through an originally opaque medium [1,2]. However, its application is limited due to harsh production conditions of the stable optical pumping and low temperature. Compared with the traditional EIT effect, plasmon-induced transparency (PIT) effect overcomes these harsh conditions [3]. Recently, many researchers have focused on various metamaterial structures to achieve PIT, which is the analog of EIT effect [4–8]. Previous studies have shown that PIT effect can be achieved mainly via bright and bright mode, bright and dark mode, and bright and quasi-dark mode [9–12]. At a PIT peak region, strong dispersion can occur, causing slow light effect which can be used in optical information processing [13–16].

In order to meet various practical applications, the properties of PIT should be able to be tuned. The tuning can be realized by changing structural parameters, using tunable materials and microelectromechanical systems (MEMS) technology. Due to the high flexibility, tunable materials-based PIT devices have become a research hotspot [17–24]. Graphene is especially widely used in the design of tunable PIT devices because of its high electron mobility, high modulation depth, tunable surface conductivity and low insertion loss characteristics. Tunable graphene-based PIT devices can realize different functions, such as the single-PIT [18–20], dual-PIT [21–23] and multi-PIT [24]. Recently, STO has also aroused the interest of researchers due to its temperature-tunable relative permittivity [25]. In 2020, Zhong proposed a tunable PIT metamaterial based on STO. When the device temperature is changed, the frequency of the single PIT window and slow light effect can be adjusted [26]. However, we can see most of the work mentioned above focused mainly on the tuning of either amplitude or frequency of the PIT peak. For example, in references [18,19,22,23], researchers proposed the metamaterials to realize the regulation of the PIT peak intensity. In references [20,21,24,26], researchers realized the regulation of the frequency of the PIT peak. To our knowledge, the tunable PIT metamaterial with independently tunable amplitude and frequency of dual-PIT effect has not been previously reported.

In this paper, we designed a metal structure to achieve the dual PIT effect and realized the modulation of the intensity and frequency of the PIT effect by changing the Fermi level of graphene strips and the temperature of STO films. The mechanisms of the tunable dual-PIT effect were analyzed using a three-harmonic oscillator model.

2. Materials and Methods

The PIT metamaterial proposed is illustrated in Figure 1. From Figure 1a, it can be observed that the metamaterial is composed of three layers, which are graphene–metal structure, STO film, and sapphire substrate. The parameters of metamaterial are shown in Figure 1b. Two sets of symmetrical half-rings named upper double symmetric semiring resonators (UDSSRs) and bottom double symmetric semiring resonators (BDSSRs) are placed on the upper and bottom sides of cut wire (CW), respectively. The thickness of the metal structure is 0.2 μm. The length of CW is L = 80 μm, and the width is w = 5 μm. For UDSSRs and BDSSRs, the gap width of splits is g = 5 μm. The outer radius and inner radius of UDSSR are 20 μm and 15 μm, respectively. The outer radius and inner radius of BDSSRs are 23 μm and 18 μm, respectively. The distance between two DSSRs and CW is S = 3.5 μm. Under the UDSSRs and BDSSRs, there are two graphene strips defined as strip 1 and strip 2, respectively. The STO film with a thickness of 100 nm is placed between the metal–graphene hybrid structure and sapphire substrate. The period of metamaterial is $P_x = P_y$ = 120 μm.

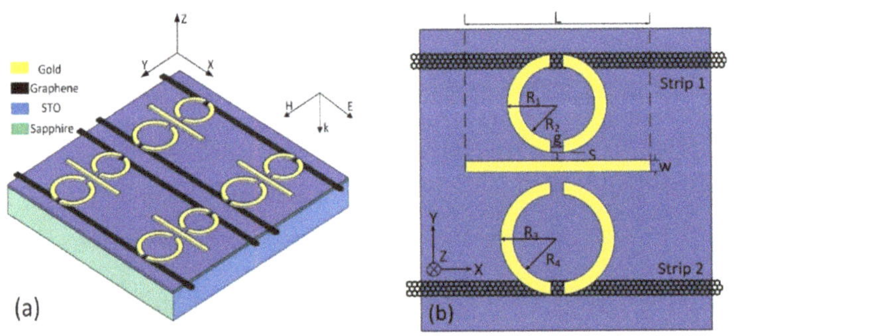

Figure 1. (**a**). Three-dimensional diagram of the hybrid metamaterial. (**b**) The top view of the unit cell.

Finite-difference time-domain (FDTD) algorithm is used for numerical simulation and calculating the near-field coupling of this design. The boundary conditions of x and y direction are periodic boundaries, and a perfect matching layer is used in the z direction. The x-polarized terahertz wave is incident vertically along the z direction. In the low frequency terahertz band, as a loss metal, gold can be represented by a static model, and its conductivity is 4.56×10^7 S/m [27]. The refractive index of the sapphire substrate is 1.78 [28].

As a 2D material, the electric-magnetic properties of graphene can be described by surface conductivity σ_g. According to random-phase approximation (RPA) theory, the conductivity σ_g can be expressed by combining intraband σ_{intra} and interband σ_{inter} [29]:

$$\sigma_g = \sigma_{intra} + \sigma_{inter} = \frac{2e^2 k_B T}{\pi \hbar^2} \frac{i}{\omega + i\tau^{-1}} ln\left[2cosh\left(\frac{E_F}{2k_B T}\right)\right] + \frac{e^2}{4\hbar}\left[\frac{1}{2} + \frac{1}{\pi} arctan\left(\frac{\hbar\omega - 2E_F}{2k_B T}\right) - \frac{i}{2\pi} ln \frac{(\hbar\omega + 2E_F)^2}{(\hbar\omega - 2E_F)^2 + 4(k_B T)^2}\right] \quad (1)$$

where e is the electron charge, k_B is the Boltzmann constant, T is the temperature in Kelvin, $\hbar$ is the reduced Planck's constant, ω is the THz frequency, τ and E_F are graphene carrier relaxation time and Fermi level.

For STO materials, the temperature-dependent relative permittivity in the THz spectral region is [30]:

$$\varepsilon_\omega = \varepsilon_\infty + \frac{f}{\omega_0 - \omega^2 - i\omega\gamma} \tag{2}$$

where ε_∞ is 9.6, representing the high-frequency bulk permittivity, f is an oscillator strength depending on temperature, with a value of 2.36×10^6 cm^2 [31]. ω, ω_0 and γ are the angular frequency, soft mode frequency and the damping factor, respectively. The formula related to temperature can be expressed as:

$$\omega_0(T)\left[\mathrm{cm}^{-1}\right] = \sqrt{31.2(T - 42.5)} \tag{3}$$

$$\gamma(T)\left[\mathrm{cm}^{-1}\right] = -3.3 + 0.094T \tag{4}$$

where T is temperature of STO. $\omega_0(T)$ can be obtained using the Cochran law and $\gamma(T)$ can be fitted by an empirical linear dependence. We can see from Equations (2)–(4) that the relative permittivity of STO under different angular frequency and temperature can be calculated.

3. Results and Discussions

In order to investigate the mechanism of double PIT transparency windows, we conducted the simulation for four arrays, composed of CW arrays, UDSSRs, BDSSRs and a combined array of them. Here, the temperature of STO was set to be 425 K. In Figure 2a, it can be seen that when the x-polarization plane wave achieves a coupling with the single CW structure, a transmission valley appears at 0.9 THz because of the localized surface plasmon resonance (LSPR) at CW. However, due to the symmetry of the isolated DSSRs structure with respect to the x-polarization incident field, the UDSSRs or BDSSRs are inactive at the same frequency [32]. Thus, the CW and the two pairs of DSSRs behave as bright and dark resonance modes, respectively. When the CW, UDSSRs and BDSSRs are combined into a unit cell, under x-polarized electric field excitation, two PIT windows arise because of the destructive interference caused by the coupling of the two LC resonance modes and LSPR mode. As shown in Figure 2b, two transparent windows at 0.87 THz and 1.016 THz can be observed and denoted as peak I and peak II.

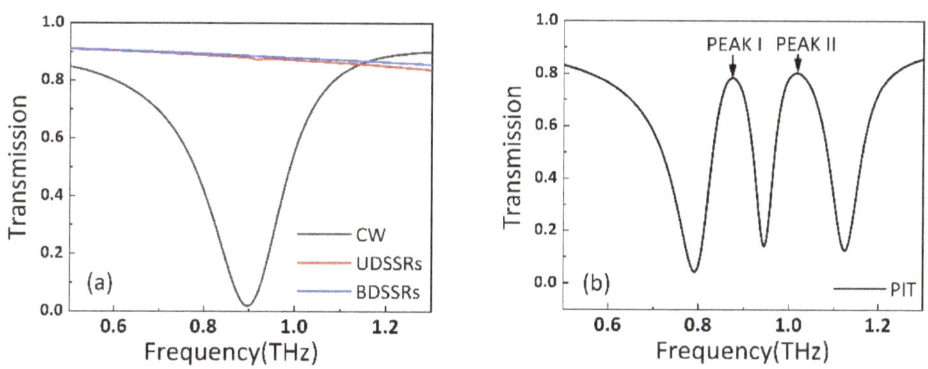

(a) **(b)**

Figure 2. (**a**) The transmission spectra of isolated CW array, BDSSRs array and UDSSRs array. (**b**) The transmission spectra of the PIT metamaterial composed of two parts of DSSRs and CW.

In order to analyze the mechanism of PIT effect, we studied the electric field and charge distribution at resonance frequency. As shown in Figures 3a and 4a, when CW is coupled with a plane wave, it can be observed that there is a strong electric field at the x-axis edges and corners of CW, and the charges are concentrated in the same position. This phenomenon, which can be excited, directly belongs to LSPR and can be described as bright resonance mode. When BDSSRs, UDSSRs and CW are placed in the arrays to achieve coupling, the electric field in 0.873 THz and 1.016 THz are shown in Figure 3b,c, the charge distribution is shown in Figure 4b,c.

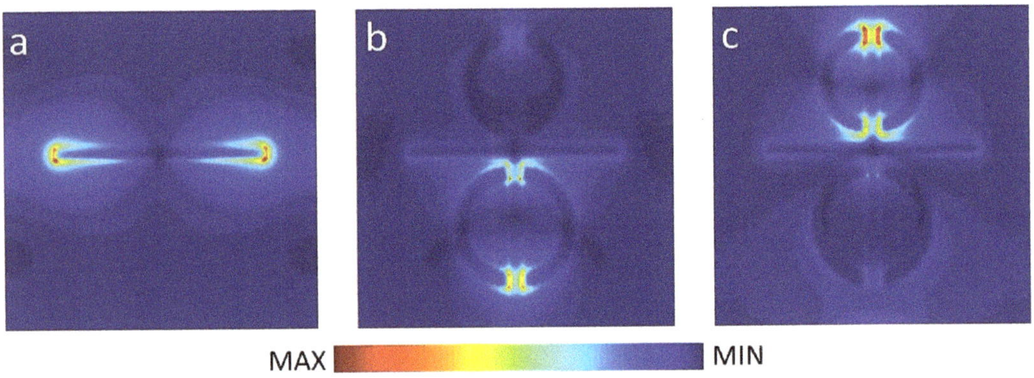

Figure 3. Electric field distributions of (**a**) the CW, (**b**) the PIT metamaterials at 0.873 THz, and (**c**) the PIT metamaterials at 1.016 THz.

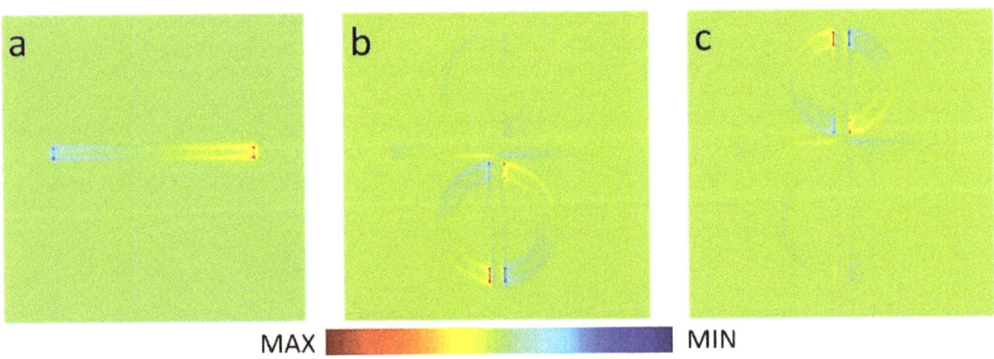

Figure 4. Charge distribution of (**a**) the CW, (**b**) the PIT metamaterials at 0.873 THz, and (**c**) the PIT metamaterials at 1.016 THz.

From Figures 3b and 4b, we can see that the enhancement of the electric field and accumulation of opposite charge transfer from the edges and corners of CW to the splits of BDSSRs. Similarly, in Figures 3c and 4c, we can see the electric field enhancement and opposite charge transfer to the splits of UDSSRs. These two resonance modes generated by indirect coupling with CW belong to the LC resonance and can be regarded as dark modes. Due to the phase difference of π between bright resonance mode and dark resonance mode, destructive interference will occur between LSPR and LC resonance, which results in the appearance of transparent windows [33].

Next, the individually tunable properties of the device are analyzed. Figure 5 shows the simulated and theoretical transmission spectrum with different Fermi levels of strip 2 and strip 1, respectively. In Figure 5a,c, it can be found that the two PIT transparency windows of this metamaterial can be achieved, and the independent on-to-off switching function at two PIT windows can be realized by tuning the graphene Fermi level. Figure 5a (top panel) is the transmission spectra when the graphene strips are absent. The amplitude of transmission of peak I and peak II are 0.7814 and 0.8017, respectively. When strip 2 is placed under the splits of the BDSSRs and the Fermi level is set to 0.2 eV, the transmission of peak I reduces to 0.424. As the graphene Fermi level increases, peak I undergoes a continuous decrease, whereas peak II changes minimally. Previous studies have shown that the graphene Fermi level can be modulated to be 1.2 eV [34]. When the Fermi level increases to 1.2 eV, peak I disappears completely, which causes an off state. In order to quantitatively describe the modulation depth of the PIT transparent windows, we introduce the formula $\Delta T = \left[(T_0 - T_g)/T_0\right] \times 100\%$, where T_0 and T_g refer to the amplitude of transmission peak without and with graphene, respectively. Finally, with the Fermi level of 1.2 eV, the transmission of peak I reduces to 0.137, correspondingly the modulation depth of peak I is calculated to be 82.4% using the formula.

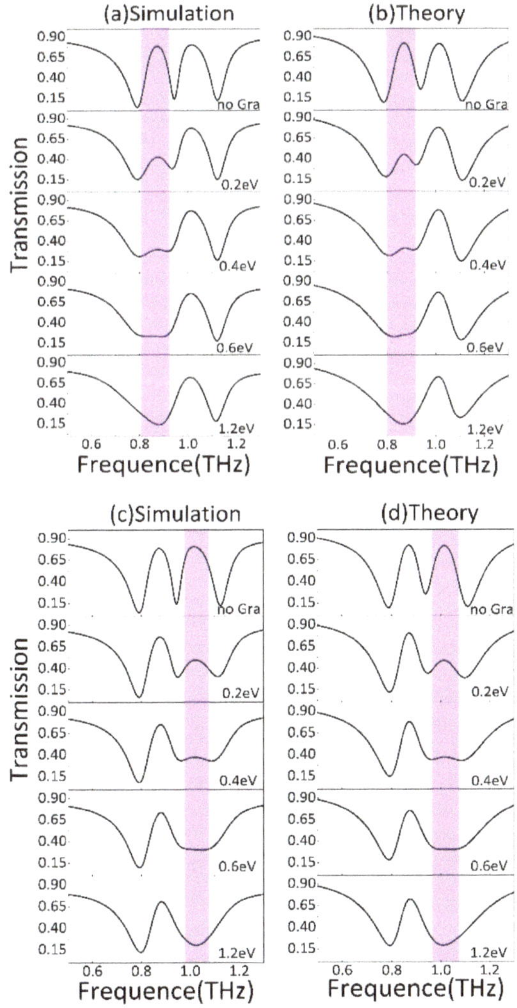

Figure 5. (**a**) The simulated and (**b**) analytical fitted transmission spectrum with different Fermi levels of strip 2. (**c**) The simulated and (**d**) analytical fitted transmission spectrum with different Fermi levels of strip 1.

In Figure 5c, it can be observed that, as the Fermi level of strip 1 increases from 0.2 eV to 1.2 eV, the transmission change of peak II is similar to that of peak I; namely, the amplitude of peak II decreases with the increase in the graphene Fermi level. When the graphene Fermi level reaches1.2 eV, the transmission of peak II is 0.2022. The modulation depth of peak II can achieve 74.7%. Therefore, this design can realize the optical switch-like regulation of peak I and peak II by adjusting the Fermi level of strip 1 and strip 2, respectively.

In order to further investigate the independent tunable mechanism of the dual-PIT transparency window by tuning the graphene Fermi level, we analyzed the interaction of the bright and two dark modes using the three-harmonic oscillator model [35]. As a bright mode, the LSPR at CW can be represented by oscillator 1 arising from direct coupling with the plane wave. As the dark modes excited through near field coupling with the bright mode, the BDSSRs and UDSSRs are represented by oscillator 2 and 3, respectively. The coupling effect between the three resonance modes is described by the following formula:

$$\ddot{x}_0(t) + \gamma_0\dot{x}_0(t) + \omega_0^2 x_0(t) + \kappa_1\dot{x}_1(t) + \kappa_2\dot{x}_2(t) = \lambda_0 E \tag{5}$$

$$\ddot{x}_1(t) + \gamma_1\dot{x}_1(t) + \omega_1^2 x_1(t) - \kappa_1\dot{x}_0(t) = 0 \tag{6}$$

$$\ddot{x}_2(t) + \gamma_2\dot{x}_2(t) + \omega_2^2 x_2(t) - \kappa_2\dot{x}_0(t) = 0 \tag{7}$$

Here, E represents the incident electromagnetic field, λ_0 describes the coupling strength of the electromagnetic field. ω_0, ω_1, ω_2 are the resonance frequencies of oscillator 1, oscillator 2 and oscillator 3, respectively. x_0 and γ_0 are the amplitude and damping of the bright resonance mode. x_1 and x_2 are the amplitudes of the dark resonance mode at BDSSRs and UDSSRs, respectively, and γ_1 and γ_2 are the damping of the dark resonance mode at BDSSRs and UDSSRs, respectively. The coupling coefficients between the two dark state modes and the bright state are κ_1 and κ_2, respectively. After solving the Equations (5)–(7) with $\omega^2 - {\omega_b}^2 \approx (\omega - \omega_b)\cdot 2\omega$, and $\lambda = \lambda_0/2\omega$, the susceptibility χ of the PIT metamaterials can be obtained as:

$$\chi(\omega) = (\chi_r + i\chi_i) \propto \frac{1}{A}\left(\omega_2 - \omega - i\frac{\gamma_1}{2}\right)\left(\omega_3 - \omega - i\frac{\gamma_2}{2}\right) \tag{8}$$

where:

$$A = \left(\omega_2 - \omega - i\frac{\gamma_1}{2}\right)\left(\omega_1 - \omega - i\frac{\gamma_0}{2}\right)\left(\omega_3 - \omega - i\frac{\gamma_2}{2}\right) - \frac{\kappa_1^2}{4}\left(\omega_3 - \omega - i\frac{\gamma_2}{2}\right) - \frac{\kappa_2^2}{4}\left(\omega_2 - \omega - i\frac{\gamma_1}{2}\right) \tag{9}$$

In Equation (8) χ_r represents the dispersion. The transmittance T can be calculated by the formula $T = 1 - \lambda_0\chi_i$, where χ_i is proportional to the energy loss [17,36].

Figure 5b,d show the theoretical results of the transmission spectrum. It is observable that they are in strong agreement with the simulation results shown in Figure 5a,c. Correspondingly, the fitting parameters are obtained and shown in Figure 6a,b. In Figure 6a, it can be found that the damping rate of the dark mode γ_1 has a significant increase from 0.025 THz for the case of no graphene to 0.65 THz for the case of Fermi level of 1.2 eV, whereas the fitting parameters γ_2, κ, and δ remain roughly unchanged. This phenomenon indicates that the increased Fermi level of strip 2 leads to an increased damping γ_1 at BDSSRs. In this design, as the Fermi level increases, the conductivity of the graphene strip connecting the two SSRs increases. When the Fermi level is 1.2 eV, the LC resonance at BDSSRs is hindered. Consequently, the destructive interference between BDSSRs and CW is weakened and peak I disappears.

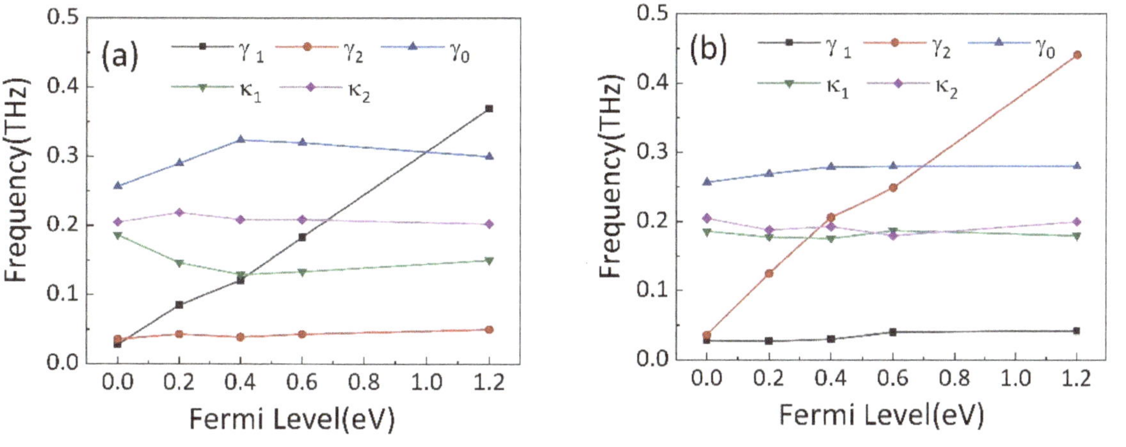

Figure 6. The variations of γ_1, γ_2, γ_0, κ_1 and κ_2 with different Fermi levels of (**a**) strip 2 and (**b**) strip 1.

On the other hand, when the Fermi level of strip 1 is changed from 0.2 eV to 1.2 eV, in Figure 6b, we can see the fitting parameters γ_1, κ and δ remain basically unchanged, whereas the damping rate γ_2 of dark mode increases significantly from 0.025 THz to 0.6 THz with the changing of Fermi level from 0.2 eV to 1.2 eV. This phenomenon can be explained by a similar principle; namely, as the Fermi level of increases, the increase in the conductivity of strip 1 reduces the intensity of LC resonance caused by the coupling of UDSSRs and CW, resulting in the weakening of destructive interference. The increase in damping rate γ_2 eventually leads to a disappearance in peak II.

In order to further explain the physical mechanism of the tunable metamaterials, in Figure 7, we present the distributions of the electric field and charge at resonance peak I and peak II. The electric field and charge distributions at peak I with different Fermi levels of strip 2 are shown in Figure 7a–f. In the absence of strip 2, as shown in Figure 7a,d, a strong electric field and accumulation of opposite charges are observed at the splits of BDSSRs. Thus, the dark mode at BDSSRs provides weak damping. When placing strip 2 under the BDSSRs and changing the Fermi level to 0.4 eV, the distribution of the electric field and charge are shown in Figure 7b,e. It is obvious that the electric field and opposite charge distribution at the splits decreases, while the electric field at CW and the opposite charge increases. This is due to the fact that the charges that accumulated opposingly at the splits are neutralized by the conductive graphene, which causes the weakening of the LC resonance at BDSSR, resulting in the weakening of the destructive interference between LC resonance at BDSSR and LSPR at CW. When the graphene Fermi level is 1.2 eV, the simulation result is shown in Figure 7c,f. We can see the electric field enhancement and the opposite charge accumulation at BDSSRs almost disappear. However, the electric field enhancement at CW is recovered and the opposite charges are re-accumulated on both edges and corners of CW. The reason is that the graphene strip almost completely neutralizes the opposite charges at the BDSSRs splits caused by the strong recombination effect of the monolayer graphene and the damping of the dark BDSSRs is too large to support LC resonance.

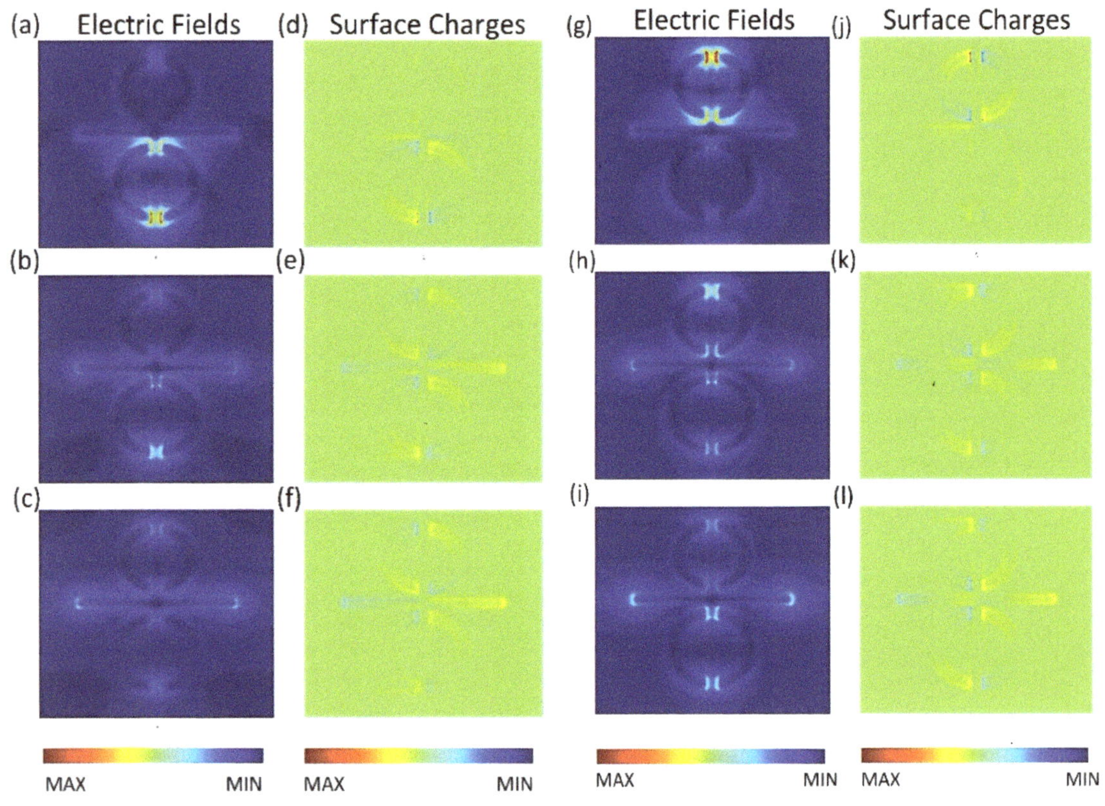

Figure 7. The distribution of (**a**–**c**) electric field and (**d**–**f**) surface charge at the PIT resonance with different strip 2 Fermi levels of No Gra, 0.4 eV and 1.2 eV. The distribution of (**g**–**i**) electric field and (**j**–**l**) surface charge at the PIT resonance with different strip 1 Fermi level of No Gra, 0.4 eV and 1.2 eV.

Therefore, it can be concluded that the damping enhancement caused by the increase in the graphene Fermi level weakens the LC resonance mode at BDSSRs and the LSPR mode, which disappears due to the destructive interference between CW and BDSSRs recovering gradually.

Figure 7g–l show the distributions of the electric field and charge at peak II with different Fermi levels of strip 1. In Figure 7g,j, when the graphene strip is not implanted into UDSSRs, a strong electric field and opposite charges are concentrated at the splits and the dark UDSSRs exhibit a low damping when placing strip 1 under the UDSSRs and changing the Fermi level to be 0.4 eV. From Figure 7h,k, it can be found that the intensity of the electric field at the UDSSRs splits, becomes weaker, and the charge density decreases, whereas the electric field intensity and charge density of CW increase gradually. With the maximum Fermi level of 1.2 eV, the simulation result is shown in Figure 7i,l, the opposite charges at UDSSRs are almost completely neutralized and the intensity of the electric field almost disappears, resulting in the disappearance of LC resonance. The intensity of the electric field and the opposite charge at CW are further recovered to achieve a strong LSPR. This is due to the fact that the high Fermi level of graphene causes the damping at UDSSRs to be too large to support the LC resonance modes, and leads to a disappearance of the destructive interference between the UDSSRs and CW. Therefore, the origin of the independent modulation of two PIT resonance can be attributed to the actively tunable Fermi level of the graphene strips under the BDSSRs and UDSSRs.

Previous studies have shown that the PIT effect is usually accompanied by the changing of dispersion properties and causes the light to slow down. Generally, group delay can be used to describe the slow light effect quantitatively [37], which can be described as:

$$t_g = \frac{d\varphi}{d\omega} \tag{10}$$

where φ and $\omega = 2\pi f$ are the transmission phase shift and frequency. The temperature of STO in this part is maintained at 425 K. Figure 8a,b shows the group delay of devices with different Fermi levels of strip 2 and strip 1, and it is evident that two parts of group delay achieve a good modulation. In the absence of graphene, the phase produces a steep jump at two transparent windows generated by PIT effect, resulting in the group delay of 1.47 ps and 1.15 ps. When strip 2 is placed under BDSSRs, the group delay change of Fermi level from 0.01 eV to 0.2 eV is shown in Figure 8a. It can be found that with the increase in Fermi level, the group delay generated at peak I gradually decreases. Finally, when the Fermi level is 0.2 eV, the group delay at peak II is only 0.45 ps. However, the group delay at peak II decreases slightly and still maintains the group delay of 0.98 ps.

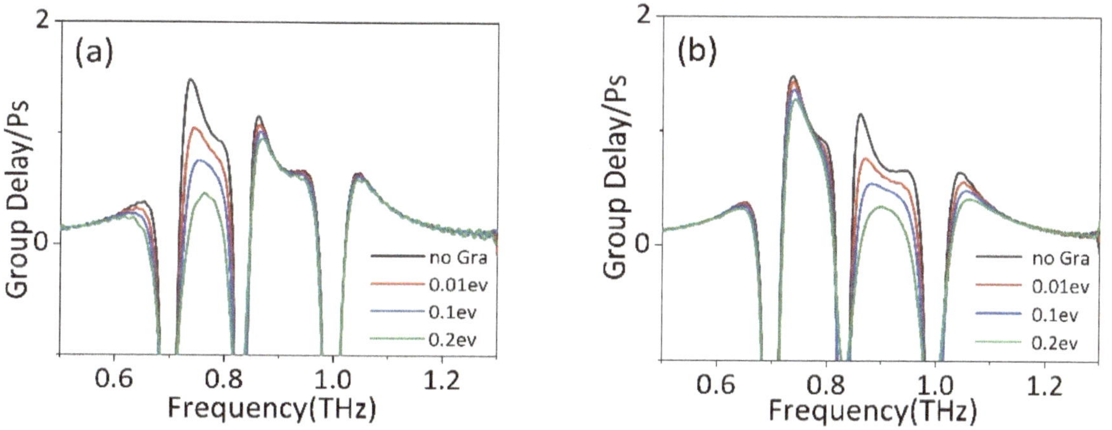

Figure 8. The group delay of PIT metamaterial with different Fermi level of (**a**) strip 2 and (**b**) strip 1.

Similarly, as shown in Figure 8b, the group delay at peak II decreases gradually by increasing the Fermi levels of strip 1 from 0.01 eV to 0.2 eV. When the Fermi level increases to 0.2 eV, the group delay at peak II gradually decreases to 0.32 ps. However, the group delay at peak I also decreases slightly and still maintains the group delay of 1.27 ps. Therefore, this design can modulate two slow light effects independently and continuously by shifting the graphene Fermi level, which is of great research significance for devices with independent tunable dual slow light.

In this section, we discuss the influence of temperature on the PIT effect. In the absence of strip 1 and strip 2, by modulating the temperature of STO, the changing of transmission spectrum is shown in Figure 9a. It can be found that peak I and peak II show blue shift. Specifically, as the temperature increases from 275 K to 425 K, the frequency of peak I moves from 0.76 THz to 0.87 THz, and the frequency of peak II moves from 0.88 THz to 1.01 THz.

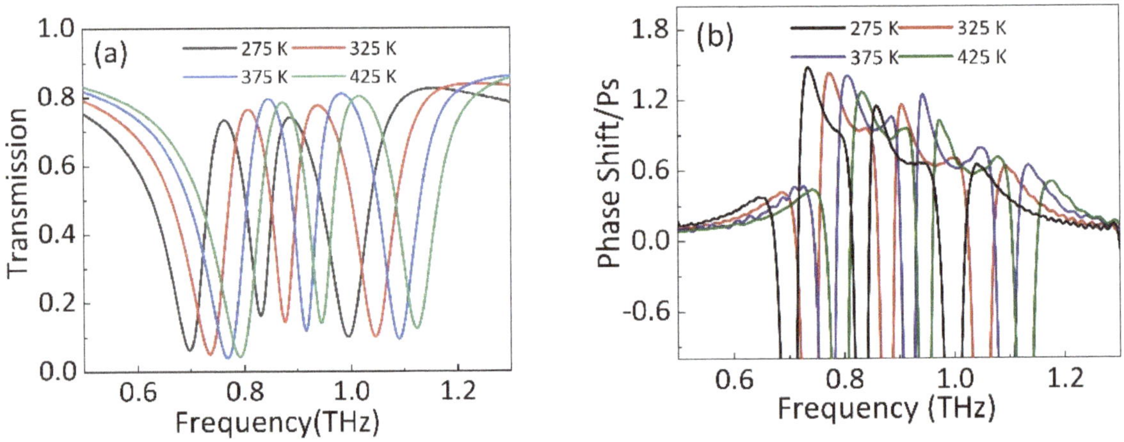

Figure 9. In the absence of graphene strips, (**a**) the transmission and (**b**) group delay of PIT metamaterial with different temperature of STO film.

Figure 9b shows the frequency change of group delay by tuning the STO temperature. When the temperature of STO film increases, the two parts of group delay caused by double PIT effect can achieve blue shift with increasing temperature. Specifically, as the temperature increases from 275 K to 425 K, the peak frequency of the two group delay moves from 0.73 THz to 0.83 THz and 0.85 THz to 0.97 THz, respectively. Therefore, the frequency selection function of double PIT windows and group delay can be realized by tuning the temperature of STO film.

Since the frequency of the PIT peak is affected by the LC resonance produced by the dark mode DSSRs, the DSSRs can be regarded as a frequent-selective surface and the resonance frequency can be estimated by [38,39]:

$$f = \frac{c}{2\pi R\sqrt{\varepsilon_{eff}}} \tag{11}$$

where R is the effective radius of the DSSRs, and R is constant because of the fixed parameters of DSSRs. ε_{eff} and c are effective dielectric constants of STO and the light speed in free space, respectively. In simulation, the ε_{eff} decreases with the temperature increasing, so the frequencies of PIT peak and group delays both cause the blue shift. Therefore, this design can not only realize the amplitude tuning of the PIT transparency windows and the slow light effect, but also realize the selection of the resonance frequency of the dual PIT transparency windows and the slow light effect.

4. Conclusions

In conclusion, we achieved the modulation of double PIT effect by integrating monolayer graphene strips and STO film into PIT metamaterials. The simulation results show that the two PIT peaks can realize the on-to-off modulation by independently shifting the Fermi level of strip 1 and strip 2. The coupling effect in the PIT metamaterial has been studied using the three-harmonic oscillator model, and the theoretical analysis shows that the recombination effect of the conductive graphene will cause the changing of dark mode damping, resulting in the tuning of the PIT peak amplitude. Through the study of the slow light effect of this metamaterial, it was found that two group delays of this metamaterial can be tuned independently. In addition, the frequency selection function of the double PIT transparency window and the double slow light is also realized by controlling the temperature of the STO layer. The multi-functional controllable metamaterial realizes the amplitude control and frequency selection of PIT transparency windows and provides a new path for future PIT control and slow light tuning devices.

Author Contributions: Conceptualization, T.W. and Y.G. (Yachen Gao); software, T.W. and Y.G. (Yang Gao); formal analysis, T.W. and G.W.; data curation, T.W. and Y.J.; writing—original draft preparation, T.W.; writing—review and editing, T.W. and Y.G. (Yachen Gao); funding acquisition, Y.G. (Yachen Gao); supervision, Y.S., C.C. and J.H. All authors have read and agreed to the published version of the manuscript.

Funding: This work was supported by the Natural Science Foundation of Heilongjiang Province (F2018027), East University of Heilongjiang Scientific Research Fund (HDFHX210110, 210111, HD-FKYTD202105) and Natural Science Foundation of Heilongjiang Province (LH2021A019).

Data Availability Statement: All content and data have been displayed in the manuscript.

Conflicts of Interest: The authors declare no conflict of interest.

References

1. Harris, S.E. Electromagnetically Induced Transparency. *Phys. Today* **1997**, *50*, 36–42. [CrossRef]
2. Miller, F.P.; Vandome, A.F.; Mcbrewster, J. Electromagnetically induced transparency: Optics in coherent media. *Rev. Mod. Phys.* **2005**, *77*, 633–673.
3. Lu, X.Q.; Shi, J.H.; Liu, R.; Guan, C. Highly-dispersive electromagnetic induced transparency in planar symmetric metamaterials. *Opt. Express* **2012**, *20*, 17581–17590. [CrossRef]
4. Liu, N.; Weiss, T.; Mesch, M.; Langguth, L.; Eigenthaler, U.; Hirscher, M.; Sonnichsen, C.; Giessen, H. Planar metamaterial analogue of electromagnetically induced transparency for plasmonic sensing. *Nano Lett.* **2010**, *10*, 1103–1107. [CrossRef]
5. Papasimakis, N.; Fedotov, V.A.; Zheludev, N.I.; Prosvirnin, S.L. Metamaterial Analog of Electromagnetically Induced Transparency. *Phys. Rev. Lett.* **2008**, *101*, 53903. [CrossRef]
6. Zhu, Z.H.; Yang, X.; Gu, J.Q.; Jiang, J.; Yue, W. Broadband plasmon induced transparency in terahertz metamaterials. *Nanotechnology* **2013**, *24*, 214003. [CrossRef]
7. Yang, Y.M.; Kravchenko, I.I.; Briggs, D.P.; Valentine, J. All-dielectric metasurface analogue of electromagnetically induced transparency. *Nat. Commun.* **2014**, *5*, 5753. [CrossRef]
8. Hu, S.; Liu, D.; Yang, H.L. Electromagnetically induced transparency in an integrated metasurface based on bright–dark–bright mode coupling. *J. Phys. D Appl. Phys.* **2019**, *52*, 175305. [CrossRef]
9. Tassin, P.; Zhang, L.; Koschny, T.; Economou, E.N.; Soukoulis, C.M. Planar designs for electromagnetically induced transparency in metamaterials. *Opt. Express* **2009**, *17*, 5595–5605. [CrossRef]
10. Singh, R.; Rockstuhl, C.; Lederer, F.; Zhang, W. Coupling between a dark and a bright eigenmode in a terahertz metamaterial. *Phys. Rev. B* **2009**, *79*, 144–147. [CrossRef]
11. Chen, H.; Zhang, H.Y.; Zhao, Y.K.; Liu, S.D.; Cao, M.Y.; Zhang, Y.P. Broadband tunable terahertz plasmon-induced transparency in Dirac semimetals. *Opt. Laser Technol.* **2018**, *104*, 210–215. [CrossRef]
12. Zhang, Y.D.; Li, J.; Li, H.Y.; Yao, C.B.; Yuan, P. Plasmon-induced-transparency in subwavelength structures. *Opt. Laser Technol.* **2013**, *49*, 202–208. [CrossRef]
13. Phillips, D.F.; Fleischhauer, A.; Mair, R.A.; Walsworth, L.; Lukin, M.D. Storage of light in atomic vapor. *Phys. Rev. Lett.* **2001**, *86*, 783–786. [CrossRef]
14. Liu, X. Electromagnetically induced transparency in terahertz plasmonic metamaterials via dual excitation pathways of the dark mode. *Appl. Phys. Lett.* **2012**, *100*, 36.
15. Niu, X.X.; Hu, X.Y.; Yan, Q.C.; Zhu, J.K.; Cheng, H.D.; Huang, Y.F.; Lu, C.C.; Fu, Y.L.; Gong, Q.H. Plasmon-induced transparency effect for ultracompact on-chip devices. *Nanophotonics* **2019**, *8*, 1125–1149. [CrossRef]
16. Zhao, Z.; Zheng, X.; Wei, P.; Zhang, J.; Shi, W. Localized terahertz electromagnetically-induced transparency-like phenomenon in a conductively coupled trimer metamolecule. *Opt. Express* **2017**, *25*, 24410. [CrossRef] [PubMed]
17. He, X.J.; Wang, Y.; Tao, M.N.; Yu, Y.Z.; Pei, Z.; Wang, B.H.; Yang, Y.Q.; Jiang, J.X.; Geng, Z.G. Dynamical switching of electromagnetically induced reflectance in complementary terahertz metamaterials. *Opt. Commun.* **2019**, *448*, 98–103. [CrossRef]
18. Chen, M.M.; Xiao, Z.Y.; Lu, X.J.; Lv, F.; Zhou, Y.J. Simulation of dynamically tunable and switchable electromagnetically induced transparency analogue based on metal-graphene hybrid metamaterial. *Carbon* **2020**, *159*, 273–282. [CrossRef]
19. He, X.; Liu, F.; Lin, F.T.; Shi, W. Graphene patterns supported terahertz tunable plasmon induced transparency. *Opt. Express* **2018**, *26*, 9931–9944. [CrossRef] [PubMed]
20. Shu, C.; Mei, J.S. Analogue of tunable electromagnetically induced transparency based on graphene-nanostrip in two perpendicular polarization directions. *Opt. Commun.* **2019**, *439*, 16–20. [CrossRef]
21. Xu, H.; Zhao, M.Z.; Zheng, M.F.; Xiong, C.; Zhang, B.; Peng, Y.; Li, H. Dual plasmon-induced transparency and slow light effect in monolayer graphene structure with rectangular defects. *J. Phys. D Appl. Phys.* **2018**, *52*, 025104. [CrossRef]

22. Liu, T.; Wang, H.; Yong, L.; Xiao, L.; Zhou, C.; Liu, Y.; Xu, C.; Xiao, S. Independently tunable dual-spectral electromagnetically induced transparency in a terahertz metal-graphene metamaterial. *J. Phys. D Appl. Phys.* **2018**, *51*, 415105. [CrossRef]
23. Liu, J.X.; Jin, K.L.; He, X.Y.; Zhang, W.J.; Lin, X.; Jin, Z.M.; Ma, G.H. Independently tunable dual-band plasmon induced transparency enabled by graphene-based terahertz metamaterial. *Appl. Phys. Express* **2019**, *12*, 075010. [CrossRef]
24. Liu, Z.M.; Zhang, X.; Zhang, Z.B.; Gao, E.D.; Zhou, F.Q.; Li, H.J.; Luo, X. Simultaneous switching at multiple frequencies and triple plasmon-induced transparency in multilayer patterned graphene-based terahertz metamaterial. *New J. Phys.* **2020**, *22*, 083006. [CrossRef]
25. Zhao, Y.; Li, B.; Lan, C.; Ke, B.; Qu, Z. Tunable silicon-based all-dielectric metamaterials with strontium titanate thin film in terahertz range. *Opt. Express* **2017**, *25*, 22158. [CrossRef] [PubMed]
26. Zhong, M. Design and modulation of the plasmon-induced transparency based on terahertz metamaterials. *Infrared Phys. Technol.* **2020**, *108*, 103377. [CrossRef]
27. Song, Z.Y.; Chen, A.; Zhang, J.H. Terahertz switching between broadband absorption and narrowband absorption. *Opt. Express* **2020**, *28*, 2037–2044. [CrossRef] [PubMed]
28. Wang, X.; Meng, H.; Deng, S.; Lao, C.; Wei, Z.; Wang, F.; Tan, C.; Huang, X. Hybrid Metal Graphene-Based Tunable Plasmon-Induced Transparency in Terahertz Metasurface. *Nanomaterials* **2019**, *9*, 385. [CrossRef]
29. Lee, S.H.; Choi, M.; Kim, T.T.; Lee, S.; Liu, M.; Yin, X.; Choi, H.K.; Lee, S.S.; Choi, C.G.; Choi, S.Y. Switching teraherz waves with gate-controlled active graphene metamaterials. *Nat. Mater.* **2012**, *11*, 936–941. [CrossRef]
30. Němec, H.; Kuzel, P.; Duvillaret, L.; Pashkin, A.; Dressel, M.; Sebastian, M.T. Highly tunable photonic crystal filter for the terahertz range. *Opt. Lett.* **2005**, *30*, 549–551. [CrossRef]
31. Kužel, P.; Kadlec, F. Tunable structures and modulators for THz light. *C. R. Phys.* **2008**, *9*, 197–214. [CrossRef]
32. Liu, T.T.; Zhou, C.B.; Cheng, L.; Jiang, X.Y.; Wang, G.Z.; Xu, C.; Xiao, S.Y. Actively tunable slow light in a terahertz hybrid metal-graphene metamaterial. *J. Opt.* **2019**, *21*, 035101. [CrossRef]
33. Ye, Z.L.; Zhang, S.; Wang, Y.; Park, Y.S.; Zentgraf, T.; Bartal, G.; Yin, X.B.; Zhang, X. Mapping the near-field dynamics in plasmon-induced transparency. *Phys. Rev. B* **2012**, *86*, 155148. [CrossRef]
34. Liu, Z.M.; Gao, E.D.; Zhang, Z.B.; Li, H.J.; Xu, H.; Zhang, X.; Luo, X.; Zhou, F.Q. Dual-mode on-to-off modulation of plasmon-induced transparency and coupling effect in patterned graphene-based terahertz metasurface. *Nanoscale Res. Lett.* **2020**, *159*, 1–9. [CrossRef]
35. Miyata, M.; Hirohata, J.; Nagasaki, Y.; Takahara, J. Multi-spectral plasmon induced transparency via in-plane dipole and dual-quadrupole coupling. *Opt. Express* **2014**, *22*, 11399–11406. [CrossRef] [PubMed]
36. Xiao, S.; Wang, T.; Liu, T.; Yan, X.; Li, Z.; Xu, C. Active modulation of electromagnetically induced transparency analogue in terahertz hybrid metal-graphene metamaterials. *Carbon* **2018**, *126*, 271–278. [CrossRef]
37. Hua, L.; Liu, X.; Dong, M. Plasmonic analog of electromagnetically induced transparency in multi-nanoresonator-coupled. *Phys. Rev. A* **2012**, *85*, 1–7.
38. Anonymous. Frequency selective surfaces: Theory and design [Book Review]. *IEEE Signal. Proc. Mag.* **2001**, *18*, 94. [CrossRef]
39. Qi, L.M.; Liu, C. Broadband multilayer graphene metamaterial absorbers. *Opt. Mater. Express* **2019**, *9*, 1298–1309. [CrossRef]

nanomaterials

MDPI

Article

Iron in Hydroxyapatite: Interstitial or Substitution Sites?

Leon Avakyan [1], Ekaterina Paramonova [2], Vladimir Bystrov [2], José Coutinho [3], Sandrine Gomes [4] and Guillaume Renaudin [4]

1 Faculty of Physics, Southern Federal University, 5 Zorge St., 344090 Rostov-on-Don , Russia
2 Institute of Mathematical Problems of Biology, Branch of Keldysh Institute of Applied Mathematics, Russian Academy of Sciences, 1 Vitkevicha St., Pushchino, 142290 Moscow, Russia; ekatp11@gmail.com (E.P.); vsbys@mail.ru (V.B.)
3 I3N, Department of Physics, Campus Santiago, University of Aveiro, 3810-193 Aveiro, Portugal; jose.coutinho@ua.pt
4 Université Clermont Auvergne, Clermont Auvergne INP, CNRS, ICCF, F-63000 Clermont-Ferrand, France; sandrine.renaudin@sigma-clermont.fr (S.G.); guillaume.renaudin@sigma-clermont.fr (G.R.)
* Correspondence: laavakyan@sfedu.ru; Tel.: +7-918-563-0854

Abstract: Iron-doped hydroxyapatite (Fe-HAp) is regarded as a promising magnetic material with innate biocompatibility. Despite the many studies reported in the literature, a detailed theoretical description of Fe inclusions is still missing. There is even no consensual view on what kind of Fe defects take place in Fe-HAp—iron interstitial or calcium substitutions? In order to address these questions, we employ modern first-principles methodologies, including hybrid density functional theory, to find the geometry, electronic, magnetic and thermodynamic properties of iron impurities in Fe-HAp. We consider a total of 26 defect configurations, including substitutional (phosphorus and calcium sites) and interstitial defects. Formation energies are estimated considering the boundaries of chemical potentials in stable hydroxyapatite. We show that the most probable defect configurations are: Fe^{3+} and Fe^{2+} substitutions of Ca(I) and Ca(II) sites under Ca-poor conditions. Conversely, Fe interstitials near the edge of the hydroxyl channel are favored in Ca-rich material. Substitutional Fe on the P site is also a probable defect, and unlike the other forms of Fe, it adopts a low-spin state. The analysis of Fe K-XANES spectra available in the literature shows that Fe-HAp usually contains iron in different configurations.

Keywords: iron doping; hydroxyapatite bioceramics; density functional theory; hybrid density functional; X-ray absorption spectroscopy

Citation: Avakyan, L.; Paramonova, E.; Bystrov, V.; Coutinho, J.; Gomes, S.; Renaudin, G. Iron in Hydroxyapatite: Interstitial or Substitution Sites? *Nanomaterials* **2021**, *11*, 2978. https://doi.org/10.3390/nano11112978

Academic Editor: Gregory M. Odegard

Received: 2 October 2021
Accepted: 1 November 2021
Published: 5 November 2021

1. Introduction

Ceramic materials based on calcium apatites are important substances in medicine, biology, ecology and catalysis [1–4]. Among the apatites, hydroxyapatite ($Ca_{10}(PO_4)_6(OH)_2$ = HAp) is singled out as the main inorganic component of vertebrate bones and teeth. Biogenic HAp differs from the synthetic compound due to the presence of foreign atomic (Mg^{2+}, Na^+, Cl^-) and molecular (CO_3^{2-}, SiO_4^{4-}) ions. Although the role of most impurities is largely unknown, some studies suggest that the presence of a small concentration of iron (below 157 ppm) prevents the loss of calcium from bones and desorption of enamel from teeth [5,6]. Iron-doped HAp (Fe-HAp) has also attracted interest due to its intrinsic magnetic properties [7]. It is a promising material for magneto-resonance imaging agents [8], heat centers for magnetic hyperthermia [4,9], sunscreen filter creams [10], antimicrobial coatings [11], and to be incorporated in drug delivery systems [12,13].

Despite nearly five decades of research on Fe-HAp and related systems [6,14–19], our knowledge regarding the atomistic and electronic structure and iron defects is still very limited. The main difficulty arises from the quality of the samples—single-phase products are hard to synthesize, iron metal and iron oxide often segregate at the surface, and the admixture of stable phosphates, in particular tricalcium phosphate ($Ca_3(PO_4)_2$ = TCP), is usually observed [16,18]. Discrepancies between the results from different authors are often attributed to different preparation conditions, including sintering temperature and atmosphere, the duration and temperature of doping treatments, the source of precursors, etc. A comprehensive review of experimental methods and results can be found in Ref. [6].

The HAp structure contains two symmetrically distinct positions of calcium, Ca(I) and Ca(II) (Figure 1), both of which are candidates for iron replacement. The pioneering study of Fe-HAp using Mössbauer (MB) spectroscopy by Ok [14] indicated the preference of $Fe_{Ca(I)}$ over $Fe_{Ca(II)}$, while subsequent MB studies [15,16] concluded on comparable amounts of $Fe_{Ca(I)}$ and $Fe_{Ca(II)}$ defects. In both cases, electron paramagnetic resonance (EPR) data indicated the presence of isolated high-spin Fe^{3+} ions with 4 or 6 neighbors, in coexistence with superparamagnetic iron oxide on the surface. The MB experiments of Boda et al. [6] suggest that $Fe^{3+}_{Ca(I)}$ defects are more probable when conventional sintering is used during sample preparation, whereas $Fe^{3+}_{Ca(II)}$ substitutions appear as a result of hot pressing.

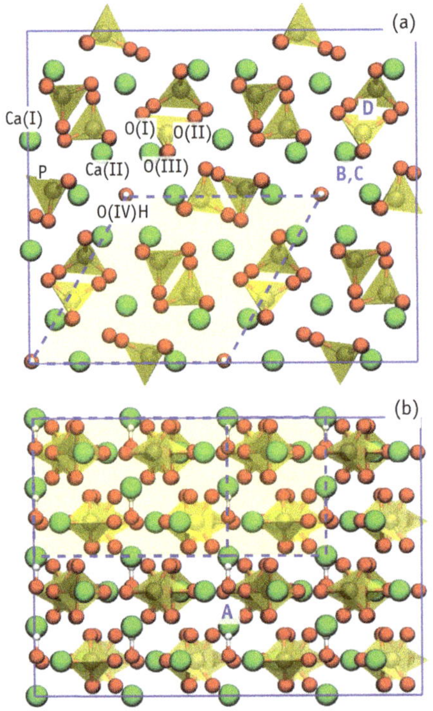

Figure 1. Top-projection (**a**) and side-view (**b**) of an orthorhombic 2 × 2 × 2 HAp supercell with the hexagonal unit cell highlighted. O, Ca and H atoms are shown in red, green and white, respectively. Each PO_4 group contains O(I), O(II) and two O(III) sites. Capital letters A-D indicate the location of the interstitial sites considered.

Interstitial iron (Fe_i) defects have been reported as well. Analysis of X-ray diffraction (XRD) data shows that iron can occupy 2*b* or 12*i* Wyckoff positions [5,18–20]. The first site corresponds to the formation of a linear O-Fe-O structure in the center of the OH channel (site A at Figure 1), while the latter involves a three-fold coordinated Fe atom displacement towards the wall of the channel (site C at Figure 1). However, from XRD and Raman scattering data, Kato et al. [19] argues on the formation of four-fold coordinated Fe. The absence of Fe_{Ca} in these experiments was postulated based on the lack of CO_3^{2-}

groups [6,16]. These are commonly present in biogenic HAp, and lead to the formation of several defects, including: $[CO_3^{2-}]_{[PO_4^{3-}]}$, $[CO_3^{2-}]_{[OH^-]}$, $[O^{2-}]_{[OH^-]}$ and, finally, $[Fe^{3+}]_{Ca^{2+}}$. Here we limit our analysis to iron-only defects, leaving carbonate-related defects outside the scope of this work.

Besides the lack of agreement between different MB studies, XRD data also holds controversy due to a low electronic contrast between Fe^{3+} and Ca^{2+} ions [18]. Hence, atomic-level simulations are regarded as a source of clarification. Using empiric interatomic potentials, Jiang et al. [15] concluded that the Fe^{2+} ion finds its favorite site at the Ca(I) position, while Fe^{3+} favors the Ca(II) site. More recently, Zilm et al. [17] investigated this problem using semi-local density functional theory, but the results were limited to the 2+ oxidation state. Despite enlightening, both studies focused on Fe_{Ca} substitutions only, with alternative substitutional and interstitial forms of iron being overlooked. Hence, we may conclude that the starting point of the present work is a largely unknown picture of Fe impurities in HAp regarding their local structure, oxidation and magnetic states, and optoelectronic properties.

Recently we demonstrated [21] that the use of hybrid density functional theory (hybrid-DFT) can describe the electronic properties of HAp with an accuracy that rivals with highly accurate Green's function (G) method with screened interaction (W), so called GW approximation [22]. The GW and hybrid-DFT methods provided band gaps of HAp close to 7 eV, but generalized gradient functional resulted in a 30 % smaller gap [23–28]. Errors of the same magnitude are therefore expected for the calculated optoelectronic properties of defects in HAp, as shown for the case of the OH vacancy [21].

In view of the above, we present a hybrid-DFT study of the most probable Fe-related defects in hydroxyapatite (substitutional and interstitial). The results are interpreted and scrutinized from the perspective of the available experimental data. The paper is organized as follows: Section 2 outlines the methodologies and computational details. Section 3 discusses the results, subdivided in Subsections pertaining the chemical phase diagram of HAp (Section 3.1), the structure and thermodynamics of neutral and charged defects (Sections 3.3 and 3.4, respectively), and a simulation of the Fe K-XANES spectra of Fe-HAp available in the literature (Section 3.5). Finally, Section 4 summarizes the results.

2. Computation Methods

HAp crystallizes in the form of an molecular ionic crystal of symmetry $P6_3/m$ (#176 in the crystallographic tables), whose hexagonal unit cell encloses two $Ca_5(PO_4)_3OH$ units [29,30]. Figure 1 illustrates the supercell used in this study, comprising 8 unit cells (16 HAp formula units). HAp primitive cell contains two and four inequivalent calcium and oxygen sites , respectively. The Ca(I) cation columns are surrounded by O(I) and O(II) from PO_4 anion groups, while mirror-symmetric O(III) sites and Ca(II) ions form a hexagonal channel enclosing the hydroxyl anions O(IV)H. Variation of alignment of OH dipoles in these channels lead to different HAp phases. Namely: (i) a hexagonal disordered phase, with random orientations of OH dipoles; (ii) a hexagonal ordered phases, where OH dipoles are all oriented along the same direction, or (iii) a monoclinic ($P2_1/b$) phase, made of two cells repeated along a basal lattice vector (**a** or **b**), with the first possessing a ...-OH-OH-... column, while the second one showing an opposite ...-HO-HO-... ordering [31]. The last phase shows anti-ferroelectric properties [32]. Given that the electronic band structure of the above polytypes is rather similar [27], the impact of OH-flipping on the problems being investigated is expected to be minor. Henceforth, we will consider the introduction of iron defects in the ordered hexagonal phase.

2.1. First-Principles Defect Energetics

The calculations were performed using density functional theory using the QUANTUM ESPRESSO package [33,34]. The many-body electronic potential was evaluated using the hybrid density functional of Heyd, Scuseria and Ernzerhof (HSE) [35,36]. The Kohn-Sham eigenstates obtained using PBE [37] exchange-correlation functional were used to initiate the self-consistent cycle. Due to severe limitations of the semi-local functionals (such as PBE) in describing the electronic structure of HAp in the band gap vicinity [21], the application of hybrid-DFT turns out to be critical in the evaluation of the stability of charged defects.

By opting for a local treatment of the electronic correlation, we expect some error in the description of the d-shell in Fe. Using a GGA+U approach would lead to a narrow gap material, and hence to an artificial hybridization between the d-levels of Fe and those from the conduction and valence band edges of HAp. Unfortunately, it was not possible to combine the non-local HSE functional, with the Hubbard-like local correction to the electronic correlation on the Fe atom.

Core electron states were described by means of optimized norm-conserving Vanderbilt pseudopotentials [38,39], while the Kohn-Sham problem was solved within the plane wave formalism with cut-off energy up to 60 and 240 Ry to expand the wave functions and semi-local potential. The exact exchange operator was evaluated on a grid which corresponded to a plane wave cut-off of 120 Ry. The occupation of states in the vicinity of the Fermi energy were smeared out with a Gaussian function of width 0.002 Ry to improve convergence.

The equilibrium lattice parameters of the unit cell were obtained $a = 9.481$ Å and $c = 6.859$ Å using HSE exchange-correlation functional. These values are particularly close to $a = 9.417$ Å and $c = 6.875$ Å from an experimental report [30].

Defects were introduced in orthorhombic supercells made up of 8 HAp primitive cells (spacegroup $P6_3/m$), containing a total of 352 atoms. The smooth dispersion of the band structure of the large HAp supercell allowed us to use a single point ($\mathbf{k} = \Gamma$) for sampling the Brillouin zone (BZ). Convergence tests showed that the total energy of such supercells obtained with Γ only sampling, differs by less than 0.1 eV (0.3 meV/atom) from a calculation where the zone was sampled over a $2\times2\times2$ grid of $\mathbf{k}$-points.

Defect containing supercells were first relaxed on a PBE level, since claculation of forces on a hybrid-DFT level is prohibitively expensive for large systems. During the relaxations, coordinates of all atoms were varied until the maximum force became less than 0.4 mRy/Bohr $\approx$ 10 meV/Å. On a second step the total energy was calculated on hybrid-DFT level for obtained structures. It was shown [40–42] that relative errors in energies obtained within this methodology usually have an order of 10 meV.

The formation energy E_f of a defect in a crystalline sample can be expressed as [43]:

$$E_\mathrm{f} = E_\mathrm{d}(q, R) - E_\mathrm{HAp} - \sum_i \Delta n_i \mu_i + q(E_\mathrm{v} + E_\mathrm{F}), \tag{1}$$

where the main contribution is the energy difference between defective (E_d) and pristine (E_HAp) crystals. R and q denotes the defect configuration and charge state, correspondingly. Last two terms account for stoichiometric and charge differences between the defect and pristine cells. In particular, μ_i is the chemical potential of species i which must be added ($\Delta n_i > 0$) or removed ($\Delta n_i < 0$) to or from the ideal crystal to create the defect, respectively. The last term in Equation (1) accounts for the exchange of electrons between the defect with charge q and an *electron reservoir* with chemical potential $\mu_e = E_\mathrm{v} + E_\mathrm{F}$. Here E_v and E_F denotes the energy of the valence band top and Fermi level, respectively. The Fermi energy can vary within the band gap ($0 \leq E_\mathrm{F} \leq E_\mathrm{g}$), where E_g is the band gap width), depending on the doping of the crystal.

The variation of chemical potentials μ_i is also limited,

$$\sum_i n_i^\phi \cdot \left(\mu_i - \mu_i^0\right) \leq \Delta H_\mathrm{f}^\phi, \tag{2}$$

with the upper limit taking place if HAp becomes unstable with respect to formation of a compound ϕ made of n_i^ϕ elements of species i and with heat of formation ΔH_f^ϕ. The chemical potentials of O, H, Ca and P in standard phases, μ_i^0, were calculated from the energy per atom in molecular oxygen, molecular hydrogen, calcium metal and black phosphorous. The molecules (O_2 in the spin-triplet state or H_2) were placed in a simulation cubic box of 20 Å size. The chemical potential of iron was found from the *bcc*-Fe phase with the BZ sampled over a $16\times16\times16$ **k**-point mesh. The resulting ground state spin density corresponded to a magnetic moment of 2.80 μ_B per cell. The deviation from the experimental value of 2.22 μB results from a known over-localization of the exchange interactions in bulk Fe when using hybrid functionals [44]. An identical result was obtained using projected augmented-wave potentials. Nevertheless, it is noted that hybrid functionals have successfully been used for the study of iron oxides and defects in oxides (see for instance [45,46]).

The heat of formation of the compounds ϕ used in Equation (2) was estimated from their hybrid-DFT total energies (E^{ϕ}) with respect to energies of their constituents in their standard phases (μ_i^0),

$$\Delta H_f^{\phi} = E^{\phi} - \sum_i n_i^{\phi} \mu_i^0. \tag{3}$$

Due to periodic boundary conditions, the calculation of charged defects ($q \neq 0$) is accompanied by a compensating uniform charge density of opposite sign [47]. The artificial interactions between the periodically repeated charged defects and the background lead to the deviation of calculated periodic total energy $\widetilde{E}_d$ from E_d, in Equation (1). The energy correction E_{corr}, so that $E_d = \widetilde{E}_d + E_{corr}$, is obtain following the recipe of Lany and Zunger [48],

$$E_{corr} \approx \frac{2}{3} \frac{\alpha_M q^2}{\epsilon L}, \tag{4}$$

where α_M is the Madelung constant of the HAp supercell with edge length L, and ϵ is the static dielectric constant ($\epsilon \approx 11$ [21]). For a singly charged defect the correction amounts to about 0.04 eV. However, this figure can grow considerably in the case of multi-ionized defects ($E_{corr} \approx 0.33$ eV for $q = +3$).

The magnetization of the Fe defects was calculated from spin-polarized electron densities $n_\uparrow(\mathbf{r})$ and $n_\downarrow(\mathbf{r})$,

$$M_z = \int_\Omega (n_\uparrow - n_\downarrow)\, d^3r, \tag{5}$$

while formal oxidation states of iron ions were deduced from the occupations of d-orbitals as proposed by Sit et al. [49].

2.2. X-ray Absorption Near Edge Structure (XANES)

In order to compare the defect structures found from the first-principles total energy calculations, with those reported in the literature, we simulated the Fe K-XANES spectra for each structure. The calculations were performed using a full-multiple scattering method and the "muffin-tin" approximation for the interatomic potential as implemented in the FDMNES code [50]. The size of the atomic cluster and spectral convolution parameters were adjusted using the spectra of magnetite Fe_3O_4 as reference. The radius of the cluster was estimated of 7 Å which gives cluster of $Fe_1Ca_{\sim30}P_{\sim15}O_{\sim70}H_{\sim5}$, with exact composition dependent on the structure. The used approach cannot reproduce pre-edge features, which is a well known limitation of a description based on dipolar transitions and the single-particle approximation [51,52].

3. Results and Discussion

3.1. Chemical Stability Diagram

The formation of the most probable iron-related defects in HAp depends most notably on the chemical potentials of the several elements involved. Each chemical potential can vary within a certain range, limited by thermodynamic stability conditions of the HAp crystal itself. We estimate the ranges using Equation (2) with respect to a set of boundary phases $\{\phi\}$. The methodology used to find the structure and energy of each phase was identical to that used for the HAp supercell. This includes the exchange-correlation functional and energy cutoffs. The first candidates for the bordering phases are those involved in the HAp synthesis. There is a variety of production routes, but we will only consider reactants whose species are common to those found in HAp, namely CaO [6], $Ca(OH)_2$ [8], H_3PO_4, H_2O and P_2O_5 [18]. Next, we considered calcium phosphates: dicalcium phosphate dihydrate $CaHPO_4 \cdot 2H_2O$ = DCPD (mineral brushite), anhydrous dicalcium phosphate $CaHPO_4$ = DCPA (mineral monetite), tricalcium phosphate $Ca_3(PO_4)_2$ = TCP and tetracalcium phosphate $Ca_4(PO_4)_2O$ = TTCP. The last two have [Ca] : [P] concentration ratio of 1.5 and 2.0, respectively, while HAp has an intermediate [Ca] : [P] ratio of $5/3 \approx 1.67$.

Table 1 presents the formation energies ΔH_f, calculated according to Equation (3) within HSE, for the materials enumerated above. The overall agreement with the reference data [53,54] (including the energy, cell volume and bulk modulus) is in line with the usual accuracy of hybrid-DFT. The phosphoric acid and water were calculated in a gas phase (single molecules in large periodic box), hence the lack of cell volume and bulk modulus.

Table 1. Comparison of calculated heats of formation ΔH_f (eV/formula unit), cell volumes Ω (Å^3) and bulk modulus B (GPa) of compounds with available reference data [53–57].

Compound	HSE			Ref.		
	ΔH_f	Ω	B	ΔH_f	Ω	B
HAp	−64.40	532.1	87	−69.44	528.7	89
CaO	−6.00	110.4	114	−6.58	110.5	116
$Ca(OH)_2$	−9.63	56.46	31	−10.23	54.78	33
P_2O_5	−14.22	330.7	64	−15.60	322.0	12–40
H_3PO_4	−11.52	–	–	−13.35	–	–
H_2O	−2.61	–	–	−2.97	–	–
TCP	−39.2	445.5	105	−42.8	445.5	100
TTCP	−45.12	796.0	79	−49.4	797.3	–
DCPD	−23.89	490.3	51	−22.33	497.7	–
DCPA	−17.41	311.9	75	–	309.3	–

From the data of Table 1 and Equation (2), we can estimate the chemical potential ranges within which HAp is thermodynamically stable. We can also reduce the number of independent chemical potentials, for instance to μ_H, μ_{Ca} and μ_P, by expressing the chemical potential of oxygen as a function of μ_i of the remaining elements (c.f. Equation (2)),

$$\mu_O - \mu_O^0 = \left[\Delta H_f^{HAp} - \sum_{i=Ca,\,P,\,H} n_i^{HAp}\left(\mu_i - \mu_i^0\right)\right] / n_O^{HAp}.$$

Let us look first at the range of hydrogen chemical potentials. In this case, phases with stoichiometry (not) satisfying $n_H^{\phi} \cdot n_O^{HAp} < n_O^{\phi} \cdot n_H^{HAp}$ give lower (upper) bounds for μ_H. In particular, the planes corresponding to CaO, TCP and pure O_2 may form lower bounds, while planes for hydrogen-containing phases may be responsible for upper bounds.

According to our calculations, the HAp stability domain is limited by CaO, $Ca(OH)_2$, P, H_2, DCPD and DCPA phase planes. Figure 2 illustrates the domain of chemical stability of HAp as a convex faceted hull with vertices P_i (with $i = 1, \ldots, 11$). These correspond to regions in phase space where three different phases coexists with HAp. Table S1 in Supplementary Materials contains the calculated values of chemical potentials at these intersection points. We note that there is no boundary with Ca metal, with the upper limit for $\mu_{Ca} - \mu_{Ca}^0$ being −1.8 eV . This means that if $\mu_{Ca} = \mu_{Ca}^0$ was assumed, formation energies of defects involving Ca substitutions would be affected by an error of $\sim$ 2 eV. In the analysis regarding substitutional iron defects, we will consider the following chemical phase space conditions:

1. Ca-rich and P-rich, points P_7 and P_8, $\mu_{Ca} - \mu_{Ca}^0 = -1.8$ eV, $\mu_P - \mu_P^0 = 0$;
2. Ca-poor, point P_3, $\mu_{Ca} - \mu_{Ca}^0 = -7.8$ eV, $\mu_P - \mu_P^0 = -7.9$ eV;

3. P-poor, points P_1 and P_2, $\mu_{Ca} - \mu^0_{Ca} = -6.0$ eV, $\mu_P - \mu^0_P \approx -10.7$ eV.

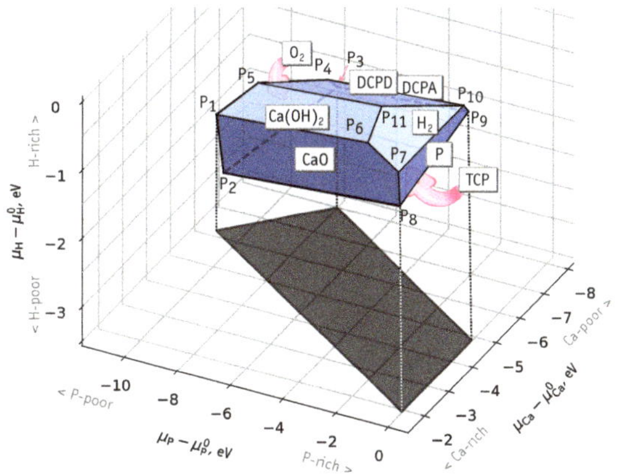

Figure 2. Domain of chemical stability of HAp represented by the limits of chemical potentials of Ca, P, H. Symbols P_1–P_{11} correspond to points where three boundary phases coexists with the HAp crystal.

3.2. Notation for Defect Structures

Before describing the structure of the iron defects let us first establish a few notation rules. We denote the defect structure as M^n_S, where M is a substituting species, most notably iron. However, this can also be a vacancy V or group of atoms (e.g., OH). S denotes the position of the defect, which can be a crystalline atomic site (Ca(I), Ca(II) or P) or an interstitial site (i(A), i(B), etc). The flipping of an hydroxyl unit is denoted as OH_{HO}. The number n is the formal oxidation state of iron, estimated from the occupation of the d-orbitals [49]. Table 2 shows the correspondence between the used notation, Kröger-Vink notation and the net charge of the defective supercell. We found a systematic correspondence between the formal oxidation state of iron atom and the defect charge state. This indicates that the gap states of the Fe defects under investigation (which can trap electrons or holes) belong to the 3d shell.

Table 2. Correspondence between the defect charge state q (in units of e), Kröger-Vink notation and that used in this work (right-most column).

q	Notation	
	Kröger-Vink	**This Work**
0	$Fe_i^{\times}$	Fe_i^0
+1	$Fe_i^{\bullet}$	Fe_i^{+}
+2	$Fe_i^{\bullet\bullet}$	Fe_i^{2+}
+3	$Fe_i^{\bullet\bullet\bullet}$	Fe_i^{3+}
0	$Fe_{Ca}^{\times}$	Fe_{Ca}^{2+}
+1	$Fe_{Ca}^{\bullet}$	Fe_{Ca}^{3+}
−2	$Fe_P^{\prime\prime}$	Fe_P^{3+}
−1	$Fe_P^{\prime}$	Fe_P^{4+}
0	$Fe_P^{\times}$	Fe_P^{5+}
+1	$Fe_P^{\bullet}$	Fe_P^{6+}

3.3. Neutral Iron Defects

We start by considering neutral iron defects based on structures previously proposed in the literature. Table 3 summarizes the relevant data collected. The first seven rows correspond to the Fe_{Ca} substitutions, while the next five to iron interstitials. There is no apparent correlation between the defect structure and sample preparation conditions (e.g., Fe concentration or synthesis temperature).

We considered Ca(I), Ca(II), as well as P substitutions, the later being unexplored in the literature. Figure 3a–c illustrate the substitutional defect structures obtained after relaxation, while Figure 3d–g depict the interstitial structures. The relaxation of the structure where iron was initially set up with the $Fe_{i(A)}$ configuration (Wyckoff position 2*b*), resulted in another structure, where iron is displaced away from the center of the channel. This new structure is denoted as $Fe_{i(C)}$ (Figure 3f) and approximately corresponds to Wyckoff position 12*i*. In order to achieve a stable $Fe_{i(A)}$ configuration, one of the nearest OH groups has to be flipped (Figure 3d). Another interstitial position with the iron ion displaced from the center of the channel is labeled as $Fe_{i(B)}$ (Figure 3e) and corresponds to Wyckoff position 2*c*. The $Fe_{i(C)}$ structure may be slightly modified by flipping of the neighboring OH group, leading to a minute ($\sim$0.1 eV) benefit in the total energy. Such small difference is lower than the flipping of isolated OH, estimated as 0.22 eV. Our result suggest that OH flipping may be stimulated by the presence of interstitial iron in the channel.

Table 3. List of proposed structures for Fe defects in Fe-HAp. Also included (when available) are the iron-oxygen interatomic distances $R_{\text{Fe-O}}$, numbers of oxygen neighbors $N_{\text{Fe-O}}$, number of Fe atoms per HAp unit cell n_{Fe}, temperatures of sample preparation (doping treatment or sintering) and detection methods.

Structure	$R_{\text{Fe-O}}$, Å	$N_{\text{Fe-O}}$	n_{Fe}	$T_{\text{prep.}}$, °C	Methods	Ref.
$Fe_{Ca(I)}$				1000	MB	[14]
$Fe_{Ca(II)}$			$\geq$0.15		XRD	[58]
$Fe_{Ca(I)}$	2.2–2.3	6	0.5	90	MD, EPR, MB	[15]
Fe_{Ca}	2.4–2.5	6	$\leq$0.5	600–1000	EPR, MB	[16]
Fe_{Ca}			0.2	40	XRD, XAS	[9]
$Fe_{Ca(II)}/Fe_{Ca(I)}$		5 or 6/6	0.5–2.0	950	XRD, XPS, MB	[6]
$Fe_{Ca(II)}/Fe_{Ca(I)}$	2.0–2.2/2.2–2.3	4 / 6	0.012	25	XRD, DFT	[17]
$Fe_{Ca(II)}$			0.3–6.0	100	XPS, XRD	[59]
FeOOH				biogenic/120	MB	[5,60]
$Fe_{i(A)}$	1.7	2	0.15–0.75	<1000	XRD	[18]
$Fe_{i(C)}$	1.80–1.85	3	0.15–0.75	1100	XRD	[18]
$Fe_{i(C)}$	1.84–1.94	3 ± 0.7	0.15	1100	EXAFS	[18]
$Fe_{i(C)}$	1.8–2.4	4	0.1–0.9	60	XRD, IR	[19]

Finally, we considered iron inserted in the region between PO_4 groups (Wyckoff position 6*g*), marked as $Fe_{i(D)}$ (Figure 3g). At this location, the negatively charged PO_4^{3-} groups are expected to screen the positive charge of the iron ion. Although we scanned the relative stability of other structures, those whose formation energy was above 7 eV were discarded and not investigated further (e.g., Fe_{OH} and Fe_i in the vicinity of the Ca(I) column).

Table 4 shows the formation energies of the most stable defects, estimated according to Equation (1) using chemical potentials at extreme points of the HAp stability diagram, namely for material grown under Ca- and P-rich (P_7 and P_8), Ca-poor (P_3) and P-poor (P_1) conditions. At Ca- and P-rich conditions, the $Fe^0_{i(A)}$ defect is likely to be the most probable as it shows the lowest formation energy. Other stable defects (within less that 1 eV above the ground state) are $Fe^0_{i(B)}$ and Fe^{2+}_{Ca} substitutions. Other interstitial sites ($Fe^0_{i(C)}$, $Fe^0_{i(D)}$) have higher formation energies. However, they still should be considered since hole or electron trapping may stabilize them.

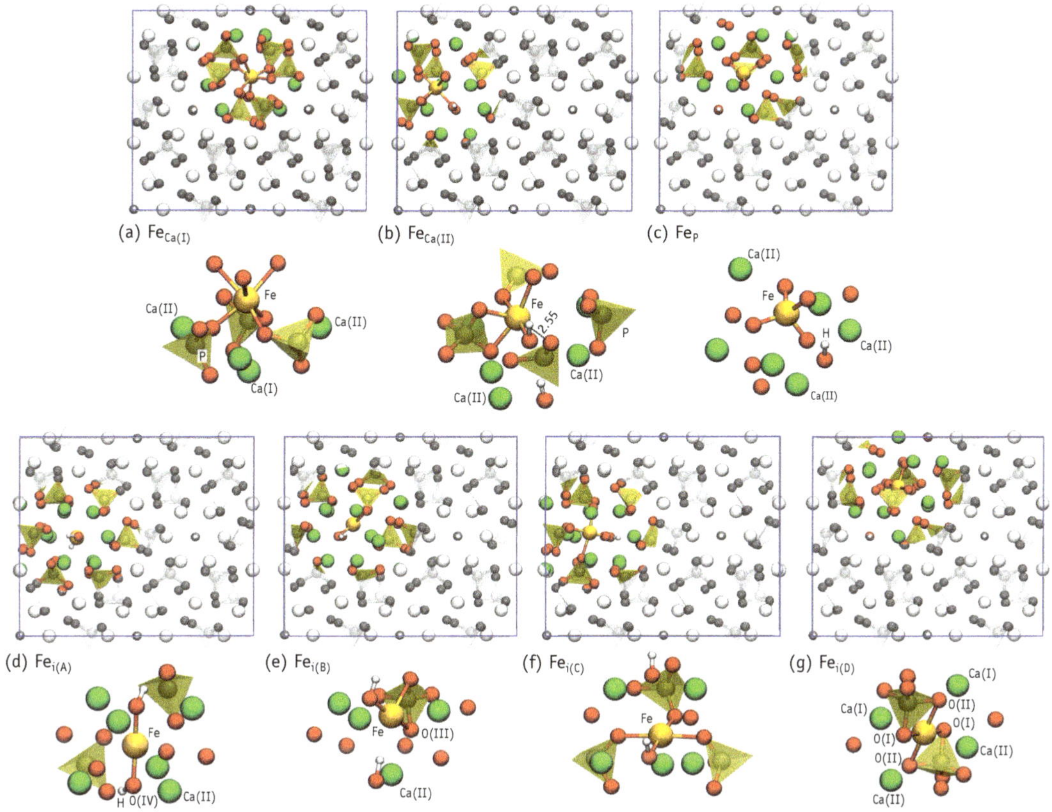

Figure 3. Illustration of neutral defect structures in Fe-HAp. For each defect, we show a top view of its location within supercell and a lateral view of its detailed atomic structure in the vicinity of ~4 Å around the iron atom.

Table 4. Formation energies (E_f in eV) under extreme growth conditions of HAp, number of nearest oxygen neighbors ($N_{\text{Fe-O}}$), Fe-O bond length ($R_{\text{Fe-O}}$ in Å) and magnetic moment (M_z in μ_B units) calculated for several defect configurations in the neutral charge state, $q = 0$.

Defect Structure		E_f		$N_{\text{Fe-O}}$	$R_{\text{Fe-O}}$	M_z
	Ca- & P-Rich	Ca-Poor	P-Poor			
$Fe^{2+}_{Ca(I)}$	5.25	−0.76	1.02	6	2.04–2.24	0.0
$Fe^{2+}_{Ca(II)}$	5.05	**−0.96**	0.82	5	1.99–2.21	0.0
Fe^{5+}_{P}	6.05	−0.09	**−3.03**	4	1.69–1.70	1.0
$Fe^{0}_{i(A)}$	**4.48**	4.48	4.48	2	1.88–1.89	2.0
$Fe^{0}_{i(B)}$	5.24	5.24	5.24	3	1.86–2.22	2.0
$Fe^{0}_{i(C)}$	6.05	6.05	6.05	4	1.94–2.18	2.0
$Fe^{0}_{i(D)}$	6.50	6.50	6.50	4	2.02–2.06	2.0

At Ca-poor conditions Fe^{2+}_{Ca} defects are the most stable, with formation energy ~5 eV below that of $Fe^{0}_{i(A)}$. A low value of μ_P at Ca-poor conditions (P_3 in Figure 2) leads to an easier depletion of phosphorus and to the stabilization of Fe^{5+}_{P} as well. Subsequently, at P-poor conditions this effect is further enhanced and Fe^{5+}_{P} becomes nearly 4 eV more stable than Fe^{2+}_{Ca}, and about 7.5 eV than the most favorable interstitial defect, $Fe^{0}_{i(A)}$.

Table 4 reports the number of oxygen atoms neighboring iron (coordination number), their respective Fe-O distances, and the magnetic moment of each defect. The six-fold coordination of $Fe_{Ca(I)}$ (Figure 3a) is in line with the results of Jiang et al. [15], Low et al. [16] and Zilm et al. [17]. The five-fold coordinated $Fe_{Ca(II)}$ structure differs: six-fold coordinated iron is proposed by Jiang et al, while four-fold – by Zilm et al. However, we may count six neighbors if we include the far oxygen neighbor at distance $R_{\text{Fe-O}} = 2.55$ Å marked with dashed line on Figure 3b. Zilm et al obtained two oxygen atoms at high distances of $R_{\text{Fe-O}} = 2.73$ and 2.97 Å, but excluded them from the neighbors count, resulting in four-fold coordination.

The structure of $Fe_{i(A)}$ with two O(IV) neighbors shows Fe-O distances ~0.2 Å longer than those observed by XRD [18]. This may result from the XRD analysis, which gives an ordered structure where oxygen atoms are fixed to crystallographic sites, or may reflect the distances of a positively charged state, where the iron cation is closer to the oxygen anions. $Fe_{i(C)}$ configurations have four-fold coordinated iron. This is at variance with the XRD study of Gomes et al. [18], where three-fold coordinated iron was found, but is in line with XRD results of Kato et al. [19]. The only three-fold coordination of Fe that we find is in the $Fe_{i(B)}$ structure (Figure 4e). However, the geometry differs from that proposed in Ref. [18], where Fe connects to one oxygen atom from a neighboring PO_4 group and two O atoms from OH groups. In general, the calculated Fe-O distances look slightly overestimated. Of course, the picture could improve when considering charged cells.

3.4. Charged Iron Defects

Iron (Fe:$4s^2 3d^6$) is not isoelectronic with respect to the species being replaced (Ca:$4s^2$ and P:$3s^2 3p^3$). Hence, Fe impurities are expected to create states within the band gap of HAp. These states are localized and may act as hole or electron traps. The description of the Fe-HAp requires the consideration these cases, which can be accounted for by changing the occupation of the highest occupied gap states of the defective supercell. In a real crystal, the capture of a hole (creation of a local positive charge) can be compensated by numerous possibilities: cation vacancies V_{Ca}^-, V_{H}^-, foreign anion interstitials like $[CO_3^{2-}]_i$, etc., thus leading to a lowering of the Fermi energy (electron chemical potential), We leave the exact mechanisms of charge compensation out of the scope of this work.

An qualitative picture of the charge states allowed for each defect can be found from the respective Kohn-Sham levels in the band gap [61]. Positive charge states (hole trapping) requires the presence of filled states in the gap, while the negatively charged defects (electron trapping) requires the presence of empty states. Figure 4 illustrates the energy of the one-electron states in the band gap obtained from spin-polarized calculations of substitutional and interstitial defects in the neutral charge state.

Substitutional iron on Ca sites, Fe_{Ca}^{2+}, have six electrons in the 3d shell, but after the first ionization all filled states move below the valence band top. Hence, further ionization would require an energy equivalent to the band gap width (~7 eV), which essentially tell us that Fe_{Ca}^{2+} defects can be single donors, but not double donors. Unoccupied states of Fe_{Ca}^{2+} are rather close to the conduction band, suggesting that they are not acceptors either (can not trap electrons).

Iron on the phosphorous site in the Fe_P^{5+} state (neutral charge state) has three filled one-electron states and 7 empty states in the gap ($3d^3$ configuration). We will show below that Fe_P^{5+} can trap one hole or up to two electrons, becoming Fe_P^{6+} or Fe_P^{4+} and Fe_P^{3+} states, respectively.

Neutral interstitial defects show only filled states in the gap. These can trap up to three holes, thus leading to Fe_i^+, Fe_i^{2+} and Fe_i^{3+}, respectively.

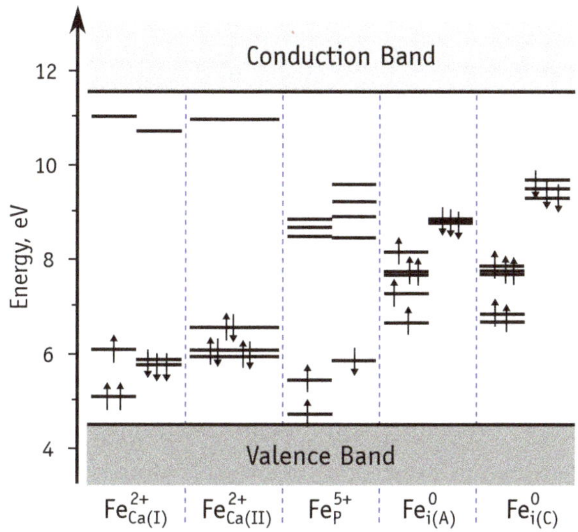

Figure 4. Energy level diagram based on the Khon-Sham states of Fe defects in HAp.

We found that the structure of the defects depend strongly on the local charge. Figure 5 illustrates some of the most remarkable structural changes of the defects induced by the capture of electrons and holes. Additional reconfigurations are depicted in Figure S1 in Supplementary Information. Table 5 presents the number of nearest oxygen neighbors to the Fe ion ($N_{\text{Fe-O}}$), the range of Fe-O bond lengths ($R_{\text{Fe-O}}$), as well as the magnetic moment of the defect (M_z). Note, that some states exhibit a high magnetization, up to $5\mu_B$ for $\text{Fe}^{3+}_{\text{Ca(I)}}$ and $\text{Fe}^{3+}_{\text{i(C)}}$. These are to be compared to the case of neutral defects, where the magnetic moment was at most $2\mu_B$.

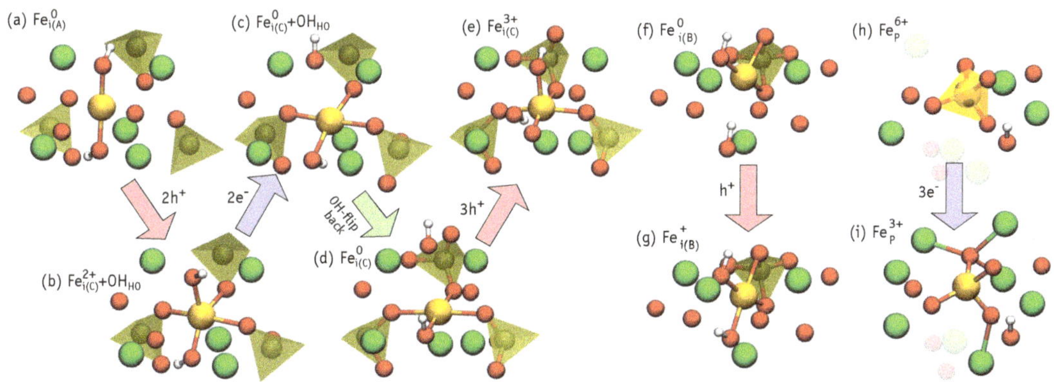

Figure 5. Illustration of the change in the local atomic structure of selected iron defects (ligands within $\lesssim 4$ Å from the Fe atom are also depicted), following the capture of electrons or holes.

Table 5. Charge state, q, Number of oxygen neighbors, $N_{\text{Fe-O}}$, and respective range of distances, $R_{\text{Fe-O}}$ (Å), and magnetization M_z (μ_B units) calculated for charged iron defects in HAp.

Defect Structure	q	$N_{\text{Fe-O}}$	$R_{\text{Fe-O}}$	M_z
$Fe^{3+}_{Ca(I)}$	+1	6	1.96–2.12	5.0
$Fe^{3+}_{Ca(II)}$	+1	5	1.90–2.01	1.0
Fe^{6+}_{P}	+1	4	1.65–1.67	0.0
Fe^{4+}_{P}	−1	4	1.74–1.75	0.0
Fe^{3+}_{P}	−2	4	1.81–1.84	1.0
$Fe^{+}_{i(A)}$	+1	2	1.84–1.85	3.0
$Fe^{+}_{i(B)}$	+1	4	1.89–2.02	3.0
$Fe^{2+}_{i(B)}$	+2	4	1.82–1.95	2.0
$Fe^{3+}_{i(B)}$	+3	4	1.78–1.88	3.0
$Fe^{+}_{i(C)}$	+1	4	1.89–2.08	3.0
$Fe^{2+}_{i(C)}$	+2	5	1.84–2.16	4.0
$Fe^{3+}_{i(C)}$	+3	5	1.72–2.12	5.0
$Fe^{+}_{i(D)}$	+1	4	1.98–1.99	3.0
$Fe^{2+}_{i(D)}$	+2	4	1.90–1.93	2.0
$Fe^{3+}_{i(D)}$	+3	6	1.95–2.06	1.0

In general, an increase of the positive charge leads to a decrease of Fe-O bond lengths. This effect is strikingly illustrated in Figure 5h,i, which showcase the Fe^{3+}_{P}–Fe^{6+}_{P} sequence of defects. This rule is understandably violated when there is a change in the coordination of the Fe ion, and therefore, a significant modification of the local electrostatics. Examples are the increases of $R_{\text{Fe-O}}$ in $Fe^{0}_{i(B)}$, $Fe^{+}_{i(C)}$, or $Fe^{2+}_{i(D)}$ upon hole capture,

$Fe^{0}_{i(B)} + h^{+} \rightarrow Fe^{+}_{i(B)}$ (Figure 5f,g),

$Fe^{+}_{i(C)} + h^{+} \rightarrow Fe^{2+}_{i(C)}$ (Figure S1p,q),

$Fe^{2+}_{i(D)} + h^{+} \rightarrow Fe^{3+}_{i(D)}$ (Figure S1y,z).

The above processes are reversible, i. e., electron trapping at (or hole emission from) the positively charged defects result in the recovery of the longer bond lengths. As expected, HAp cations repel the iron ion. In the case of $Fe^{3+}_{i(C)}$ without OH flipping, this effect leads to migration of a proton from OH to a close PO_4 group (see Figure 5d,e). We found that the high-symmetry defect configuration $Fe^{2+}_{i(A)}$ is not stable and spontaneously transforms to $Fe^{2+}_{i(C)}$. Hence, upon hole capture by (or electron emission from) $Fe^{+}_{i(A)}$, the iron moves toward the edge of the OH channel (see Figure 5a,b),

$$Fe^{+}_{i(A)} + h^{+} \rightarrow Fe^{2+}_{i(C)} + OH_{HO}.$$

The reverse process involves overcoming a barrier (not calculated), and $Fe^{+}_{i(A)}$ is not recovered spontaneously (see Figure 5b,c).

The formation energy of a charged defect is a function of the Fermi energy (c.f. Equation (1)). This dependence is clearly illustrated in Figure 6a,b, which show the results for Ca- and P-rich material and for Ca-poor HAp, respectively. The red shadow area on both diagrams indicates the whole range for the formation energy of Fe_P. The upper bound corresponds to P-rich conditions, while the lower bound to P-poor material. The solid red line in Figure 6b shows the formation energy of Fe_P under Ca-poor conditions. Thick dashed-dotted lines correspond to formation energies of Fe_{Ca} defects, and thick solid line and other non-solid lines correspond to interstitial defects (see legend).

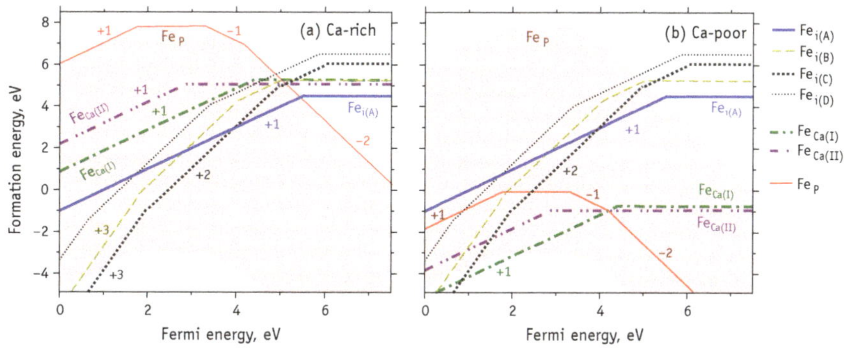

Figure 6. Formation energies of Fe-HAp as a function of the Fermi energy at extreme points of HAp stability diagram: (**a**) Ca-rich and P-rich, (**b**) Ca-poor. The horizontal segments correspond to neutral simulation cells (Table 4), and slanted – to non-neutral with charge marked by numbers from −2 to +3.

According to our results, the phosphorous substitutions, especially Fe_P^{3+} (charge state $q = -2$) is a rather stable species when there is abundance of electrons in the material (n-type HAp). This is even more evidenced in Ca-poor and P-poor conditions, where in p-type material we expect Fe_P^{4+}, Fe_P^{5+} and even Fe_P^{6+} to become more stable and compete with other species, namely Fe_i and Fe_{Ca}.

However, the phosphorous substitutions were not previously considered in the literature, since the replacing of phosphorus by iron cation requires the breaking of pretty stable P–O bonds of PO_4 group. Alternatively, the replacement of whole PO_4 group by FeO_4 one will provide the same structural result. The obtained most probable Fe_P configuration is tetrahedral coordinated ferric cation ($[FeO_4]^{5-}$ group) is not unusual and can be found in magnetite [52] or in iron-phosphate glass [62].

Regarding substitutional Fe at calcium sites, we find that $Fe_{Ca(I)}^{3+}$ is the most probable form under Ca-poor and p-type conditions (Figure 6b). We note that despite $Fe_{Ca(II)}^{2+}$ having lower formation energy than $Fe_{Ca(I)}^{2+}$ in n-type HAp, the Fe_P^{3+} species is even more stable, thus making the phosphorus substitution more probable than a replacement of calcium.

In contrast, under Ca- and P-rich conditions (Figure 6a) the interstitial defects are expected to prevail, especially in p-type and intrinsic HAp. Depending on the Fermi level location, the most probable states are $Fe_{i(C)}^{3+}$, $Fe_{i(C)}^{2+}$ and $Fe_{i(A)}^{+}$ defects. The two-fold coordinated $Fe_{i(A)}^{0}$ defect, is more favorable in n-type HAp, yet again, Fe_P^{3+} is more stable and more likely to form.

We can compare the obtained local atomic structure of iron defects in HAp (see Tables 4 and 5) with those reported in the literature (Table 3). Iron substitutions with long distances $R_{Fe\text{-}O} \geq 2.2$ Å and 6 oxygen neighbors [15,16] are best described by $Fe_{Ca(I)}^{2+}$. The slightly more favorable $Fe_{Ca(II)}^{2+}$ substitution has 5 oxygen neighbors, which could explain the result of Boda et al. [6]. The iron with only two neighbors in the study of Gomes et al. [18] could be described by $Fe_{i(A)}^{+}$, although the small Fe-O distances of 1.7 Å are only reproduced by Fe_P^{6+}, which is rather unstable. Iron with three [18] or four [19] oxygen neighbors at distances $R_{Fe\text{-}O} \geq 1.8$ Å can be accounted for by $Fe_{i(C)}^{3+}$ or Fe_P^{3+} defects.

In summary, we find that Fe-HAp can contain both substitutional and interstitial defects depending on the preparation conditions. The phosphorus substitutions have iron in low-spin states ($M_z \leq \mu_B$,), making them less useful for many applications envisaged for Fe-HAp. To avoid those defects the synthesis should be performed closer to P-rich and p-type conditions. In that case we expect the formation of high-spin defects with $M_z = 5\mu_B$ ($Fe_{Ca(I)}^{3+}$ and $Fe_{i(C)}^{3+}$) and $M_z = 4\mu_B$ ($Fe_{i(C)}^{2+}$).

3.5. Fe K-XANES of Fe-HAp

The near-edge structure of X-ray absorption (XANES) spectra is particularly sensitive to the details of local atomic structure of the absorbing element [51]. X-ray absorption spectroscopy has been applied to materials without long range order, and that includes the Fe-HAp. We consider the experimental spectra of Fe-HAp published by Gomes et al. [18]. We also keep the notation of the original study regarding the experimental conditions, i.e., 15Fe-500 and 15Fe-1100 which correspond to 15 mol % of Fe per HAp unit cell of samples sintered at 500 °C and 1100 °C, respectively.

The top three (black colored) curves in Figure 7 shows the experimental data of Gomes et al. [18] for Fe-HAp and for magnetite (Fe_3O_4). The latter was used as a reference for the alignment of the theoretical energy scale to the experimental one. Greek letters mark the main spectral features: pre-edge (α), the main peak (β), its satellite peaks (β' and β''), and a more distant peak (γ). The vertical dashed line in Figure 7 provides guidance for the relative positions of the minimum between features β and γ.

In order to determine the types of Fe defects in the 15Fe-500 and 15Fe-1100 samples we simulated the Fe K-XANES spectra for each structure considered in Sections 3.3 and 3.4. Figure 7 shows the simulated spectra of the most probable defect structures. Figure S2 in Supplementary Information shows all calculated spectra. The comparison of experimental and simulated spectra for magnetite Fe_3O_4 (dotted curves in Figure 7) reveals the main insufficiencies of the simulation method, including (i) the lack of pre-edge features (α) and (ii) a poor reproduction of the β' satellite. However, the intensities and energy positions of the main features (β, β'' and γ) of the magnetite spectrum are correctly reproduced. All spectral features of another reference iron oxide, hematite Fe_2O_3, could be reproduced (not shown) using exact the same calculation scheme. In this case, the differences between experimental and simulated spectra show qualitatively the same insufficiencies as for magnetite.

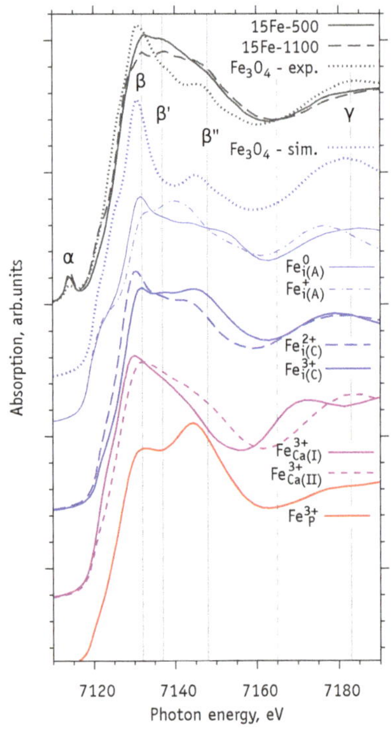

Figure 7. Comparison between experimental Fe K-XANES spectra of Fe-HAp published in Ref. [18] (top three curves) and simulated spectra (all other curves) as described in the text.

The simulated spectra for Fe-HAp show good sensitivity with respect to their atomistic structures and charge states. In most cases the increase of the iron oxidation state shifts the position of the main peak (β) to higher energies. This follows from an increase of the local positive charge and binding energy of K-electrons. An exception is seen for the $Fe^{3+}_{i(D)}$ spectrum (Figure S2d), which is explained by significant changes in its local atomic structure (coordination change from 4 to 6) upon ionization.

The relative intensities I_β and $I_{\beta''}$, corresponding to features β and β'' from the simulated spectra, are sensitive to the defect type: phosphorus substitutions have $I_\beta < I_{\beta''}$, calcium substitutions show $I_\beta > I_{\beta''}$, and most interstitial defects have $I_\beta \sim I_{\beta''}$. Additionally, the spectra show considerable differences concerning the position and shape of the γ feature.

The spectrum of the 15Fe-1100 sample shows comparable β and β'' intensities ($I_\beta \sim I_{\beta''}$), suggesting that most Fe defects are of interstitial character. Conversely, the higher intensity of peak β in the spectrum of sample 15Fe-500 may indicate the presence of substitutional Fe on calcium sites or $Fe^{n}_{i(A)}$ defect ($n = 0, 1+$). The latter scenario is in line with the conclusions of Gomes et al. [18], where the formation of the high-symmetry $Fe_{i(A)}$ structure was proposed in the 15Fe-500 and 15Fe-800 samples, whereas $Fe_{i(C)}$-type defects were suggested for the high-temperature treated 15Fe-1100 sample.

4. Conclusions

We performed a hybrid-DFT study of iron defects in hydroxyapatite. The most favorable structures (and respective charge states) were used to simulate Fe K-XANES spectra of Fe-HAp. These were compared to experimental data available in the literature. The results allow us to draw the following conclusions:

1. The chemical stability of HAp ($Ca_{10}(PO_4)_6(OH)_2$) is limited by a thermodynamic equilibrium with O_2, CaO, $Ca(OH)_2$, H_2, P, DCPD ($CaHPO_4 \cdot 2H_2O$), DCPA ($CaHPO_4$) and TCP ($Ca_3(PO_4)_2$) phases. Formation of these compounds limit the range of values of Ca and P chemical potentials, and we could identify three extreme cases: Ca- and P-rich, Ca-poor, P-poor.
2. Under P-poor conditions, phosphorous substitutions are the most favorable, resulting in the formation of low-spin Fe defect states. However, p-type HAp may contain high spin iron interstitials $Fe^{3+}_{i(C)}$.
3. Under Ca-poor conditions, the high-spin calcium substitution, $Fe^{3+}_{Ca(I)}$, is the most favorable species. Fe^{2+}_{Ca} defects have higher formation energy comparing to Fe^{3+}_{P} and $Fe^{3+}_{Ca(I)}$.
4. Under Ca- and P-rich conditions, interstitial iron atoms in the OH channel are the most prominent. Depending on the position of Fermi level the most favorable are $Fe^{3+}_{i(C)}$, $Fe^{2+}_{i(C)}$ and $Fe^{+}_{i(A)}$. When compared to $Fe_{i(C)}$, the last structure involves the flipping of a nearby hydroxyl unit. The OH flipping does not introduce a significant change to the formation energy of $Fe^{3+}_{i(C)}$ and $Fe^{2+}_{i(C)}$ defects.
5. High-spin iron defects are $Fe^{3+}_{Ca(I)}$ and $Fe^{3+}_{i(C)}$. These are both expected in p-type HAp. Such configurations are expected to be most useful for materials targeting magnetic hyperthermia or magnetic resonance imaging applications.
6. The comparison of Fe K-XANES spectra of theoretically predicted defect structures with experimental data [18] confirms the interstitial character of iron defects in samples sintered at high (1100 °C) temperature, but does not exclude the substitution defects for samples sintered at lower temperatures.

Supplementary Materials: The following are available online at https://www.mdpi.com/article/10.3390/nano11112978/s1, Table S1: Chemical potentials in eV at selected points from the phase stability diagram, Figure S1: Illustration of the local atomic structure of iron atom ($\lesssim$4 Å) for all considered Fe-HAp configurations, Figure S2: Simulated Fe K-XANES spectra for the considered

Fe-HAp structures. Pseudopotential files, example input script and quantum espresso output file. Atomic coordinates of Fe-HAp structures in POSCAR format.

Author Contributions: Conceptualization, J.C. and V.B.; methodology, L.A. and E.P.; software, E.P. and J.C.; validation, S.G., G.R. and J.C.; investigation, L.A. and E.P.; resources, V.B. and J.C.; writing—original draft preparation, L.A.; writing—review and editing, S.G., G.R. and J.C.; visualization, L.A.; funding acquisition, V.B. All authors have read and agreed to the published version of the manuscript.

Funding: This research was funded by Russian Science Foundation (RSF) grant number 21-12-00251. J.C. thanks the FCT in Portugal for support through Projects UIDB/50025/2020, UIDP/50025/2020.

Institutional Review Board Statement: Not applicable.

Informed Consent Statement: Not applicable.

Data Availability Statement: The atomic coordinates of optimized Fe-HAp structures are available in POSCAR format through Supplementary Materials.

Acknowledgments: Authors are grateful to the I3N at the University of Aveiro and to the Institute of Mathematical Problems of Biology by providing access to the HPC facilities. Authors acknowledge SLS for provision of synchrotron radiation facilities and would like to thank Maarten Nachtegaal for assistance in using beamline SuperXAS.

Conflicts of Interest: The authors declare no conflict of interest.

Abbreviations

The following abbreviations are used in this manuscript:

HAp	Hydroxyapatite
Fe-HAp	Iron-doped HAp
DFT	Density functional theory
HSE	Heyd, Scuseria and Ernzerhof / Hybrid Screened Exchange
XANES	X-ray absorption near-edge structure

References

1. Epple, M.; Ganesan, K.; Heumann, R.; Klesing, J.; Kovtun, A.; Neumann, S.; Sokolova, V. Application of calcium phosphate nanoparticles in biomedicine. *J. Mater. Chem.* **2010**, *20*, 18–23. [CrossRef]
2. Bystrov, V. Computational studies of the hydroxyapatite nanostructures, peculiarities and properties. *Math. Biol. Bioinform.* **2017**, *12*, 14–54. [CrossRef]
3. Duminis, T.; Shahid, S.; Hill, R.G. Apatite Glass-Ceramics: A Review. *Front. Mater.* **2017**, *3*, 59. [CrossRef]
4. Degli Esposti, L.; Carella, F.; Adamiano, A.; Tampieri, A.; Iafisco, M. Calcium phosphate-based nanosystems for advanced targeted nanomedicine. *Drug Dev. Ind. Pharm.* **2018**, *44*, 1223–1238. [CrossRef]
5. Mayer, I.; Diab, H.; Felner, I. Ferric iron in synthetic carbonate apatites: A Mössbauer effect study. *J. Inorg. Biochem.* **1992**, *45*, 129–133. [CrossRef]
6. Boda, S.K.; V., A.A.; Basu, B.; Sahoo, B. Structural and Magnetic Phase Transformations of Hydroxyapatite-Magnetite Composites under Inert and Ambient Sintering Atmospheres. *J. Phys. Chem. C* **2015**, *119*, 6539–6555. [CrossRef]
7. Mondal, S.; Manivasagan, P.; Bharathiraja, S.; Santha Moorthy, M.; Kim, H.H.; Seo, H.; Lee, K.D.; Oh, J. Magnetic hydroxyapatite: A promising multifunctional platform for nanomedicine application. *Int. J. Nanomed.* **2017**, *12*, 8389–8410. [CrossRef]
8. Laranjeira, M.S.; Moço, A.; Ferreira, J.; Coimbra, S.; Costa, E.; Santos-Silva, A.; Ferreira, P.J.; Monteiro, F.J. Different hydroxyapatite magnetic nanoparticles for medical imaging: Its effects on hemostatic, hemolytic activity and cellular cytotoxicity. *Colloids Surf. B Biointerfaces* **2016**, *146*, 363–374. [CrossRef]
9. Tampieri, A.; D'Alessandro, T.; Sandri, M.; Sprio, S.; Landi, E.; Bertinetti, L.; Panseri, S.; Pepponi, G.; Goettlicher, J.; Bañobre-López, M.; et al. Intrinsic magnetism and hyperthermia in bioactive Fe-doped hydroxyapatite. *Acta Biomater.* **2012**, *8*, 843–851. [CrossRef]
10. Piccirillo, C.; Rocha, C.; Tobaldi, D.M.; Pullar, R.C.; Labrincha, J.A.; Ferreira, M.O.; Castro, P.M.L.; Pintado, M.M.E. A hydroxyapatite–Fe2O3 based material of natural origin as an active sunscreen filter. *J. Mater. Chem. B* **2014**, *2*, 5999–6009. [CrossRef]
11. Gambardella, A.; Bianchi, M.; Kaciulis, S.; Mezzi, A.; Brucale, M.; Cavallini, M.; Herrmannsdoerfer, T.; Chanda, G.; Uhlarz, M.; Cellini, A.; et al. Magnetic hydroxyapatite coatings as a new tool in medicine: A scanning probe investigation. *Mater. Sci. Eng. C* **2016**, *62*, 444–449. [CrossRef] [PubMed]
12. Lin, K.; Chen, L.; Liu, P.; Zou, Z.; Zhang, M.; Shen, Y.; Qiao, Y.; Liu, X.; Chang, J. Hollow magnetic hydroxyapatite microspheres with hierarchically mesoporous microstructure for pH-responsive drug delivery. *CrystEngComm* **2013**, *15*, 2999. [CrossRef]
13. Gu, L.; He, X.; Wu, Z. Mesoporous Fe 3 O 4 /hydroxyapatite composite for targeted drug delivery. *Mater. Res. Bull.* **2014**, *59*, 65–68. [CrossRef]

14. Ok, H.N. Mössbauer Studies of Natural Apatite. *Phys. Rev.* **1969**, *185*, 477–482. [CrossRef]
15. Jiang, M.; Terra, J.; Rossi, A.M.; Morales, M.A.; Saitovitch, E.M.B.; Ellis, D.E. Fe2+/Fe3+substitution in hydroxyapatite: Theory and experiment. *Phys. Rev. B* **2002**, *66*, 224107. [CrossRef]
16. Low, H.R.; Phonthammachai, N.; Maignan, A.; Stewart, G.A.; Bastow, T.J.; Ma, L.L.; White, T.J. The Crystal Chemistry of Ferric Oxyhydroxyapatite. *Inorg. Chem.* **2008**, *47*, 11774–11782. [CrossRef]
17. Zilm, M.E.; Chen, L.; Sharma, V.; McDannald, A.; Jain, M.; Ramprasad, R.; Wei, M. Hydroxyapatite substituted by transition metals: Experiment and theory. *Phys. Chem. Chem. Phys.* **2016**, *18*, 16457–16465. [CrossRef]
18. Gomes, S.; Kaur, A.; Grenèche, J.M.; Nedelec, J.M.; Renaudin, G. Atomic scale modeling of iron-doped biphasic calcium phosphate bioceramics. *Acta Biomater.* **2017**, *50*, 78–88. [CrossRef]
19. Kato, S.; Ikeda, S.; Saito, K.; Ogasawara, M. Fe incorporation into hydroxyapatite channels by Fe loading and post-annealing. *J. Solid State Chem.* **2018**, *265*, 411–416. [CrossRef]
20. Renaudin, G.; Gomes, S.; Nedelec, J.M. First-row transition metal doping in calcium phosphate bioceramics: A detailed crystallographic study. *Materials* **2017**, *10*, 92. [CrossRef]
21. Avakyan, L.A.; Paramonova, E.V.; Coutinho, J.; Öberg, S.; Bystrov, V.S.; Bugaev, L.A. Optoelectronics and defect levels in hydroxyapatite by first-principles. *J. Chem. Phys.* **2018**, *148*, 154706,
22. Martin, R.; Reining, L.; Ceperley, D. *Interacting Electrons*; Cambridge University Press: Cambridge, UK, 2016.
23. Calderin, L.; Stott, M.; Rubio, A. Electronic and crystallographic structure of apatites. *Phys. Rev. B.* **2003**, *67*, 134106. [CrossRef]
24. Rulis, P.; Ouyang, L.; Ching, W.Y. Electronic structure and bonding in calcium apatite crystals: Hydroxyapatite, fluorapatite, chlorapatite, and bromapatite. *Phys. Rev. B* **2004**, *70*, 155104. [CrossRef]
25. Rulis, P.; Yao, H.; Ouyang, L.; Ching, W.Y. Electronic structure, bonding, charge distribution, and X-ray absorption spectra of the (001) surfaces of fluorapatite and hydroxyapatite from first principles. *Phys. Rev. B* **2007**, *76*, 245410. [CrossRef]
26. Matsunaga, K.; Kuwabara, A. First-principles study of vacancy formation in hydroxyapatite. *Phys. Rev. B* **2007**, *75*, 014102. [CrossRef]
27. Slepko, A.; Demkov, A.A. First-principles study of the biomineral hydroxyapatite. *Phys. Rev. B* **2011**, *84*, 134108. [CrossRef]
28. Bystrov, V.S.; Coutinho, J.; Bystrova, A.V.; Dekhtyar, Y.D.; Pullar, R.C.; Poronin, A.; Palcevskis, E.; Dindune, A.; Alkan, B.; Durucan, C.; Paramonova, E.V. Computational study of hydroxyapatite structures, properties and defects. *J. Phys. D Appl. Phys.* **2015**, *48*, 195302. [CrossRef]
29. Kay, M.; Young, R.; Posner, A. Crystal Structure of Hydroxyapatite. *Nature* **1964**, *204*, 1050–1052. [CrossRef] [PubMed]
30. Hughes, J.M.; Rakovan, J. The Crystal Structure of Apatite, Ca5(PO4)3(F,OH,Cl). *Rev. Mineral. Geochem.* **2002**, *48*, 1–12. [CrossRef]
31. Elliott, J.; Mackie, P.; Young, R. Monoclinic hydroxyapatite. *Science* **1973**, *180*, 1055–1057. [CrossRef] [PubMed]
32. Ikoma, T.; Yamazaki, A.; Nakamura, S.; Akao, M. Preparation and dielectric property of sintered monoclinic hydroxyapatite. *J. Mater. Sci. Lett.* **1999**, *18*, 1225–1228. [CrossRef]
33. Giannozzi, P.; Baroni, S.; Bonini, N.; Calandra, M.; Car, R.; Cavazzoni, C.; Ceresoli, D.; Chiarotti, G.L.; Cococcioni, M.; Dabo, I.; et al. QUANTUM ESPRESSO: A modular and open-source software project for quantum simulations of materials. *J. Phys. Condens. Matter* **2009**, *21*, 395502. [CrossRef] [PubMed]
34. Giannozzi, P.; Andreussi, O.; Brumme, T.; Bunau, O.; Nardelli, M.B.; Calandra, M.; Car, R.; Cavazzoni, C.; Ceresoli, D.; Cococcioni, M.; et al. Advanced capabilities for materials modelling with Quantum ESPRESSO. *J. Phys. Condens. Matter* **2017**, *29*, 465901. [CrossRef] [PubMed]
35. Heyd, J.; Scuseria, G.E.; Ernzerhof, M. Hybrid functionals based on a screened Coulomb potential. *J. Chem. Phys.* **2003**, *118*, 8207–8215. [CrossRef]
36. Krukau, A.V.; Vydrov, O.A.; Izmaylov, A.F.; Scuseria, G.E. Influence of the exchange screening parameter on the performance of screened hybrid functionals. *J. Chem. Phys.* **2006**, *125*, 224106. [CrossRef]
37. Perdew, J.P.; Burke, K.; Ernzerhof, M. Generalized Gradient Approximation Made Simple. *Phys. Rev. Lett.* **1996**, *77*, 3865–3868. [CrossRef]
38. Hamann, D.R. Optimized norm-conserving Vanderbilt pseudopotentials. *Phys. Rev. B* **2013**, *88*, 085117. 085117. [CrossRef]
39. Schlipf, M.; Gygi, F. Optimization algorithm for the generation of ONCV pseudopotentials. *Comput. Phys. Commun.* **2015**, *196*, 36–44. [CrossRef]
40. Coutinho, J.; Torres, V.J.; Demmouche, K.; Öberg, S. Theory of the carbon vacancy in 4 H-SiC: Crystal field and pseudo-Jahn-Teller effects. *Phys. Rev. B* **2017**, *96*, 174105. [CrossRef]
41. Bystrov, V.S.; Avakyan, L.A.; Paramonova, E.V.; Coutinho, J. Sub-Band Gap Absorption Mechanisms Involving Oxygen Vacancies in Hydroxyapatite. *J. Phys. Chem. C* **2019**, *123*, 4856–4865. [CrossRef]
42. Gouveia, J.D.; Coutinho, J. Can we rely on hybrid-DFT energies of solid-state problems with local-DFT geometries? *Electron. Struct.* **2019**, *1*, 015008. [CrossRef]
43. Qian, G.X.; Martin, R.M.; Chadi, D.J. First-principles study of the atomic reconstructions and energies of Ga- and As-stabilized GaAs(100) surfaces. *Phys. Rev. B* **1988**, *38*, 7649–7663. [CrossRef]
44. Jang, Y.R.; Yu, B.D. Structural, Magnetic, and Electronic Properties of Fe: A Screened Hybrid Functional Study. *J. Magn.* **2011**, *16*, 201–205. [CrossRef]
45. Meng, Y.; Liu, X.W.; Huo, C.F.; Guo, W.P.; Cao, D.B.; Peng, Q.; Dearden, A.; Gonze, X.; Yang, Y.; Wang, J.; et al. When Density Functional Approximations Meet Iron Oxides. *J. Chem. Theory Comput.* **2016**, *12*, 5132–5144. [CrossRef] [PubMed]

46. Ingebrigtsen, M.E.; Varley, J.B.; Kuznetsov, A.Y.; Svensson, B.G.; Alfieri, G.; Mihaila, A.; Badstübner, U.; Vines, L. Iron and intrinsic deep level states in Ga2O3. *Appl. Phys. Lett.* **2018**, *112*, 042104. [CrossRef]
47. Makov, G.; Payne, M.C. Periodic boundary conditions in ab initio calculations. *Phys. Rev. B* **1995**, *51*, 4014–4022. [CrossRef] [PubMed]
48. Lany, S.; Zunger, A. Assessment of correction methods for the band-gap problem and for finite-size effects in supercell defect calculations: Case studies for ZnO and GaAs. *Phys. Rev. B* **2008**, *78*, 235104. [CrossRef]
49. Sit, P.H.L.; Car, R.; Cohen, M.H.; Selloni, A. Simple, Unambiguous Theoretical Approach to Oxidation State Determination via First-Principles Calculations. *Inorg. Chem.* **2011**. *50*, 10259–10267. [CrossRef]
50. Bunau, O.; Joly, Y. Self-consistent Aspects of X-ray Absorption Calculations. *J. Phys. Condens. Matter* **2009**, *21*, 345501. [CrossRef]
51. Lamberti, C.; van Bokhoven, J.A., X-ray Absorption and X-ray Emission Spectroscopy: Theory and Applications. In *X-ray Absorption and X-ray Emission Spectroscopy*; Chapter X-ray Absorption and Emission Spectroscopy for Catalysis; van Bokhoven, J.A., Lamberti, C., Eds.; John Wiley & Sons, Ltd.: Hoboken, NJ, USA, 2016; pp. 351–383. [CrossRef]
52. Soldatov, M.A.; Göttlicher, J.; Kubrin, S.P.; Guda, A.A.; Lastovina, T.A.; Bugaev, A.L.; Rusalev, Y.V.; Soldatov, A.V.; Lamberti, C. Insight from X-ray Absorption Spectroscopy to Octahedral/Tetrahedral Site Distribution in Sm-Doped Iron Oxide Magnetic Nanoparticles. *J. Phys. Chem. C* **2018**, *122*, 8543–8552,
53. Lide, D.R. (Ed.) *CRC Handbook of Chemistry and Physics*, 88th ed.; CRC Press: Boca Raton, FL, USA, 2005; p. 2661.
54. Gražulis, S.; Daškevič, A.; Merkys, A.; Chateigner, D.; Lutterotti, L.; Quiros, M.; Serebryanaya, N.R.; Moeck, P.; Downs, R.T.; Le Bail, A. Crystallography Open Database (COD): An open-access collection of crystal structures and platform for world-wide collaboration. *Nucleic Acids Res.* **2012**, *40*, D420–D427. [CrossRef] [PubMed]
55. Fukui, H.; Ohtaka, O.; Fujisawa, T.; Kunisada, T.; Suzuki, T.; Kikegawa, T. Thermo-elastic property of Ca(OH)2portlandite. *High Press. Res.* **2003**, *23*, 55–61. [CrossRef]
56. Brazhkin, V.V.; Gromnitskaya, E.L.; Danilov, I.V.; Katayama, Y.; Lyapin, A.G.; Popova, S.V. High pressure behavior of P2O5crystalline modifications: compressibility, elastic properties and phase transitions. *Mater. Res. Express* **2015**, *2*, 025201. [CrossRef]
57. Zhai, S.; Liu, X.; Shieh, S.R.; Zhang, L.; Ito, E. Equation of state of -tricalcium phosphate, -Ca3(PO4)2, to lower mantle pressures. *Am. Mineral.* **2009**, *94*, 1388–1391. [CrossRef]
58. Hughes, J.; Fransolet, A.; Schreyer, W. The atomic arrangement of iron-bearing apatite. In *Neues Jahrbuch Fur Mineralogie-Monatshefte*; Schweizerbart Science Publishers: Stuttgart, Germany, 1993 ; p. 504–510.
59. Carrera, K.; Huerta, V.; Orozco, V.; Matutes, J.; Fernández, P.; Graeve, O.; Herrera, M. Formation of vacancy point-defects in hydroxyapatite nanobelts by selective incorporation of Fe3+ ions in Ca(II) sites. A CL and XPS study. *Mater. Sci. Eng. B* **2021**, *271*, 115308. [CrossRef]
60. Bauminger, E.; Ofer, S.; Gedalia, I.; Horowitz, G.; Mayer, I. Iron uptake by teeth and bones: A mossbauer effect study. *Calcif. Tissue Int.* **1985**, *37*, 386–389. [CrossRef] [PubMed]
61. Tsuneda, T.; Song, J.W.; Suzuki, S.; Hirao, K. On Koopmans' theorem in density functional theory. *J. Chem. Phys.* **2010**, *133*, 174101. [CrossRef]
62. Stoch, P.; Szczerba, W.; Bodnar, W.; Ciecinska, M.; Stoch, A.; Burkel, E. Structural properties of iron-phosphate glasses: Spectroscopic studies and ab initio simulations. *Phys. Chem. Chem. Phys.* **2014**, *16*, 19917–19927. [CrossRef] [PubMed]

nanomaterials

MDPI

Article

Quantitative Assessment of Chirality of Protein Secondary Structures and Phenylalanine Peptide Nanotubes

Alla Sidorova [1,*], Vladimir Bystrov [2], Aleksey Lutsenko [1], Denis Shpigun [1], Ekaterina Belova [1] and Ilya Likhachev [2]

1 Faculty of Physics, Lomonosov Moscow State University, 119991 Moscow, Russia; aleksluchrus@yandex.ru (A.L.); denish.den@mail.ru (D.S.); ev.malyshko@physics.msu.ru (E.B.)
2 Institute of Mathematical Problems of Biology, The Branch of Keldysh Institute of Applied Mathematics, RAS, 142290 Pushchino, Russia; vsbys@mail.ru (V.B.); ilya_lihachev@mail.ru (I.L.)
* Correspondence: sky314bone@mail.ru

Abstract: In this study we consider the features of spatial-structure formation in proteins and their application in bioengineering. Methods for the quantitative assessment of the chirality of regular helical and irregular structures of proteins are presented. The features of self-assembly of phenylalanine (F) into peptide nanotubes (PNT), which form helices of different chirality, are also analyzed. A method is proposed for calculating the magnitude and sign of the chirality of helix-like peptide nanotubes using a sequence of vectors for the dipole moments of individual peptides.

Keywords: helical structures; peptide nanotubes; phenylalanine; self-assembly; molecular modeling; dipole moments; polarization; chirality; protein secondary structure

Citation: Sidorova, A.; Bystrov, V.; Lutsenko, A.; Shpigun, D.; Belova, E.; Likhachev, I. Quantitative Assessment of Chirality of Protein Secondary Structures and Phenylalanine Peptide Nanotubes. *Nanomaterials* **2021**, *11*, 3299. https://doi.org/10.3390/nano11123299

Academic Editor: Eunji Lee

Received: 30 October 2021
Accepted: 2 December 2021
Published: 5 December 2021

1. Introduction

There exists a sophisticated understanding of the relationship between amino acid sequences and the structure of various types of protein elements. This understanding has expanded the possibilities of managing the assembly of both natural proteins and artificial structures in the field of protein engineering, materials science, etc. Due to the biocompatibility of molecular recognition properties and availability for production, biomolecular nanostructures are attractive for use in various fields of biomedicine, biotechnology, and bioengineering. Artificial peptides, such as natural ones, can be targeted for self-assembly to perform a specific function. For example, some of the earliest artificially created peptides in tissue engineering demonstrated that self-organizing peptides are capable of supporting cell attachment and proliferation [1,2]. Other studies have shown that the use of peptides can promote the regeneration of axons and restoration of the brain of animals [3], cultivation of stem cells [4], coordination of lanthanide ions [5] and DNA binding [6]. Peptide nanotubes allow numerous chemical modifications and assist in exploiting the specificity of biological systems. For example, they are used to study the ability of very short aromatic peptides to form ordered amyloid fibrils, which have similar biophysical and structural properties and are a hallmark of a diverse group of diseases (Alzheimer's disease, type 2 diabetes, prion diseases). The spatial structure and forces of interaction of aromatic fragments provide the direction and energy necessary for these ordered structures' formation [7]. Therefore, many works studies have been conducted to study the three-dimensional structure of proteins in the context of determining the structural and functional features of regular and irregular protein secondary structures.

Chirality occupies a valuable space in studies of artificial structures, as it is used as a control characteristic of stratification in hierarchies of biomacromolecule structures and, as a consequence, their functional features [8,9]. During the formation of complex protein structures, a chirality sign alternation was identified, from left-handed (L) to right-handed (D), and during the transition between hierarchical levels. However, this pattern requires confirmation in the form of a qualitative and quantitative assessment of chirality for various protein structural levels.

One of the main objectives of protein engineering is to improve protein stability, and this task is associated with the chirality of protein structures. Syndiotactic chains have an enormous ensemble of available conformations; therefore, L- and D-amino acids are often used in bioengineering [10]. Heterochirality is not a characteristic feature of most biological systems. Homochiral amino acid chains, which have significantly fewer possible conformations than heterochiral ones, promote the formation of regular secondary structures since protein isotacticity reduces the entropic component of folding and, accordingly, increases the stability of protein structures [11]. Amino acids of different chirality have different effects on the self-assembly of proteins, and the substitution of enantiomers can alter the kinetics, morphology, and the mechanical properties of self-assembly of the peptides [12–21]. Thus, a substitution for a D-amino acid is able to disrupt the structure of the helix or β-sheet and destabilize the process of peptide self-assembly [22–25].

In [26,27], a new self-assembly mode based on the use of helical peptides with a chiral center, where chirality determines the self-assembly of helical structures, is proposed.

A study of the effect of the chirality of amino acids on the structures of diphenylalanine (FF) and its derivatives showed that switching the chirality of one Phe in FF derivatives changed the morphology of their self-assembly but retained the ability to self-assemble into nanotubes, and heterochirality made nanotubes more stable [13,28,29]. At the same time, it was shown that the hydrogel formed by the racemic ferrocene-diphenylalanine mixture was mechanically weaker than the enantiopure hydrogels [30]. Thus, there are differences in self-assembly between racemic mixtures and pure enantiomers. The introduction of D-amino acids into self-organizing L-peptides is widely used to increase the enzymatic stability of structures and affect their biological functions [19].

The left-handed helix of polyproline II (PPII) belongs to the trans-isomers (steric more favorable) of peptide bonds, and the more compact right-handed helix of polyproline I (PPI) belongs to the cis isomers. PPII helices are involved in signal transduction and in the assembly of the protein complex, transcription, protein self-assembly and elasticity, the regulation of many intracellular signaling complexes, and they play a significant structural role in amyloidogenic proteins [31,32].

The formation of sign-alternating chiral structures of different scales can also be observed in cholesteric liquid crystals formed by chiral molecules. In each layer, the molecules are predominantly oriented along the director, and upon passing to the neighbouring layer, the director rotates around the cholesteric axis (rotation is associated with stereospecific molecular restrictions). A helix is formed that is opposite in sign to the chirality of the molecules. "Left" cholesterol defines the dextrorotation of the director [33]. In the cholesteric phase of DNA, a change in the sign of chirality is also observed during the transition to the next level of organization [34]. The cholesteric phase comprises a standard organized parallel layers of DNA molecules. Each layer is rotated relative to the previous layer by a small angle. Right-handed DNA forms layers, which in turn form a left-handed helix.

The combination of flexibility and rigidity within one protein molecule is possibly associated with the aperiodicity of the protein structure crystals [35]. Orientational symmetry is broken in the aperiodic arrangement of secondary structural elements, and the folded structures are nematic droplets. At certain values of the introduced nematic order parameter P_2, the arrangement of structural elements can withstand mechanical forces. This approach is found to be valid when considering the relationship between the three-dimensional organization and the nematic order of protein allostery.

In this article, we discuss methods for determining the chirality sign of regular (helical) and irregular (turns and loops) protein secondary structures, as well as the possibility of their application for helix-like peptide nanotubes based on amino acids.

2. Models and Computational Methods

2.1. Objects of Study

2.1.1. Protein Secondary Structures

A particular manifestation of chirality is the helicity of structures. The helix boundaries are determined by a set of amino acids, the sequence of which is encoded in the DNA [8,36]. In this article, three types of regular secondary structures are considered–α, 3_{10}, and π-helices.

α-helices in natural proteins are more stable and resistant to mutations than β-sheets [37]. The most common protein-regular secondary structure is the right-handed α-helix [38].

The third most common structures after α-helices and β-sheets in globular proteins are 3_{10}-helices [39]. These short helices are located at the sites of turns of α-helices or their ends [40]. The 3_{10}-helices have three residues per turn and are less-stable structures than α-helices (possibly due to a slightly different structure of hydrogen bonds) [41]. Similar to α-helices, 3_{10}-helices are mainly observed in the right-handed conformation.

π-helices are formed as a result of the exclusion or addition of one amino acid residue in the α-helix [42]; they are found in 15% of protein structures [42] and, as a rule, are located near the functional sites of proteins [42]. Thus, 32% of 6-residue π-helices are involved in ligand binding or constitute an active site, and 77% have conserved residues among homologous proteins [43]. The overwhelming majority of natural π-helices consist of seven residues and at least two consecutive π-type H-bonds [44].

Irregular secondary protein structures (turns and loops) act as a link between regular secondary structures [45] and are important elements of molecular recognition in protein folding [46]. The formation of irregular regions is largely determined by the primary amino acid sequence of the polypeptide chain [47]. Turns and coils account for approximately 30% [48] to 50% of the total secondary structure of globular proteins [49]. These structures are often present in the active sites of proteins, facilitating specific interactions between molecules [50] and, as a rule, are located on the surface of a globular protein.

Turns are sufficiently stable structures, since these isolated elements should actively promote the folding and maintenance of the globular form of the protein [51]. Depending on the number of residues separating the pair connected via the hydrogen bond, the turns are divided into δ-, γ-, β-, α-, and π-turns, consisting of 2 to 6 amino acid residues, respectively [52]. In this article, we consider such irregular secondary structures as β- and α-turns, and Ω-loops.

For β- and α-turns, the distance between the first and last α-carbons of the turn is less than 10 Å, and it has no hydrogen bond [52].

A β-turn (of 4 residues) is the most frequent type of turn [52], accounting for 63% of the residues in loops between regular secondary structures and around 25–30% of all protein residues [53]. Even for relatively small peptides, it is believed that the β-turn conformation is bioactive (the rate of β-turns formation is ten times lower than the rate of α-helices formation) [54].

The α-turn is not part of the α-helix. Turns that are not hydrogen bonded contain more hydrophobic residues at i and $i + 4$ positions and can provide (and stabilize) hydrophobic interactions between turns. The most common structural motif with an α-turn is a β-hairpin. According to the hypothesis, one of the mechanisms of α-turn initiation is the development of an initial β-turn into an α-turn [55]. The β-turn is important for understanding protein folding mechanisms.

Ω-loops consist of 6–16 amino acid residues, where the lower limit of length serves to exclude reverse turns. As a rule, they are located on the surface of globular proteins, connect membrane α-helices on the cytoplasmic or extracellular surface, and are often involved in recognition processes [56]. On average, a protein molecule contains around four Ω-loops, the distance between the ends is less than the α-α carbon separation in a loop, the twisting angles of the main chain are not repeated, and there are fewer hydrogen bonds of the main chain [57]. The hydrogen bond in the main chain of the loop is irregular, which favors the packing of side chains within long loops [51]. Therefore, depending on the three-dimensional shape, Ω-loops exhibit different degrees of flexibility during protein folding [58] and affect the function of protein structures [59]. Furthermore, Ω-loops can be used in bioengineering since their replacement or elimination affects the stability and enzymatic activity of the protein.

It is clear that the identification of irregular structures in proteins is necessary for understanding their structure and functions since they connect secondary structural elements, change the direction of the polypeptide chain, and often contain residues of active sites.

2.1.2. Peptide Nanotubes

One example of the self-assembly of complex biomolecular structures is the formation of a helical nanotube-type structure based on a phenylalanine amino acid (F or Phe) [60]. It is known that, based on such a phenylalanine amino acid, dipeptides, diphenylalanine (FF or $(Phe)_2$), are also formed, which are then easily assembled into peptide nanotubes [61–63]. Diphenylalanine dipeptide and peptide nanotubes (PNT), based on this process (FF PNT), are currently well studied, as they are of considerable interest due to their special structural and physical properties, which are important in various applications. However, single phenylalanine molecules can also form nanotubes and nanofibrils [64,65]. In [60], the modeling and assembly of a set of $(Phe)_{48}$ molecules into a helix-like tubular structure of a phenylalanine peptide nanotube (F48 PNT) using the molecular dynamics method (MD manipulator) was considered. Data on the formation of nanotubes of the "right" (D-F48 PNT) chirality from the initial L-F peptides and nanotubes of the "left" (L-F48 PNT) chirality from the D-F peptides were obtained.

In this work, using the approach described in [66], based on the values of the dipole moments of individual peptides, we calculated the sign and magnitude of chirality for such a phenylalanine spiral nanotube [60].

2.2. Methods for Evaluating the Chirality of Regular Helical and Irregular Protein Secondary Structures

According to M. Petitjean, the measure of chirality should be of a continuous characteristic and should be determined for a space of any dimension, and the chirality index should not depend on the method of selecting the mirror image [67]. At present, there are a significant number of works in which various methods for evaluating the chirality of helical structures are proposed. The main methods are as follows: a "connectivity index" for alkanes depending on physical parameters [68–70], binary code for benzenoids in 2D space [71], the deviation of a chiral set from a reference achiral set [72], chirality as overlapping the initial set with its mirror image [73], a measure of continuous symmetry based on determining the distance from a distorted molecular shape to a selected symmetry element [74,75], pseudoscalar measures of electronic chirality for molecular systems using the rotational polarizability of molecules [76], calculating the "degree of chirality" based on the overlap and an infinite hierarchy of pseudoscalar parameters for spiral ribbons, using a two-dimensional plane [77]. The described methods are often highly specialized and, as a rule, provide an estimate of symmetry, not chirality, and are fairly difficult to calculate, even if they are not based on experiments. The folding of peptide chains by rotating the planes of peptide bonds, relative to their bonds with α-carbon atoms, is described by dihedral angles, the values of which provide the basis for assessing the chirality of proteins in the Ramachandran method, which allows for the observation of the predominant conformation of elements of the secondary structure of biomolecules [78]. Since its inception, this method has become the main method for characterizing proteins. In [79], the structural RP-analysis of proteins was extended from a two-dimensional map to a three-dimensional one, and a quantitative degree of chirality was added as a continuous measure of chirality (CCM), covering all bond angles and bond lengths of the amino acid residue at each point of the map. The addition of chirality to the Ramachandran plots made it possible to emphasize the sensitivity of the protein structure to minor conformational changes. It has been shown that points with higher values of chirality are special transition points in a protein, such as bends of the α-helix, twisting of β-chains [80]. The authors of [81] presented a method based on the consideration of a peptide framework as a helix with axial (d) and angular (θ) displacements, which were obtained based on the co-ordinates of the peptide framework and dihedral angles. The method is used to characterize each region of the Ramachandran plot for both cis ($\omega = 0°$) and trans ($\omega = 180°$) peptides.

There are several main approaches used in recognizing and modeling turns and loops. In [82], the parameters of loops and turns that are similar in the length and geometry of their endpoints are used. The mechanisms for searching for structural motifs are described in [83]: the extension of the DeepView/Swiss-PdbViewer algorithms allows for the determination of structural motifs in a large databases of protein structures. The knob-socket model serves to determine the role of coils and turns within a tertiary structure. In the works of various authors for β-turns, the dihedral angles φ, ψ are calculated on the Ramachandran plots [47,84–89]. Several statistical and computational methods for determining the structure of γ-turns are described. In particular, in [90], a method for recognizing γ-turns using neural networks was proposed. The determination of the structural characteristics of the loops seems to be especially difficult since their residues do not correspond to the pattern of dihedral angles or hydrogen bonds. The average prediction accuracy is primarily limited by the accuracy of the energy function, not by the degree of conformational sampling [91]. Attempts to determine rotations by including the interaction with the participation of atoms of the side chains in the analysis were made [92,93].

We considered the estimated chirality of the helices in [94,95]. In this paper, we describe in more detail the mathematical aspects of the solution. According to the mentioned method for the assessment of the chirality sign of helical structures, the mutual arrangement of α-carbons-reference points in the helices provides a sufficient condition [66,94,95]. The chirality sign of secondary helical structures can be assessed by the sum of mixed products for triples of vectors ($\mathbf{v}_i$) built between successive reference points Cα as follows:

$$\chi_{total} = \sum_{i=1}^{n-3} ([\mathbf{v}_i, \mathbf{v}_{i+1}], \mathbf{v}_{i+2}), \tag{1}$$

where the original vectors are calculated by taking into account the Cα coordinates presented in the PDB (Figure 1)

$$([\mathbf{v}_1, \mathbf{v}_2], \mathbf{v}_3) = (y_1 z_2 - y_2 z_1) x_3 + (z_1 x_2 - z_2 x_1) y_3 + (x_1 y_2 - x_2 y_1) z_3. \tag{2}$$

In [94], we proposed a model for the normalized value of the chirality of protein helical structures as follows:

$$\chi_{norm} = \sum_{i=1}^{n-3} \frac{([\mathrm{v}_i, \mathrm{v}_{i+1}], \mathrm{v}_{i+2})}{\mathrm{C}_i}, \tag{3}$$

where the normalization factor is calculated as $\mathrm{C}_i = \left(\frac{1}{3}\sum_{j=0}^{2} |\mathrm{v}_{i+j}|\right)^k$.

The chirality normalization was calculated based on considering the chirality value of the helix as a physical object, which should tend to a certain value with an increase in the number of reference points (that is, with an increase in the density of points). Therefore, each mixed product is normalized to the power k of the average length of the vectors.

To find the value of k, we establish the behavior of the chirality characteristic χ when the length of the vectors connecting the reference points changes. To execute this, consider a helix of radius R and height H. Let there be N reference points on this turn. We then performed all calculations in cylindrical coordinates. The origin of the coordinates coincides with the middle of the turn (Figure 2).

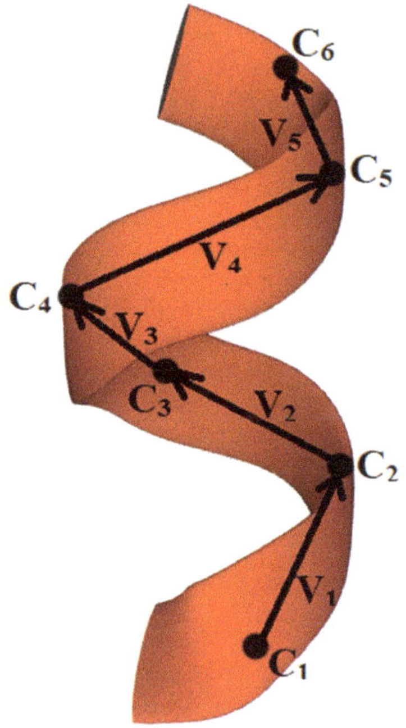

Figure 1. Graphic construction of vectors $\mathbf{v}_i$ for calculating the mixed vector product of helical protein structure (1L63 [96], α-helix, residues 45–50). C_i–atoms of α-carbons, reference points in helix.

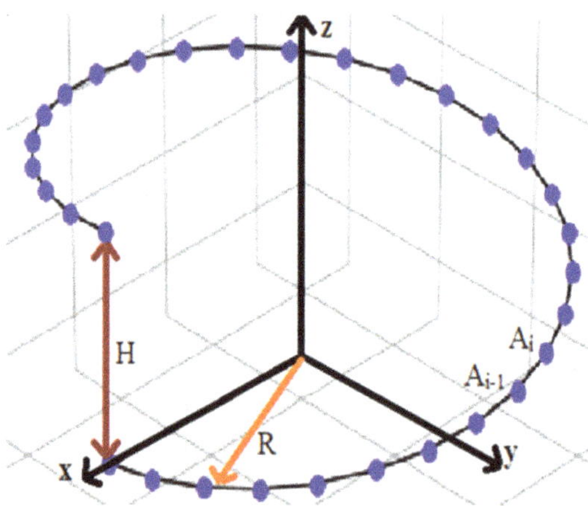

Figure 2. Normalization of chirality. The blue dots are reference points for this structure.

Then, the coordinates of the i-th point of the turn (denote it A_i):

$$A_i\ :(R\cos\varphi_i; R\sin\varphi_i; ih),\ where\ i = 0\ldots N,\ h = \frac{H}{N}, \varphi_i = i\varphi\left(\varphi = \frac{2\pi}{N}\right). \tag{4}$$

In accordance with this, the coordinates of the $\vec{r_i}$ vector connecting the points A_{i-1} and A_i have the following form:

$$\vec{r_i}\ :\ \{R\cos\varphi_i - R\cos\varphi_{i-1}; R\sin\varphi_i - R\sin\varphi_{i-1}; h\}. \tag{5}$$

Consider separately the x and y coordinates of the $\vec{r_i}$ vector:

$$x_i = R(\cos\varphi_i - \cos\varphi_{i-1}) = -2\sin[\frac{1}{2}(\varphi_i + \varphi_{i-1})]\sin[\frac{1}{2}(\varphi_i - \varphi_{i-1})] = -2\sin[(2i-1)\varphi]\sin\frac{\varphi}{2}, \tag{6}$$

$$y_i = R(\sin\varphi_i - \sin\varphi_{i-1}) = 2\sin[\frac{1}{2}(\varphi_i - \varphi_{i-1})]\cos[\frac{1}{2}(\varphi_i + \varphi_{i-1})] = 2\sin\varphi\cos[(2i-1)\varphi]. \tag{7}$$

Using the obtained values for the coordinates, we calculate the modulus (length) of the $\vec{r_i}$ vector:

$$\begin{aligned}|\vec{r_i}| &= \sqrt[2]{x_i{}^2 + y_i{}^2 + z_i{}^2} = \sqrt[2]{h^2 + R^2(\cos\varphi_i - \cos\varphi_{i-1})^2 + R^2(\sin\varphi_i - \sin\varphi_{i-1})^2} \\ &= \sqrt{h^2 + 4(\sin\frac{\varphi}{2})^2 R^2\left(\sin\left[\frac{(2i-1)\varphi}{2}\right]^2 + \cos\left[\frac{(2i-1)\varphi}{2}\right]^2\right)} = \sqrt{h^2 + 4(\sin\frac{\varphi}{2})^2 R^2}.\end{aligned} \tag{8}$$

The chirality characteristic, as already mentioned, is calculated using the following formula:

$$\chi = \sum_{i=1}^{N-3}([\vec{r_i}, \vec{r_{i+1}}], \vec{r_{i+2}}). \tag{9}$$

Consider the vector product in (6) and take into account that the coordinates of the vectors are calculated by Formulas (3) and (4). Then

$$\begin{aligned}[\vec{r_i}, \vec{r_{i+1}}] &= \begin{vmatrix} i & j & k \\ x_i & y_i & z_i \\ x_{i+1} & y_{i+1} & z_{i+1}\end{vmatrix} \\ &= \vec{i}\,hR\left(2\sin\tfrac{\varphi}{2}\cos\tfrac{2i-1}{2}\varphi - 2\sin\tfrac{\varphi}{2}\cos\tfrac{2i+1}{2}\varphi\right) \\ &+ \vec{j}\,hR\left(2\sin\tfrac{\varphi}{2}\sin\tfrac{2i-1}{2}\varphi - 2\sin\tfrac{\varphi}{2}\sin\tfrac{2i+1}{2}\varphi\right) \\ &+ \vec{k}\,4R^2\left(\sin\tfrac{\varphi}{2}\sin\tfrac{2i+1}{2}\varphi\sin\tfrac{\varphi}{2}\cos\tfrac{2i-1}{2}\varphi - \sin\tfrac{\varphi}{2}\cos\tfrac{2i+1}{2}\varphi\sin\tfrac{\varphi}{2}\sin\tfrac{2i-1}{2}\varphi\right).\end{aligned} \tag{10}$$

Substituting Formula (7) into Equation (6), we obtain:

$$\begin{aligned}([\vec{r_i}, \vec{r_{i+1}}], \vec{r_{i+2}}) &= 4hR^2(\sin\frac{\varphi}{2})^2[-\left(\cos\frac{2i-1}{2}\varphi - \cos\frac{2i+1}{2}\varphi\right)\sin\frac{2i+3}{2}\varphi \\ &+\left(\sin\frac{2i-1}{2}\varphi - \sin\frac{2i+1}{2}\varphi\right)\cos\frac{2i+3}{2}\varphi + (\sin\frac{2i+1}{2}\varphi\cos\frac{2i-1}{2}\varphi \\ &-\cos\frac{2i+1}{2}\varphi\sin\frac{2i-1}{2}\varphi)].\end{aligned} \tag{11}$$

Let us consider separately several terms in (11), taking into account trigonometric transformations:

$$\cos\frac{2i-1}{2}\varphi\sin\frac{2i+3}{2}\varphi = \frac{1}{2}(\sin(2i+1)\varphi + \sin 2\varphi), \tag{12}$$

$$\cos\frac{2i+1}{2}\varphi\sin\frac{2i+3}{2}\varphi = \frac{1}{2}(\sin(2i+1)\varphi + \sin 2\varphi), \tag{13}$$

$$\sin\frac{2i-1}{2}\varphi\cos\frac{2i+3}{2}\varphi = \frac{1}{2}(\sin(2i+1)\varphi - \sin 2\varphi), \tag{14}$$

$$\sin\frac{2i+1}{2}\varphi\cos\frac{2i+3}{2}\varphi = \frac{1}{2}(\sin(2i+2)\varphi - \sin\varphi), \tag{15}$$

$$\sin\frac{2i+1}{2}\varphi\cos\frac{2i-1}{2}\varphi = \frac{1}{2}(\sin 2i\varphi + \sin\varphi), \tag{16}$$

$$\sin\frac{2i-1}{2}\varphi\cos\frac{2i+1}{2}\varphi = \frac{1}{2}(\sin 2i\varphi - \sin\varphi). \tag{17}$$

By opening the brackets and substituting Formulas (12)–(17) into (11), we obtain an expression for the mixed product:

$$\begin{aligned}([\vec{r_i}, \vec{r_{i+1}}], \vec{r_{i+2}}) = {} & 2hR^2(\sin\frac{\varphi}{2})^2[\sin(2i+2)\varphi + \sin\varphi - \sin(2i-1)\varphi \\ & - \sin 2\varphi + \sin(2i+1)\varphi - \sin 2\varphi - \sin(2i+2)\varphi + \sin\varphi + \sin 2i\varphi + \sin\varphi - \sin 2i\varphi \\ & + \sin\varphi] = 2hR^2(\sin\frac{\varphi}{2})^2[4\sin\varphi - 2\sin 2\varphi].\end{aligned} \tag{18}$$

The final expression for characterizing chirality is obtained by a summation of all the mixed products:

$$\chi = \sum_{i=1}^{N-3}([\vec{r_i}, \vec{r_{i+1}}], \vec{r_{i+2}}) = 16(N-3)HR^2(\sin\frac{\varphi}{2})^4\sin\varphi. \tag{19}$$

Consider the behavior of χ (19) and $|\vec{r}|$ (8) as the number of reference points tends to infinity:

$$\lim_{N\to\infty}\chi = \lim_{N\to\infty}16(N-3)HR^2(\sin\frac{\pi}{N})^4\sin\frac{2\pi}{N} = \lim_{N\to\infty}\frac{32(N-3)HR^2\pi^5}{N^5} \sim \frac{1}{N^5}, \tag{20}$$

$$\lim_{N\to\infty}|\vec{r}| = \lim_{N\to\infty}\sqrt{\frac{H}{N^2}^2 + 4(\sin\frac{\pi}{N})^2R^2} = \lim_{N\to\infty}\frac{1}{N}\sqrt{H^2+4\pi^2R^2} \sim \frac{1}{N}. \tag{21}$$

Thus, as $N\to\infty$ $\chi \sim |\vec{r}|^5$, therefore,

$$\chi_{norm} = \frac{\chi}{|\vec{r}|^5}. \tag{22}$$

Thus, to preserve the finite nonzero chirality with an unlimited increase in the number of points, one should take $k = 5$.

The assessment of the chirality of irregular protein secondary structures is similar to that of regular structures. In this article, the chirality of irregular protein secondary structures was assessed using the example of β-, α-turns and Ω-loops (Figures 3–5). A sufficient condition is provided by the relative position of the α-carbon reference points in turns and loops.

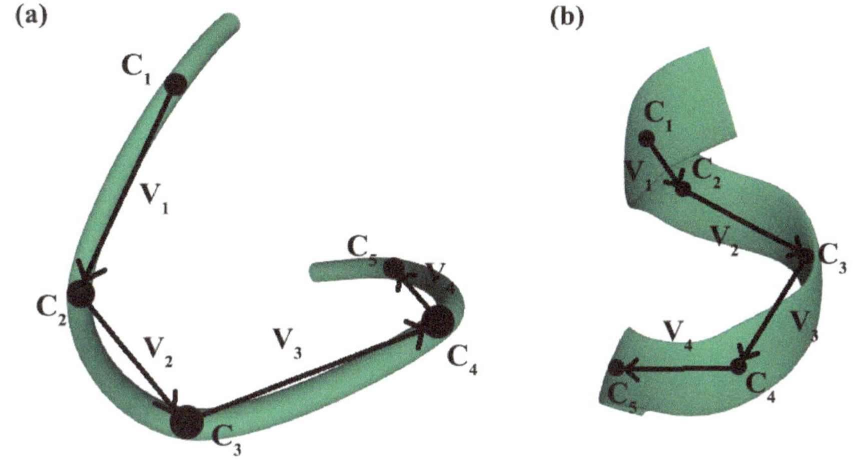

Figure 3. Graphic construction of $\mathbf{v}_i$ vectors for calculating the mixed vector product, α-turns of the 2FOX protein [97]: (**a**) α-turn, type g4 [98]; (**b**) α-turn, type g5 [98] (the studied types of α-turns are discussed in the Results section). C_i –atoms of α-carbons, reference points in α-turns.

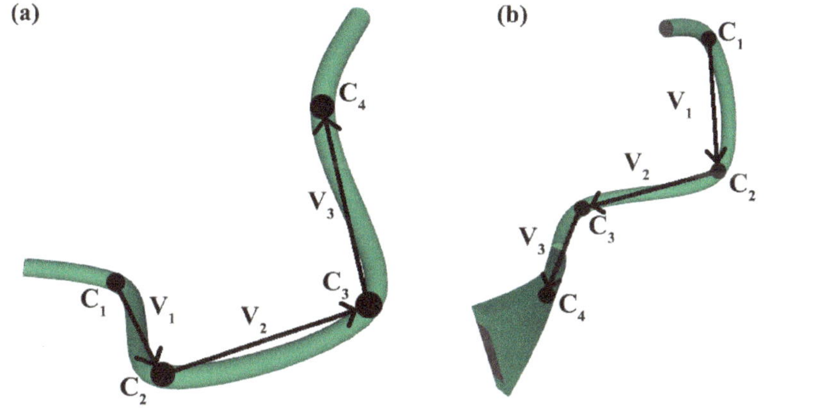

Figure 4. Graphic construction of $\mathbf{v}_i$ vectors for calculating the mixed vector product, β-turns of the 1A4G protein [99]: (**a**) β-turn, type AG [89]; (**b**) β-turn, type Dd [89] (the studied types of β-turns are discussed in the Results section). C_i –atoms of α-carbons, reference points in β-turns.

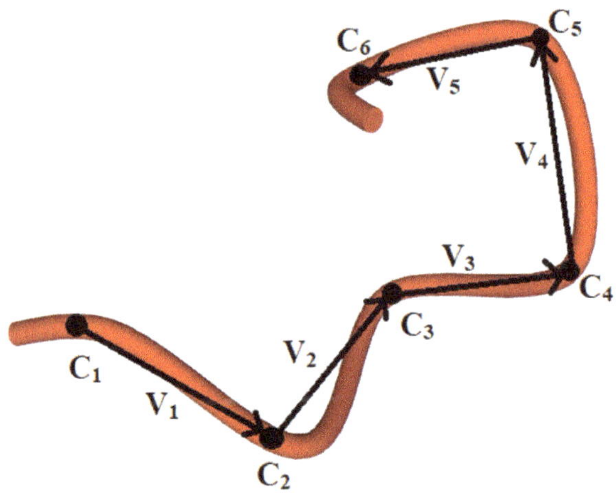

Figure 5. Graphic construction of $\mathbf{v}_i$ vectors for calculating the mixed vector product, Ω-loop of the 2ACT protein [100] (residues 8–13). C_i –atoms of α-carbons, reference points in the Ω-loop.

The results of calculating the chirality of regular (helical) and irregular (turns and loops) protein secondary structures are presented in the Results section.

2.3. Method for Calculating the Chirality of Phenylalanine (F) Helical Peptide Nanotubes (PNT) from Successive Dipole Moments of Their Constituent Phenylalanine Molecules

Recently, in [60], a new approach was proposed for modeling various molecular nanostructures, determining the implementation of the molecular dynamics simulation (MDS) run trajectory, and forming the final structure using the so-called molecular dynamics manipulator (MDM). This approach is a type of MDS, developed based on the PUMA-CUDA software package [101,102], using the physics of the PUMA software package [103,104]. Using this tool allows for the exploration of the formation process of helical structures from a linear sequence of any amino acids. It was used in [60] to assemble nanotubes from linear phenylalanine chains of different chirality (L-F and D-F) by including additional force effects in the molecular dynamics simulation program for these structures.

In this work, using the obtained helical structures of phenylalanine nanotubes of different chirality [60], we calculate the magnitude and sign of their chirality using a method similar to that developed in [94,95] and applied in [66] to diphenylalanine helical structures, based on the procedure of the mixed product of three consecutive vectors in a coil of a spiral structure. To achieve this, we select one coil of the helix from each nanotube of different chirality and apply this calculation method to them based on the mixed product of vectors of dipole moments from a number of successive phenylalanine molecules that form this coil of the helix of a phenylalanine peptide nanotube (PNT), taking into account its chirality.

Helical-like PNT structures, based on phenylalanine of different chirality (L and D), were obtained as a result of MD simulation (MDS) and their assembly [60], and consist of 48 F molecules and 4 coils in such structures of nanotubes of each chirality L-F48 and D-F48 (with coordinates of all atoms in standard *.pdb format) (Figure 6, Video S1).

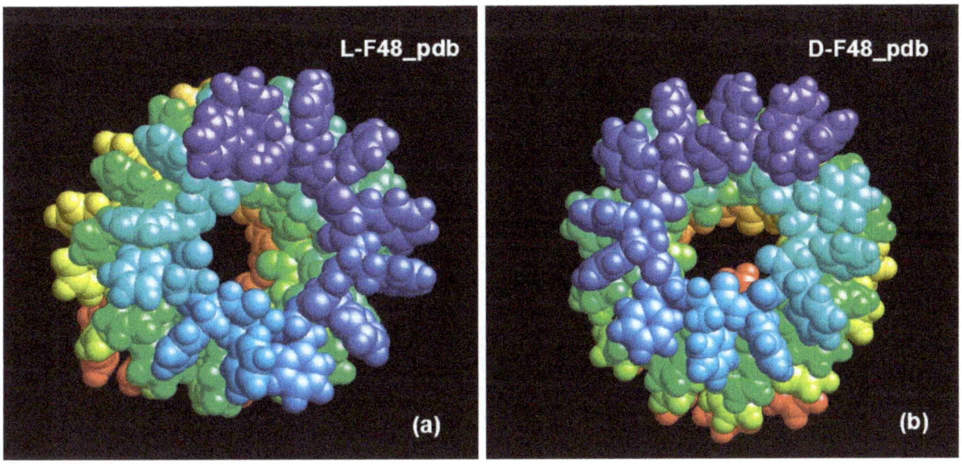

Figure 6. The obtained results of MDS and self-assembly of phenylalanine helical-like nanostructures: (**a**) L-F48_pdb; (**b**) D-F48_pdb (images were obtained from *.pdb files using the RasMol program http://www.openrasmol.org/ (accessed on 30 September 2021)).

We transferred dipole moments to the HyperChem [105] workspace (in *.hin format with Cartesian x, y, z coordinates for all atoms) for analysis (Figure 7).

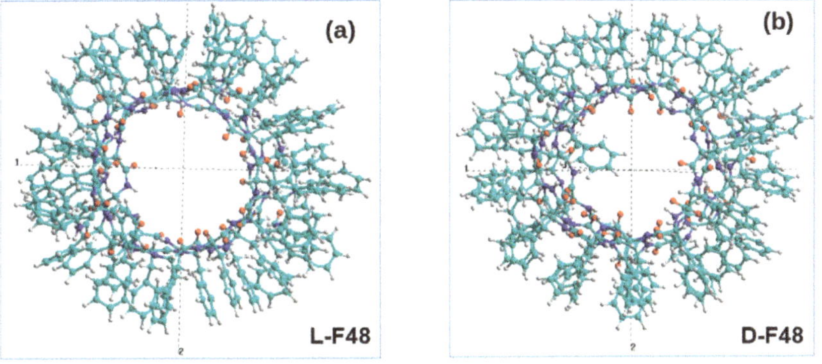

Figure 7. The obtained results of MDS and self-assembly of phenylalanine helical-like PNT nanostructures, transferred to HyperChem workspace (in Z-projection): (**a**) F48L PNT; (**b**) F48D-PNT.

We selected an individual coil at sequentially from each PNT helix consisting of four coils. When calculating the dipole moments of these coils we found that they have opposite directions to the vectors of the total dipole moments $\mathbf{D_{L\text{-}F}}$ for each L-F and $\mathbf{D_{D\text{-}F}}$ for each D-F coils (Figure 8b,d).

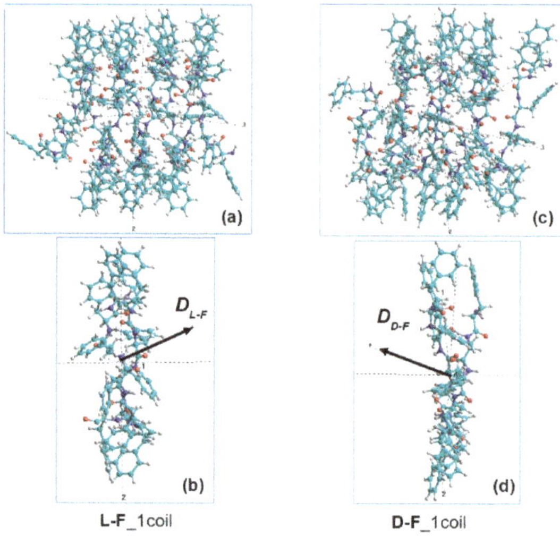

Figure 8. Selection of the one coil from the four in PNT helix-like structures (in X-projection): (**a**, **b**)–for the L-F PNT; (**c**, **d**)–for the D-F PNT.

We then selected each consequential phenylalanine F molecule from corresponding coil, containing 12 F molecules, and calculated its dipole moment $\mathbf{D}_i$ using various methods (from HyperChem package [105,106]), including the following: (1) quantum-chemical semi-empirical RM1 RHF [106]; (2) classical molecular mechanical Amber method [105]. This procedure schematically shown in Figure 9.

Similarly to the calculation of diphenylalanine nanotubes [66], we use a similar successive set of F molecules for phenylalanine nanotubes. The origin of $\mathbf{D}_i$ vectors is obtained relative to the center of mass of the corresponding molecules. The absolute value of each dipole moment $\mathbf{D}_i$ is

$$\mathbf{D}_i = |\mathbf{D}_i| = \sqrt{D_{x,i}^2 + D_{y,i}^2 + D_{z,i}^2}, \tag{23}$$

where $D_{x,i}$, $D_{y,i}$, and $D_{z,i}$ are the components of the i-th vector $\mathbf{D}_i$ in the Cartesian coordinates. Similar to Equation (1) [94,95], the sum of the scalar triple products of the dipole moments related to the PNT's chirality can be written as:

$$c_{total} = \sum_{i=1}^{n-2} ([\mathbf{D}_i, \mathbf{D}_{i+1}], \mathbf{D}_{i+2}), \tag{24}$$

It is necessary to note that the summation here has taken over i in the range from 1 to n-2, whereas in Equation (1), the i range is from 1 to n-3. Now n = 12. This is because in supramolecular helixes i numerates the individual molecules instead of the Cα atoms in proteins. The c_{total} can be normalized over the average value of the total dipole momentum of the PNT's coil, $D_{av} = \frac{1}{12}\sum_{i=1}^{12} D_i$, to obtain a universal measure of the chirality:

$$c_{norm} = \frac{c_{total}}{D_{av}^3}. \tag{25}$$

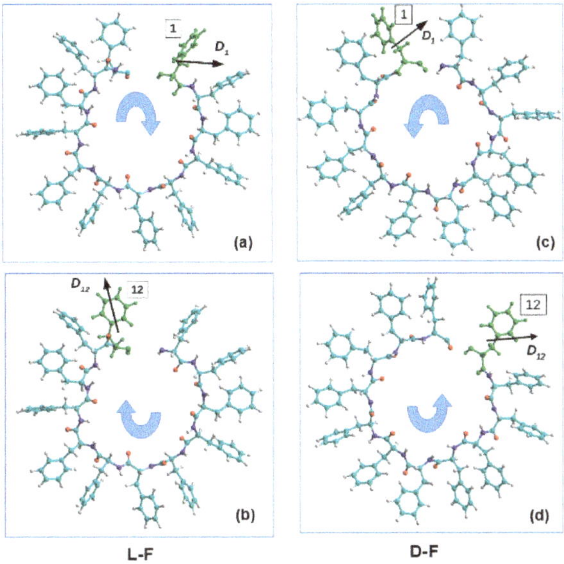

Figure 9. Schematic of the selection procedure of each consequential phenylalanine F molecule (from 1 up 12) from one corresponding coil of helix and the calculation of its dipole moment $\mathbf{D}_i$ (for i = 1, ..., 12) (in Z-projections): (**a**,**b**)–for the L-F PNT; (**c**,**d**)–for the D-F PNT, correspondingly.

Individual dipole moments of F molecules in one coil of helical PNTs were calculated using the semi-empirical quantum-mechanical method PM3 in the restricted Hartree–Fock approximation (RHF) and molecular force field method Amber from the HyperChem package [105,106]. The results for L-F and D-F are shown below in the Results section. A schematic representation of the spatial arrangement of F of the individual dipole moments $\mathbf{D}_i$ in two PNT coils is presented in Figure 9a–d for L-F and for D-F PNT. The obtained results of calculating the magnitude and sign of chirality by Formulas (23)–(25) for each case (L and D) are presented in the Results section.

3. Results

3.1. Helical Protein Secondary Structures

Using the developed method, files from PDB with the data of 983 proteins of various classes were considered. The chirality of helical structures was calculated, and data for α- and 3_{10}-helices (oxidoreductase, transferase, hydrolase, lyase, isomerase, ligase, translocase, chaperones, viruses, structural proteins, endo- and exocytosis proteins, electron transport-protein data of chirality) are presented in [95], as well as for π-helices (all π-helices are taken from [44]). In accordance with Formula (3), a map of normalized chirality for the considered helical protein structures is presented (Figure 10).

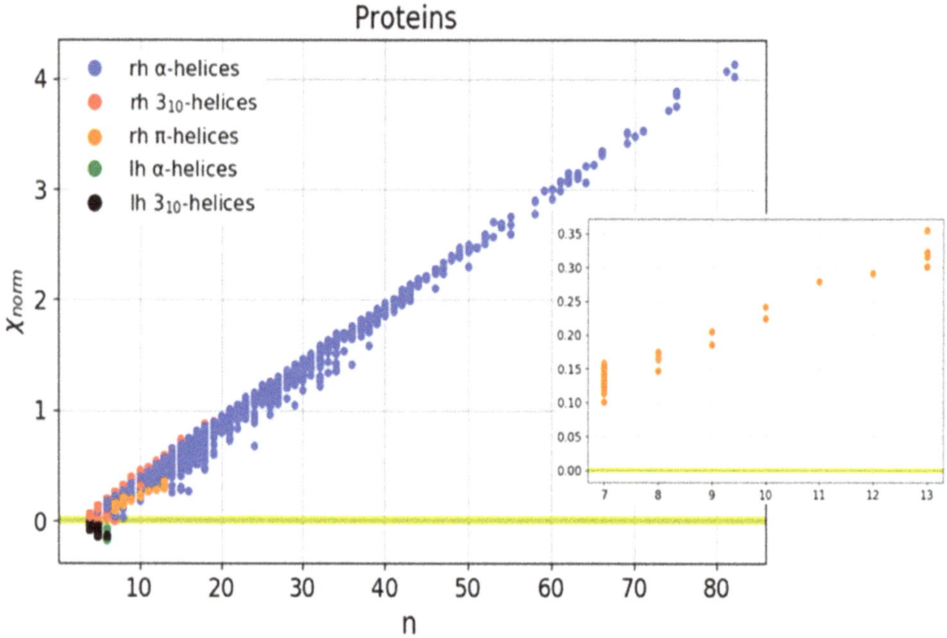

Figure 10. Normalized chirality map of protein helical structures. The horizontal axis is the length of the secondary structure in amino acid residues, the vertical axis is the normalized chirality value. The inset shows a map of π-helices; rh–right-handed structures, lh–left-handed structures.

The results obtained showed that for all the considered regular protein structures (helices), the developed measure of chirality (χ_{norm}) linearly depends on the number of atoms in the helix. Secondly, the results obtained are consistent with the literature data on the predominant conformation of right-handed helical structures.

The calculated data for α and 3_{10}-helices were presented previously in our work [95]. The calculated parameters of π-helices are presented in Table A1.

3.2. Irregular Protein Secondary Structures

We estimated the chirality of α-turns (Figure 3) on the basis of the data presented in [98], where all α-turns are classified depending on the value of the torsion angles (φ, ψ) for various amino acid residues that comprise the turns. Based on this, the authors distinguish 9 types of α-turns, including two types of F1 and F2, called families, as well as seven less common types g1, g2, g3, g4, g5, g6, g7, termed groups. In addition, the two turns identified were not included in any of the groups. We examined 78 α-turns identified by the authors of [98], calculated the chirality for each turn using our described method, and obtained the mean values and standard deviations of chirality for each type. It follows that most of these types have certain chirality values characteristic of each type of α-turn. The calculation results are presented in Table 1. The classes for α-turns are presented in Table A2.

Table 1. Mean values of chirality and standard deviations for α-turns of various types [98], calculated using the method for evaluating the chirality of regular helical and irregular protein secondary structures.

Type of α-Turn	Number	Mean Chirality Value	Standard Deviation
F1	46	0.08628	0.01473
F2	8	0.06922	0.013
g1	5	−0.03274	0.00846
g2	2	−0.04165	0.00864
g3	4	0.02838	0.02813
g4	4	−0.06686	0.00568
g5	3	0.10221	0.00191
g6	3	−0.00376	0.00415
g7	2	0.09378	0.02175
Other	2	0.0323	0.05972

For β-turns (Figure 4), we performed similar calculations using the database presented in [89]. Article [89] presents a new classification of β-turns based on an algorithm for their identification and recognition, implemented in the form of a computer program. The authors divided all β-turns into 18 types, in addition to those that were not included in any of the newly formed groups, based on the following criteria: the distance between the first and last residues of turn, the values of torsion angles (φ, ψ) for the second and third amino acid residues, and the conformation of these residues relative to peptide bonds (cis/trans). Using the computer program of the authors of [89], we distinguished 850 uniquely determined β-turns from 20 proteins taken from the PDB [107] and calculated the mean values and standard deviations of chirality for each type (Table 2).

Table 2. Mean values and standard deviations of chirality for β-turns of various types (according to [89]), calculated using the method for evaluating the chirality of regular helical and irregular protein secondary structures. The proteins with β-turns are hydrolases.

Type of β-Turn	Number	Mean Chirality Value	Standard Deviation
AD	481	0.05041	0.01634
Pd	61	0.02132	0.01371
Pa	36	0.00392	0.02101
ad	60	−0.05916	0.00556
AB1	17	0.00647	0.01820
AZ	16	0.04043	0.01255
AB2	3	0.04561	0.00151
pD	30	−0.01805	0.02312
AG	6	0.06884	0.00086
BcisP	10	0.03700	0.01090
dD	5	0.05961	0.01223
PcisD	10	0.00591	0.01596
dN	6	0.05235	0.00206
Dd	5	−0.06710	0.00107
cisDA	6	0.06506	0.00022
pG	7	0.00086	0.02898
cisDP	3	0.04490	0.02358
other	88	0.02855	0.03472

The calculation results showed that, for most types of β-turns, there are certain ranges of chirality values, which also confirms the correctness of such a classification of these structures s found in [89]. Based on the calculations performed (Tables 1, 2 and A2), a chirality map was obtained for the considered α- and β-turns (Figure 11).

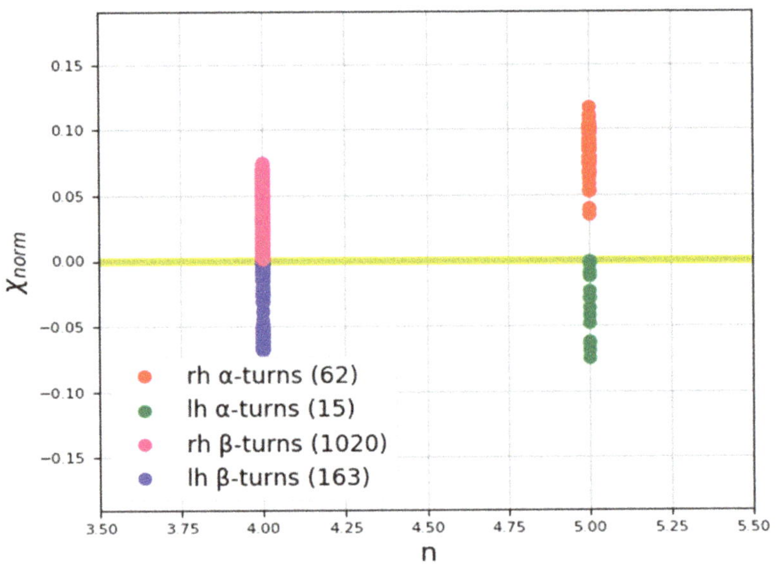

Figure 11. Normalized chirality map of α- and β-turns (mean values of various types presented in Tables 1 and 2). The horizontal axis is the length of the secondary structure in amino acid residues, the vertical axis is the normalized chirality value; rh–right-handed structures, lh–left-handed structures.

An analysis of the normalized chirality map of α- and β-turns (Figure 11) indicates that, for all the considered proteins, the measure of chirality (χ_{norm}) of turns linearly depends on the number of atoms in these structures (see data in Tables 1 and 2). However, since all considered β-turns consist of 4 amino acid residues, and α-turns consist of 5 residues, they are located parallel to the Y-axis on the map.

To calculate the chirality of the Ω-loops (Figure 5), the data in [57] were utilized. The chirality calculation data for 190 Ω-loops are presented on the chirality map (Figure 12) and in Table A3.

The spatial orientation of the loops clearly affects the quantitative values of the chirality of these structures. Let us consider a stepwise change in chirality in the process of calculating the total chirality of the loop structure using the example of a loop from the 2ACT protein (Figure 5). Depending on the number of consecutive residues used for calculation, the chirality can be altered either upwards or downwards (Table 3).

Table 3. Stages of calculating the chirality value of the Ω-loop, 2ACT protein (residues 8–13) [100].

Number of Residues	Number of Vectors	Number of Mixed Products	Addition to the Chirality Value at This Step	Total Chirality Value at This Step
1	–	–	0	0
2	1	–	0	0
3	2	–	0	0
4	3	1	0.0554	0.0554
5	4	2	−0.0631	−0.0077
6	5	3	0.0585	0.0508

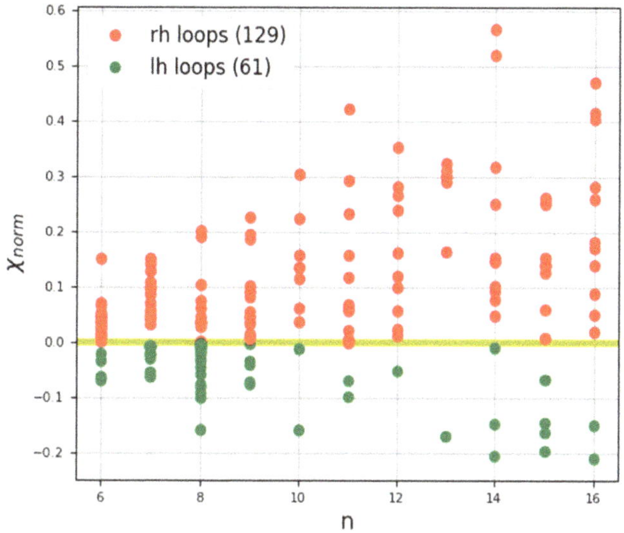

Figure 12. Normalized chirality map of Ω-loops. The horizontal axis is the length of the secondary structure in amino acid residues, the vertical axis is the normalized chirality value; rh–right-handed structures, lh–left-handed structures.

3.3. Phenylalanine (F) Helix-like Peptide Nanotubes

Following the method for calculating the chirality of phenylalanine (F) helix-like peptide nanotubes (PNT), the results of calculating the dipole moments for the sequence of individual amino acids of phenylalanine F from the turns of spiral nanotubes of different chirality are presented. Nanotubes were obtained by coiling a linear sequence of amino acid F, initially with a different chirality of L-F and D-F (by the molecular dynamics method according to [60]). The obtained results of calculating the magnitude and sign of chirality by Formulas (23)–(25) for each case (L and D) are presented in Tables 4–6.

Table 4. Values of dipole moments for a coil of the helix-like L-F PNT computed using RM1 (RHF) and Amber (after RM1) methods. All values of dipole moments are given in Debye units.

i	RM1 RHF				Amber			
	Di	*Dx*	*Dy*	*Dz*	*Di*	*Dx*	*Dy*	*Dz*
1	2.730	2.625	−0.697	0.282	2.915	2.740	−0.955	−0.282
2	3.400	2.884	0.360	−1.765	2.937	2.232	1.075	−1.578
3	2.488	1.645	1.503	1.106	2.624	0.984	1.348	2.025
4	2.615	1.710	1.869	0.650	2.671	2.086	1.603	0.463
5	2.558	−0.760	2.203	1.054	2.844	−1.325	1.618	1.928
6	2.449	−0.956	2.224	0.370	2.564	−0.554	2.429	−0.608
7	2.997	−1.096	0.851	−2.656	2.372	−1.185	−0.194	−2.046
8	2.258	−1.456	0.790	−1.534	1.631	−0.631	0.534	−1.406
9	2.436	−1.265	−1.928	−0.785	2.691	−2.137	−1.636	−0.026
10	2.611	−1.789	−1.795	0.630	1.504	−1.081	−0.999	0.310
11	1.887	−0.942	−0.771	−1.442	1.165	−1.135	−0.254	0.068
12	2.201	−0.996	−1.873	0.586	1.771	−0.819	−1.533	−0.338
Dsum	30.630	−0.396	3.433	−3.504	27.692	2.111	3.036	−2.138
Dcoil	9.995	6.624	2.820	−6.934	6.3004	4.2597	2.0522	−4.164
Dav	2.553				2.308			

RM1: $(Dav)^3$ = 16.640 Debye^3; Amber: $(Dav)^3$ = 12.294 Debye^3.

Table 5. Values of dipole moments for a coil of the helix-like D-F PNT computed using RM1 (RHF) and Amber (after RM1) methods. All values of dipole moments are given in Debye units.

i	RM1 RHF				Amber			
	Di	*Dx*	*Dy*	*Dz*	*Di*	*Dx*	*Dy*	*Dz*
1	2.992	−2.037	2.152	0.414	3.041	−2.109	1.870	1.142
2	3.676	−3.516	−0.872	−0.630	2.980	−2.724	−0.862	−0.846
3	2.864	−2.084	−1.822	−0.735	2.727	−1.612	−1.660	−1.443
4	2.943	0.259	−2.654	1.244	2.797	−0.790	−2.055	1.726
5	2.746	0.852	−2.590	−0.325	2.780	1.098	−2.542	0.245
6	2.821	1.415	−2.392	−0.483	3.192	1.285	−2.357	−1.726
7	3.781	1.444	−3.251	1.282	3.691	1.197	−2.894	1.953
8	2.503	1.317	2.048	0.578	2.387	2.070	1.060	−0.539
9	2.888	1.720	1.513	1.758	2.483	1.376	1.563	1.353
10	3.282	−1.128	2.525	1.766	2.732	−0.748	2.323	1.230
11	3.762	−3.239	1.798	0.657	3.072	−2.448	1.774	0.547
12	2.667	−2.229	0.793	−1.231	2.836	−2.774	0.589	0.034
Dsum	36.925	−7.224	−2.752	4.294	34.718	−6.179	−3.191	3.675
Dcoil	9.599	−6.395	−5.676	4.362	8.234	−4.333	−5.081	4.818
Dav	3.077				2.893			

RM1: $(Dav)^3$ = 29.134 Debye^3; Amber: $(Dav)^3$ = 24.217 Debye^3.

Table 6. Magnitudes and signs of the chirality obtained for L-F and D-F PNTs for various calculating methods.

Type of PNT	L-F		D-F	
Calculating method	RM1 RHF	Amber	RM1 RHF	Amber
c_{total}, Debye^3	20.266	18.171	−19.647	−26.204
c_{norm}	1.219	1.479	−0.674	−1.082
Chirality sign	positive	positive	negative	negative
Chirality symbol	D	D	L	L

Table 6 shows the magnitude and sign of chirality, calculated using the formulas for the mixed product of dipole moments (23)–(25).

The obtained results (Table 6) reveal a characteristic change in the sign of chirality during transition to a higher level of organization, which is observed in the structures of biomacromolecules [8,9]. The calculated chirality of a spiral nanotube, based on the L-F initial amino acid, was found to have a positive sign-D type, and the chirality of the D-F-based nanotube has a negative sign corresponding to the L chirality type.

Note that the data presented here for the cube of average values for the absolute value of the total dipole moment $D_{av} = \frac{1}{12}\sum_{i=1}^{12} D_i$ of each of the F amino acids, represent the average volume built on three consecutive vectors of the mixed product of these vectors. For each group of three such vectors, the calculated value of the magnitude of their mixed product changes, corresponding to the volume on these three vectors. hereby normalizing for the average value $(D_{av})^3$ according to Formula (25), we obtain the relative change in volume, with a slight change around "1". Moreover, for the left-handed and right-handed triplets of vectors, the corresponding volumes have different signs, which leads to a change in sign in this case.

4. Discussion

A chirality analysis was conducted for 26,150 helical structures, namely, 21,702 α-helices, 4360 3_{10}-helices, and 88 π-helices (all studied π-helices were taken from [44]). Research has shown that most of the helical structures are right-handed. Among the structures studied, we found 21,689 right-handed α-helices, 4160 right-handed 3_{10}-helices, and 88 right-handed π-helices. To study the chirality of π-helices, 84 proteins were analyzed, including 23 oxidoreductases, 22 hydrolases, 8 lyases, 7 transferases, 5 isomerases, 4 binding proteins, 2 toxins, 2 photosynthetic proteins, 2 signal proteins, 1 electron transport protein, 1 luminescent protein, 1 viral protein, 1 protein of endo- and exocytosis, 1 oxygen transport protein, 1 ion transport protein, 1 antibiotic, 1 adhesion protein and 1 iron transport protein (Table A1).

The obtained results show that, for regular helical protein secondary structures (the data for the chirality of α– and 3_{10}– helices are presented in [95]), the data for π-helices are in Table A1), the developed measure of chirality (χ_{norm}) linearly depends on the number of atoms in the helix (Figure 10). For irregular protein secondary structures (β-turns-Table 2, α-turns-Table A2, Ω-loops-Table A3), a different picture emerges. Since all of the considered β-turns consist of 4 amino acid residues, and α-turns consist of 5 residues, they were found to be located parallel to the Y-axis on the chirality map (Figure 11). The spatial orientation of Ω-loops, in contrast to turns, consists of different numbers of amino acid residues, but unlike regular helices, they are characterized by different spatial orientations. These features affect the quantitative values of the chirality of the loops, whereby, depending on the number of consecutive residues taken for the calculation, the chirality can be altered either upwards to or downwards (Figure 12, Table 3).

Irregular protein secondary structures connect regular protein secondary structures and play a key role in the formation of a protein globule [56]. Given the frequency of occurrence of α-, β-turns, and Ω-loops, this study acts as a useful tool for studying the structure of proteins, both natural and artificial, as well as for protein design and materials science.

The results obtained for calculating the magnitude and sign of chirality, for L- and D-nanotubes based on phenylalanine (Tables 4–6), are similar to the data of other works on modeling peptide and dipeptide nanotubes of different chirality and experimental data [61–63,108]. These data also fully comply with the regularity of the change in the chirality sign of molecular structures with the complication of their hierarchical level of organization [8,9]. Therefore, this method for calculating the magnitude and sign of chirality by the mixed product method of Sidorova et al. [94,95], using the values of dipole moments in the sequence of individual peptides and dipeptides (or amino acids) as vectors, is found to be suitable, and can be successfully applied for assessments on the magnitude and sign of chirality of complex self-organizing helix-like nanostructures based on amino acids, as well as peptides and dipeptides.

Supplementary Materials: The following are available online at https://www.mdpi.com/article/10.3390/nano11123299/s1, Video S1: Assembly of a left-handed nanotube based on 48 phenylalanine D-monomers.

Author Contributions: Conceptualization, A.S. and V.B.; methodology, A.S., A.L., D.S., I.L. and V.B.; software, A.L., D.S. and I.L.; validation, A.L., D.S. and I.L.; formal analysis, A.S., V.B. and E.B.; writing—original draft preparation, A.S. and V.B.; writing—review and editing, A.S. and E.B.; visualization, A.L., D.S., I.L. and V.B.; supervision, A.S. and V.B. All authors have read and agreed to the published version of the manuscript.

Funding: This research was funded by the Russian Foundation for Basic Research, grant n. 20-51-53014_GFEN_a.

Institutional Review Board Statement: Not applicable.

Informed Consent Statement: Not applicable.

Data Availability Statement: The data presented in this study are available on request from the corresponding author.

Acknowledgments: The authors are grateful to the Russian Foundation for Basic Research, grant n. 19-01-00519-a. The authors are grateful to V.A. Tverdislov for useful discussions.

Conflicts of Interest: The authors declare no conflict of interest.

Appendix A

Table A1. The calculated parameters of π-helices.

Protein PDB ID	Chain/Residues	Length of Helix	Chirality Value	Chirality Sign	Protein Class
1A8E	A/124-130	7	0.1150	Right-handed	Iron transport
1A8I	A/488-495	8	0.1632	Right-handed	Transferase
1BDM	A/217-223	7	0.1350	Right-handed	Oxidoreductase
1BG6	A/297-303	7	0.1373	Right-handed	Oxidoreductase
1D3G	A/37-43	7	0.1283	Right-handed	Oxidoreductase
1DK8	A/242-249	8	0.1711	Right-handed	Signaling protein
1DYS	A/112-118	7	0.1250	Right-handed	Hydrolase
1DZ4	A/150-156	7	0.1273	Right-handed	Oxidoreductase
1E3A	A/138-146	9	0.2050	Right-handed	Hydrolase
1EGU	A/292-298	7	0.1323	Right-handed	Lyase
1EGU	A/441-447	7	0.1229	Right-handed	Lyase
1EI5	A/177-183	7	0.1299	Right-handed	Hydrolase
1EK6	A/105-111	7	0.1502	Right-handed	Isomerase
1EL4	A/44-51	8	0.1665	Right-handed	Luminescent protein
1ELK	A/95-101	7	0.1421	Right-handed	Endocytosis/Exocytosis
1EOK	A/111-118	8	0.1670	Right-handed	Hydrolase
1EVY	A/257-263	7	0.1543	Right-handed	Oxidoreductase
1F3A	A/126-132	7	0.1235	Right-handed	Transferase
1F24	A/214-220	7	0.1336	Right-handed	Oxidoreductase
1FQA	A/279-285	7	0.1304	Right-handed	Sugar binding protein
1KVE	B/177-183	7	0.1271	Right-handed	Toxin
1MRO	A/313-324	12	0.2911	Right-handed	Transferase
1MUC	A/70-76	7	0.1270	Right-handed	Isomerase
1NCI	A/40-46	7	0.1125	Right-handed	Cell adhesin protein
1QGW	C/105-111	7	0.1287	Right-handed	Photosynthesis
1QH3	A/154-160	7	0.1316	Right-handed	Hydrolase
1QH8	A/63-72	10	0.2242	Right-handed	Oxidoreductase
1QLM	A/88-94	7	0.1216	Right-handed	Hydrolase
1QMG	A/349-355	7	0.1503	Right-handed	Oxidoreductase
1QMG	A/490-502	13	0.3543	Right-handed	Oxidoreductase
1SMD	A/27-33	7	0.1276	Right-handed	Hydrolase
1SUR	A/131-137	7	0.1303	Right-handed	Oxidoreductase
1THG	A/424-430	7	0.1360	Right-handed	Hydrolase
2SCP	A/56-62	7	0.1344	Right-handed	Binding protein
1B16	A/104-110	7	0.1277	Right-handed	Oxidoreductase
1B25	A/479-485	7	0.1247	Right-handed	Oxidoreductase
1BDB	A/112-118	7	0.1379	Right-handed	Oxidoreductase
1BXK	A/98-104	7	0.1439	Right-handed	Lyase
1C3P	A/97-103	7	0.1447	Right-handed	Lyase
1C3W	A/213-219	7	0.1543	Right-handed	Ion transport
1C7S	A/641-647	7	0.0941	Right-handed	Hydrolase
1C7S	A/801-807	7	0.1273	Right-handed	Hydrolase
1CB8	A/267-273	7	0.1317	Right-handed	Lyase
1COJ	A/26-33	8	0.1745	Right-handed	Oxidoreductase
1CXP	C/291-297	7	0.1234	Right-handed	Oxidoreductase
1CYD	A/104-110	7	0.1336	Right-handed	Oxidoreductase
1D3Y	A/253-259	7	0.1317	Right-handed	Isomerase
1D8D	A/343-349	7	0.1146	Right-handed	Transferase
1DC1	A/98-104	7	0.1341	Right-handed	Hydrolase
1DEK	A/137-145	9	0.1858	Right-handed	Transferase
1DJ0	A/81-87	7	0.1546	Right-handed	Lyase
1DOZ	A/265-274	10	0.2409	Right-handed	Lyase
1DQA	A/733-740	8	0.1738	Right-handed	Oxidoreductase
1DQS	A/142-148	7	0.1262	Right-handed	Lyase
1DXR	C/277-283	7	0.1231	Right-handed	Photosynthesis
1DXR	H/27-33	7	0.1234	Right-handed	Photosynthesis
1DXR	L/129-135	7	0.1276	Right-handed	Photosynthesis
1DXR	M/156-162	7	0.1229	Right-handed	Photosynthesis
1EA5	A/396-402	7	0.1241	Right-handed	Hydrolase
1EA5	A/522-528	7	0.1313	Right-handed	Hydrolase
1EWF	A/181-187	7	0.1379	Right-handed	Antibiotic
1EYZ	A/119-125	7	0.1426	Right-handed	Transferase
1F24	A/140-146	7	0.1259	Right-handed	Oxidoreductase
1FDS	A/111-117	7	0.1373	Right-handed	Oxidoreductase

Table A1. *Cont.*

Protein PDB ID	Chain/Residues	Length of Helix	Chirality Value	Chirality Sign	Protein Class
1FP3	A/273-279	7	0.1241	Right-handed	Isomerase
1FRP	A/276-282	7	0.1304	Right-handed	Hydrolase
1FSW	A/174-180	7	0.1318	Right-handed	Hydrolase
1FUR	A/155-161	7	0.1348	Right-handed	Hydrolase
1FUR	A/383-389	7	0.1302	Right-handed	Hydrolase
1G8K	A/181-187	7	0.1299	Right-handed	Oxidoreductase
1G8K	A/242-248	7	0.1418	Right-handed	Oxidoreductase
1GAI	A/150-156	7	0.1346	Right-handed	Hydrolase
1HFE	S/71-77	7	0.1228	Right-handed	Oxidoreductase
1HVB	A/183-189	7	0.1282	Right-handed	Hydrolase
1I0H	A/26-32	7	0.1402	Right-handed	Oxidoreductase
1LML	A/155-161	7	0.1243	Right-handed	Hydrolase
1LST	A/126-132	7	0.1449	Right-handed	Amino-acid binding protein
1LST	A/165-171	7	0.1207	Right-handed	Amino-acid binding protein
1MTY	B/140-150	11	0.2786	Right-handed	Oxidoreductase
1MTY	B/297-304	8	0.1463	Right-handed	Oxidoreductase
1MTY	D/185-191	7	0.1139	Right-handed	Oxidoreductase
1MTY	D/202-214	13	0.3008	Right-handed	Oxidoreductase
1MTY	D/306-318	13	0.3226	Right-handed	Oxidoreductase
1MTY	D/379-385	7	0.1590	Right-handed	Oxidoreductase
1ONE	A/67-73	7	0.1322	Right-handed	Lyase
1PHN	A/107-113	7	0.1335	Right-handed	Electron transport
1QOY	A/26-32	7	0.1315	Right-handed	Toxin
1SVF	A/171-177	7	0.1381	Right-handed	Viral protein
1UOK	A/393-399	7	0.1330	Right-handed	Hydrolase
1YAC	A/114-120	7	0.1410	Right-handed	Hydrolase
1YGE	A/261-267	7	0.1353	Right-handed	Oxidoreductase
1YGE	A/494-506	13	0.3152	Right-handed	Oxidoreductase
1YGE	A/684-690	7	0.1398	Right-handed	Oxidoreductase
2EBN	A/257-263	7	0.1327	Right-handed	Hydrolase
2HMQ	A/101-107	7	0.1429	Right-handed	Oxygen transport
2OLB	A/301-308	8	0.1504	Right-handed	Binding protein
4PAN	A/325-331	7	0.1580	Right-handed	Signaling protein
5CSM	A/233-239	7	0.1380	Right-handed	Isomerase
7A3H	A/146-152	7	0.1293	Right-handed	Hydrolase
9GAF	A/186-192	7	0.1005	Right-handed	Hydrolase

Table A2. The calculated parameters of α-turns.

Family or Group	Protein PDB ID	Chain/Residues	Chirality Value	Chirality Sign	Protein Class
F1	1AAP	A/24-28	0.0917	Right-handed	Proteinase Inhibitor (Trypsin)
	1ACX	A/82-86	0.0735	Right-handed	Antibacterial Protein
	2AK3	A/149-153	0.0694	Right-handed	Transferase (Phosphotransferase)
	3COX	A/391-395	0.0852	Right-handed	Oxidoreductase (Oxygen Receptor)
	1DRF	A/152-156	0.0837	Right-handed	Oxidoreductase (Ch-Nh(D)-Nad Or Nadp (A))
	1ECA	A/38-42	0.0884	Right-handed	Oxygen Transport
	1GD1	O/47-51	0.0668	Right-handed	Oxidoreductase (Aldehyde(D)-Nad (A))
	1GOX	A/345-349	0.1173	Right-handed	Oxidoreductase (Oxygen(A))
	1MBA	A/43-47	0.0861	Right-handed	Oxygen Storage
	1OMD	A/2-6	0.0911	Right-handed	Calcium Binding Protein
	1OVA	A/277-281	0.0764	Right-handed	Serpin
	1OVA	A/318-322	0.1015	Right-handed	Serpin
	1RDG	A/6-10	0.0552	Right-handed	Electron Transfer (Iron-Sulfur Protein)
	1RDG	A/14-18	0.0976	Right-handed	Electron Transfer (Iron-Sulfur Protein)
	1RDG	A/39-43	0.0585	Right-handed	Electron Transfer (Iron-Sulfur Protein)
	2SN3	A/7-11	0.0626	Right-handed	Toxin
	2SN4	A/31-35	0.0847	Right-handed	Toxin
	1THB	A/113-117	0.0937	Right-handed	Oxygen Transport
	2ACT	A/85-89	0.1009	Right-handed	Hydrolase (Proteinase)
	2AZA	A/40-44	0.0952	Right-handed	Electron Transport Protein (Cuproprotein)
	2CA2	A/34-38	0.1028	Right-handed	Lyase (Oxo-Acid)
	2CPP	A/77-81	0.0883	Right-handed	Oxidoreductase (Oxygenase)
	2CPP	A/328-332	0.1000	Right-handed	Oxidoreductase (Oxygenase)
	2CSC	A/59-63	0.0674	Right-handed	Lyase
	2CYP	A/58-62	0.0673	Right-handed	Oxidoreductase (H2O2(A))
	2ER7	E/240-244	0.0926	Right-handed	Hydrolase/Hydrolase Inhibitor
	2FCR	A/94-98	0.0923	Right-handed	Electron Transport
	2FCR	A/148-152	0.0841	Right-handed	Electron Transport
	2LTN	A/54-58	0.0776	Right-handed	Lectin

Table A2. *Cont.*

Family or Group	Protein PDB ID	Chain/Residues	Chirality Value	Chirality Sign	Protein Class
	2LTN	A/125-129	0.0523	Right-handed	Lectin
	2LTN	A/167-171	0.0852	Right-handed	Lectin
	2RHE	A/93-97	0.0986	Right-handed	Immunoglobulin
	2RSP	A/46-50	0.1031	Right-handed	Hydrolase (Aspartyl Proteinase)
	2RSP	A/219-223	0.0798	Right-handed	Hydrolase (Serine Proteinase)
	2TRX	A/59-63	0.0993	Right-handed	Electron Transport
	3BLM	A/50-54	0.0742	Right-handed	Hydrolase
	3CLA	A/97-101	0.0833	Right-handed	Transferase (Acyltransferase)
	3CLA	A/194-198	0.0880	Right-handed	Transferase (Acyltransferase)
	4FGF	A/67-71	0.0920	Right-handed	Growth Factor
	3GRS	A/164-168	0.0968	Right-handed	Oxidoreductase (Flavoenzyme)
	4ENL	A/102-106	0.0738	Right-handed	Carbon-Oxygen Lyase
	5FD1	A/35-39	0.0986	Right-handed	Electron Transport(Iron-Sulfur)
	5CPA	A/3-7	0.1063	Right-handed	Hydrolase (C-Terminal Peptidase)
	5CPA	A/29-33	0.0905	Right-handed	Hydrolase (C-Terminal Peptidase)
	5P21	A/145-149	0.0840	Right-handed	Oncogene Protein
	6LDH	A/181-185	0.1114	Right-handed	Oxidoreductase(Choh(D)-Nad(A))
F2	2AK3	A/137-141	0.0723	Right-handed	Transferase (Phosphotransferase)
	3COX	A/453-457	0.0649	Right-handed	Oxidoreductase (Oxygen Receptor)
	1FKF	A/87-91	0.0400	Right-handed	Isomerase
	1GD1	O/129-133	0.0791	Right-handed	Oxidoreductase (Aldehyde(D)-Nad(A))
	1GD2	O/267-271	0.0761	Right-handed	Oxidoreductase (Aldehyde(D)-Nad(A))
	2TEC	E/261-265	0.0750	Right-handed	Complex(Serine Proteinase-Inhibitor)
	2TRX	A/49-53	0.0799	Right-handed	Electron Transport
	5CPA	A/89-93	0.0664	Right-handed	Hydrolase (C-Terminal Peptidase)
g1	4GCR	A/47-51	−0.0291	Left-handed	Eye Lens Protein
	4GCR	A/136-140	−0.0283	Left-handed	Eye Lens Protein
	2CYP	A/35-39	−0.0422	Left-handed	Oxidoreductase (H2O2(A))
	2FBJ	A/48-52	−0.0411	Left-handed	Immunoglobulin
	2PRK	A/212-216	−0.0230	Left-handed	Serine Proteinase
g2	2FBJ	H/100-104	−0.0355	Left-handed	Immunoglobulin
	2RHE	A/50-54	−0.0478	Left-handed	Immunoglobulin
g3	1FKF	A/82-86	0.0384	Right-handed	Isomerase
	1OVA	A/69-73	0.0523	Right-handed	Serpin
	3GRS	A/55-59	0.0351	Right-handed	Oxidoreductase (Flavoenzyme)
	4BP2	A/25-29	−0.0123	Left-handed	Carboxylic Ester Hydrolase Zymogen
g4	2ACT	A/188-192	−0.0636	Left-handed	Hydrolase (Proteinase)
	2LZM	A/27-31	−0.0613	Left-handed	Hydrolase (O-Glycosyl)
	3APR	E/11-15	−0.0684	Left-handed	Hydrolase/Hydrolase Inhibitor
	2FOX	A/56-60	−0.0741	Left-handed	Electron Transport
g5	1FX1	A/71-75	0.1042	Right-handed	Electron Transfer (Flavoprotein)
	1YPI	A/25-29	0.1020	Right-handed	Isomerase (Intramolecular Oxidoreductase)
	2FOX	A/40-44	0.1004	Right-handed	Electron Transport
g6	1GD1	O/300-304	−0.0084	Left-handed	Oxidoreductase (Aldehyde(D)-Nad(A))
	2LTN	A/100-104	−0.0026	Left-handed	Lectin
	4PEP	A/9-13	−0.0003	Left-handed	Hydrolase (Acid Proteinase)
g7	2AK3	A/129-133	0.0784	Right-handed	Transferase (Phosphotransferase)
	1FX1	A/72-76	0.1092	Right-handed	Electron Transfer (Flavoprotein)
Other	1RBP	A/63-67	−0.0099	Left-handed	Retinol Transport
	2FBJ	L/166-170	0.0745	Right-handed	Immunoglobulin

Table A3. The calculated parameters of Ω-loops.

Protein PDB ID	Chain/Residues	Length of Loop	Chirality Value	Chirality Sign	Protein Class
1ABE	A/93-99	7	−0.0280	Left-handed	Binding Protein
1ABE	A/142-148	7	0.1496	Right-handed	Binding Protein
1ABE	A/203-208	6	0.0231	Right-handed	Binding Protein
1ABE	A/236-248	13	0.1632	Right-handed	Binding Protein
1ABE	A/289-294	6	0.0232	Right-handed	Binding Protein
1ABE	A/299-304	6	0.0464	Right-handed	Binding Protein
2ACT	A/8-13	6	0.0508	Right-handed	Hydrolase (Proteinase)
2ACT	A/58-64	7	0.0462	Right-handed	Hydrolase (Proteinase)
2ACT	A/89-103	15	−0.1453	Left-handed	Hydrolase (Proteinase)
2ACT	A/139-144	6	0.0709	Right-handed	Hydrolase (Proteinase)
2ACT	A/141-156	16	0.1703	Right-handed	Hydrolase (Proteinase)
2ACT	A/182-192	11	0.0027	Right-handed	Hydrolase (Proteinase)
2ACT	A/198-205	8	−0.0151	Left-handed	Hydrolase (Proteinase)
2ACT	A/203-209	7	−0.0605	Left-handed	Hydrolase (Proteinase)

Table A3. *Cont.*

Protein PDB ID	Chain/Residues	Length of Loop	Chirality Value	Chirality Sign	Protein Class
8ADH	A/14-21	8	−0.0071	Left-handed	Oxidoreductase (Nad(A)-Choh(D))
8ADH	A/100-112	13	0.3005	Right-handed	Oxidoreductase (Nad(A)-Choh(D))
8ADH	A/115-122	8	−0.0038	Left-handed	Oxidoreductase (Nad(A)-Choh(D))
8ADH	A/122-128	7	−0.0191	Left-handed	Oxidoreductase (Nad(A)-Choh(D))
8ADH	A/282-287	6	0.0022	Right-handed	Oxidoreductase (Nad(A)-Choh(D))
3ADK	A/133-142	10	0.0617	Right-handed	Transferase(Phosphotransferase)
2ALP	A/217-224	8	−0.0460	Left-handed	Hydrolase (Serine Proteinase)
3APP	A/41-55	15	0.2606	Right-handed	Hydrolase (Acid Proteinase)
3APP	A/129-136	8	0.0022	Right-handed	Hydrolase (Acid Proteinase)
3APP	A/139-149	11	0.4213	Right-handed	Hydrolase (Acid Proteinase)
3APP	A/184-192	9	0.0463	Right-handed	Hydrolase (Acid Proteinase)
2APR	A/8-17	10	−0.1572	Left-handed	Hydrolase (Aspartic Proteinase)
2APR	A/18-31	14	−0.2057	Left-handed	Hydrolase (Aspartic Proteinase)
2APR	A/43-58	16	0.0879	Right-handed	Hydrolase (Aspartic Proteinase)
2APR	A/61-69	9	0.1019	Right-handed	Hydrolase (Aspartic Proteinase)
2APR	A/76-83	8	−0.0071	Left-handed	Hydrolase (Aspartic Proteinase)
2APR	A/90-103	14	−0.1480	Left-handed	Hydrolase (Aspartic Proteinase)
2APR	A/129-138	10	0.0383	Right-handed	Hydrolase (Aspartic Proteinase)
2APR	A/189-197	9	0.0337	Right-handed	Hydrolase (Aspartic Proteinase)
2APR	A/203-211	9	−0.0328	Left-handed	Hydrolase (Aspartic Proteinase)
2APR	A/216-226	11	0.0677	Right-handed	Hydrolase (Aspartic Proteinase)
2APR	A/227-232	6	0.1507	Right-handed	Hydrolase (Aspartic Proteinase)
2APR	A/233-248	16	0.0514	Right-handed	Hydrolase (Aspartic Proteinase)
2APR	A/243-250	8	−0.0071	Left-handed	Hydrolase (Aspartic Proteinase)
2APR	A/261-273	13	−0.1703	Left-handed	Hydrolase (Aspartic Proteinase)
2APR	A/280-287	8	−0.0761	Left-handed	Hydrolase (Aspartic Proteinase)
2APR	A/291-299	9	−0.0724	Left-handed	Hydrolase (Aspartic Proteinase)
1AZU	A/9-15	7	−0.0069	Left-handed	Electron Transport (Copper Binding)
1AZU	A/35-46	12	0.0569	Right-handed	Electron Transport (Copper Binding)
1AZU	A/67-72	6	0.0376	Right-handed	Electron Transport (Copper Binding)
1AZU	A/73-83	11	0.0646	Right-handed	Electron Transport (Copper Binding)
1AZU	A/84-92	9	−0.0023	Left-handed	Electron Transport (Copper Binding)
1AZU	A/112-118	7	0.0420	Right-handed	Electron Transport (Copper Binding)
1CYO	A/32-47	16	0.4146	Right-handed	Electron Transport
1BP2	A/23-30	8	0.0360	Right-handed	Hydrolase
1BP3	A/25-39	15	−0.0675	Left-handed	Hydrolase
1BP4	A/56-66	11	0.2318	Right-handed	Hydrolase
2BP2	A/23-30	8	0.0360	Right-handed	Hydrolase Zymogen
2BP3	A/25-39	15	−0.0675	Left-handed	Hydrolase Zymogen
2BP4	A/61-68	8	−0.0056	Left-handed	Hydrolase Zymogen
256B	A/16-25	10	0.1574	Right-handed	Electron Transport
256B	A/47-58	12	0.1626	Right-handed	Electron Transport
351C	A/16-25	10	−0.0117	Left-handed	Electron Transport
351C	A/51-62	12	0.0243	Right-handed	Electron Transport
155C	A/21-28	8	−0.1010	Left-handed	Electron Transport
155C	A/47-54	8	0.0381	Right-handed	Electron Transport
155C	A/83-95	13	0.3003	Right-handed	Electron Transport
155C	A/128-133	6	0.0115	Right-handed	Electron Transport
2C2C	A/18-33	16	−0.2105	Left-handed	Electron Transport Protein (Cytochrome)
2C2C	A/30-43	14	0.1534	Right-handed	Electron Transport Protein (Cytochrome)
2C2C	A/41-56	16	0.2603	Right-handed	Electron Transport Protein (Cytochrome)
2C2C	A/74-89	16	0.4033	Right-handed	Electron Transport Protein (Cytochrome)
2CAB	A/6-12	7	0.0716	Right-handed	Hydro-Lyase
2CAB	A/17-24	8	0.1911	Right-handed	Hydro-Lyase
2CAB	A/78-87	10	0.1359	Right-handed	Hydro-Lyase
2CAB	A/98-104	7	0.0950	Right-handed	Hydro-Lyase
2CAB	A/108-114	7	−0.0536	Left-handed	Hydro-Lyase
2CAB	A/128-140	13	0.2908	Right-handed	Hydro-Lyase
2CAB	A/197-204	8	−0.0225	Left-handed	Hydro-Lyase
2CAB	A/230-240	11	−0.0683	Left-handed	Hydro-Lyase
1CA2	A/5-16	12	0.1204	Right-handed	Lyase (Oxo-Acid)
1CA3	A/17-23	7	0.1274	Right-handed	Lyase (Oxo-Acid)
1CA4	A/98-103	6	0.0470	Right-handed	Lyase (Oxo-Acid)
1CA5	A/108-114	7	−0.0629	Left-handed	Lyase (Oxo-Acid)
1CA6	A/128-140	13	0.3094	Right-handed	Lyase (Oxo-Acid)
1CA7	A/166-172	7	0.0573	Right-handed	Lyase (Oxo-Acid)

Table A3. *Cont.*

Protein PDB ID	Chain/Residues	Length of Loop	Chirality Value	Chirality Sign	Protein Class
1CA8	A/197-204	8	−0.0111	Left-handed	Lyase (Oxo-Acid)
1CA9	A/232-239	8	−0.0806	Left-handed	Lyase (Oxo-Acid)
2CHA	B/70-78	9	0.0557	Right-handed	Hydrolase (Serine Proteinase)
2CHA	B/94-102	9	0.0806	Right-handed	Hydrolase (Serine Proteinase)
2CHA	B/114-119	6	0.0369	Right-handed	Hydrolase (Serine Proteinase)
2CHA	C/217-224	8	−0.0060	Left-handed	Hydrolase (Serine Proteinase)
3CNA	A/13-21	9	0.0381	Right-handed	Lectin (Agglutinin)
3CNA	A/97-104	8	−0.0102	Left-handed	Lectin (Agglutinin)
3CNA	A/116-123	8	0.0756	Right-handed	Lectin (Agglutinin)
3CNA	A/147-155	9	0.0144	Right-handed	Lectin (Agglutinin)
3CNA	A/160-165	6	0.0127	Right-handed	Lectin (Agglutinin)
3CNA	A/199-209	11	0.0003	Right-handed	Lectin (Agglutinin)
3CNA	A/222-235	14	0.1464	Right-handed	Lectin (Agglutinin)
3CNA	A/229-237	9	0.0108	Right-handed	Lectin (Agglutinin)
3CPA	A/128-141	14	0.0767	Right-handed	Hydrolase (C-Terminal Peptidase)
3CPA	A/142-156	15	0.2551	Right-handed	Hydrolase (C-Terminal Peptidase)
3CPA	A/156-166	11	0.1575	Right-handed	Hydrolase (C-Terminal Peptidase)
3CPA	A/205-213	9	0.0898	Right-handed	Hydrolase (C-Terminal Peptidase)
3CPA	A/231-237	7	−0.0073	Left-handed	Hydrolase (C-Terminal Peptidase)
3CPA	A/244-250	7	−0.0284	Left-handed	Hydrolase (C-Terminal Peptidase)
3CPA	A/272-285	14	−0.0102	Left-handed	Hydrolase (C-Terminal Peptidase)
5CPV	A/18-23	6	0.0092	Right-handed	Calcium Binding
5CPV	A/64-77	14	0.2505	Right-handed	Calcium Binding
1CRN	A/33-44	12	−0.0524	Left-handed	Plant Protein
1CTX	A/1-15	15	0.0597	Right-handed	Toxin
1CTX	A/26-35	10	0.1348	Right-handed	Toxin
1CYC	A/18-32	15	−0.1967	Left-handed	Electron Transport
1CYC	A/30-43	14	0.0486	Right-handed	Electron Transport
1CYC	A/40-54	15	0.1398	Right-handed	Electron Transport
1CYC	A/70-84	15	0.1522	Right-handed	Electron Transport
3CYT	O/18-32	15	−0.1637	Left-handed	Electron Transport (Heme Protein)
3CYT	O/34-43	10	0.1153	Right-handed	Electron Transport (Heme Protein)
3CYT	O/40-54	15	0.1250	Right-handed	Electron Transport (Heme Protein)
3CYT	O/70-84	15	0.2509	Right-handed	Electron Transport (Heme Protein)
1ECD	A/33-42	10	0.3042	Right-handed	Oxygen Transport
1ECD	A/41-49	9	0.0899	Right-handed	Oxygen Transport
1EST	A/69-80	12	0.0984	Right-handed	Hydrolase
1EST	A/94-104	11	0.1166	Right-handed	Hydrolase
1EST	A/112-118	7	−0.0091	Left-handed	Hydrolase
1EST	A/165-178	14	0.3179	Right-handed	Hydrolase
1EST	A/216-226	11	0.0211	Right-handed	Hydrolase
7FAB	L/24-29	6	0.0311	Right-handed	Immune System
7FAB	L/122-132	11	−0.0979	Left-handed	Immune System
7FAB	L/168-173	6	−0.0200	Left-handed	Immune System
7FAB	L/182-187	6	0.0160	Right-handed	Immune System
7FAB	H/72-77	6	0.0406	Right-handed	Immune System
7FAB	H/99-105	7	−0.0067	Left-handed	Immune System
1DUR	A/12-23	12	0.2808	Right-handed	Electron Transport
1DUR	A/30-41	12	0.0120	Right-handed	Electron Transport
1DUR	A/39-50	12	0.2652	Right-handed	Electron Transport
5NLL	A/54-61	8	−0.0590	Left-handed	Electron Transport
2GCH	F/70-78	9	0.1866	Right-handed	Hydrolase (Serine Proteinase)
2GCH	F/94-101	8	0.0378	Right-handed	Hydrolase (Serine Proteinase)
2GCH	F/112-118	7	−0.0220	Left-handed	Hydrolase (Serine Proteinase)
2GCH	G/165-176	12	0.3527	Right-handed	Hydrolase (Serine Proteinase)
2GCH	G/217-224	8	−0.0086	Left-handed	Hydrolase (Serine Proteinase)
1GPD	G/47-52	6	0.0284	Right-handed	Oxidoreductase
1GPD	G/76-82	7	0.0872	Right-handed	Oxidoreductase
1GPD	G/121-129	9	0.0052	Right-handed	Oxidoreductase
1GPD	G/183-198	16	−0.1486	Left-handed	Oxidoreductase
3GRS	A/83-89	7	−0.0102	Left-handed	Oxidoreductase (Flavoenzyme)
3GRS	A/139-147	9	−0.0760	Left-handed	Oxidoreductase (Flavoenzyme)
3GRS	A/162-172	11	0.0576	Right-handed	Oxidoreductase (Flavoenzyme)
3GRS	A/239-245	7	−0.0094	Left-handed	Oxidoreductase (Flavoenzyme)
3GRS	A/256-261	6	−0.0633	Left-handed	Oxidoreductase (Flavoenzyme)
3GRS	A/268-274	7	−0.0219	Left-handed	Oxidoreductase (Flavoenzyme)

Table A3. *Cont.*

Protein PDB ID	Chain/Residues	Length of Loop	Chirality Value	Chirality Sign	Protein Class
3GRS	A/300-307	8	−0.0017	Left-handed	Oxidoreductase (Flavoenzyme)
3GRS	A/315-320	6	−0.0687	Left-handed	Oxidoreductase (Flavoenzyme)
3GRS	A/331-337	7	0.0505	Right-handed	Oxidoreductase (Flavoenzyme)
3GRS	A/404-415	12	0.2396	Right-handed	Oxidoreductase (Flavoenzyme)
3GRS	A/465-472	8	0.0467	Right-handed	Oxidoreductase (Flavoenzyme)
1HIP	A/20-26	7	0.1417	Right-handed	Electron Transfer (Iron-Sulfur Protein)
1HIP	A/28-41	14	0.0932	Right-handed	Electron Transfer (Iron-Sulfur Protein)
1HIP	A/43-49	7	0.0998	Right-handed	Electron Transfer (Iron-Sulfur Protein)
1HIP	A/44-59	16	0.1820	Right-handed	Electron Transfer (Iron-Sulfur Protein)
6LDH	A/173-188	16	0.2820	Right-handed	Oxidoreductase (Choh(D)-Nad(A))
6LDH	A/192-200	9	0.0461	Right-handed	Oxidoreductase (Choh(D)-Nad(A))
6LDH	A/203-218	16	0.0187	Right-handed	Oxidoreductase (Choh(D)-Nad(A))
6LDH	A/212-225	14	0.1010	Right-handed	Oxidoreductase (Choh(D)-Nad(A))
6LDH	A/219-226	8	0.1033	Right-handed	Oxidoreductase (Choh(D)-Nad(A))
6LDH	A/239-246	8	−0.0174	Left-handed	Oxidoreductase (Choh(D)-Nad(A))
6LDH	A/275-285	11	0.0584	Right-handed	Oxidoreductase (Choh(D)-Nad(A))
1LH1	A/41-53	13	0.3229	Right-handed	Oxygen Transport
1LH2	A/47-54	8	0.0285	Right-handed	Oxygen Transport
2LHB	A/46-59	14	0.5669	Right-handed	Oxygen Transport
2LHB	A/55-64	10	0.2227	Right-handed	Oxygen Transport
7LYZ	A/18-25	8	−0.0331	Left-handed	Hydrolase (O-Glycosyl)
7LYZ	A/36-42	7	0.0318	Right-handed	Hydrolase (O-Glycosyl)
7LYZ	A/44-52	9	−0.0407	Left-handed	Hydrolase (O-Glycosyl)
7LYZ	A/60-75	16	0.1404	Right-handed	Hydrolase (O-Glycosyl)
2LZM	A/134-139	6	0.0657	Right-handed	Hydrolase (O-Glycosyl)
1MBN	A/40-47	8	0.2021	Right-handed	Oxygen Storage
1MBS	A/37-50	14	0.5199	Right-handed	Oxygen Transport
1MBS	A/49-54	6	0.0358	Right-handed	Oxygen Transport
1MBS	A/78-84	7	0.1077	Right-handed	Oxygen Transport
2MHB	A/40-48	9	0.2250	Right-handed	Oxygen Transport
2MHB	B/39-54	16	0.4710	Right-handed	Oxygen Transport
2MHB	B/47-57	11	0.2937	Right-handed	Oxygen Transport
1NXB	A/6-13	8	−0.1576	Left-handed	Neurotoxin (Post-Synaptic)
2PAB	A/49-54	6	−0.0652	Left-handed	Transport (Thyroxine,Retinol) In Serum
9PAP	A/8-13	6	0.0535	Right-handed	Hydrolase (Sulfhydryl Proteinase)
9PAP	A/60-67	8	−0.0304	Left-handed	Hydrolase (Sulfhydryl Proteinase)
9PAP	A/86-100	15	0.0073	Right-handed	Hydrolase (Sulfhydryl Proteinase)
9PAP	A/138-153	14	0.1452	Right-handed	Hydrolase (Sulfhydryl Proteinase)
9PAP	A/175-185	11	0.0051	Right-handed	Hydrolase (Sulfhydryl Proteinase)
9PAP	A/191-198	8	0.0624	Right-handed	Hydrolase (Sulfhydryl Proteinase)
9PAP	A/198-203	6	−0.0338	Left-handed	Hydrolase (Sulfhydryl Proteinase)
1PLC	A/6-13	8	−0.0920	Left-handed	Electron Transport
1PLC	A/41-55	15	0.1393	Right-handed	Electron Transport
1PLC	A/63-68	6	0.0171	Right-handed	Electron Transport
1PLC	A/84-92	9	0.1941	Right-handed	Electron Transport

References

1. Zhang, S.; Holmes, T.; DiPersio, C.; Hynes, R.O.; Su, X.; Rich, A. Self-complementary oligopeptide matrices support mammalian cell attachment. *Biomaterials* **1995**, *16*, 1385–1393. [CrossRef]
2. Boyle, A.L. Applications of de novo designed peptides. In *Book Peptide Applications in Biomedicine, Biotechnology and Bioengineering*; Elsevier: Amsterdam, The Netherlands, 2018; pp. 51–86. [CrossRef]
3. Ellis-Behnke, R.G.; Liang, Y.-X.; You, S.-W.; Tay, D.K.C.; Zhang, S.; So, K.-F.; Schneider, G.E. Nano neuro knitting: Peptide nanofiber scaffold for brain repair and axon regeneration with functional return of vision. *Proc. Natl. Acad. Sci. USA* **2006**, *103*, 5054–5059. [CrossRef] [PubMed]
4. Gelain, F.; Bottai, D.; Vescovi, A.; Zhang, S. Designer Self-Assembling Peptide Nanofiber Scaffolds for Adult Mouse Neural Stem Cell 3-Dimensional Cultures. *PLoS ONE* **2006**, *1*, e119. [CrossRef] [PubMed]
5. Welch, J.T.; Kearney, W.R.; Franklin, S.J. Lanthanide-binding helix-turn-helix peptides: Solution structure of a designed metallonuclease. *Proc. Natl. Acad. Sci. USA* **2003**, *100*, 3725–3730. [CrossRef]
6. Kovacic, R.T.; Welch, J.T.; Franklin, S.J. Sequence-Selective DNA Cleavage by a Chimeric Metallopeptide. *J. Am. Chem. Soc.* **2003**, *125*, 6656–6662. [CrossRef]
7. Reches, M.; Porat, Y.; Gazit, E. Amyloid Fibril Formation by Pentapeptide and Tetrapeptide Fragments of Human Calcitonin. *J. Biol. Chem.* **2002**, *277*, 35475–35480. [CrossRef]

8. Tverdislov, V.A.; Malyshko, E.V. On regularities in the spontaneous formation of structural hierarchies in chiral systems of nonliving and living matter. *Phys. Uspekhi* **2019**, *62*, 354–363. [CrossRef]
9. Tverdislov, V.A.; Malyshko, E.V. Chiral Dualism as an Instrument of Hierarchical Structure Formation in Molecular Biology. *Symmetry* **2020**, *12*, 587. [CrossRef]
10. Guichard, G.; Benkirane-Jessel, N.; Zeder-Lutz, G.; van Regenmortel, M.H.; Briand, J.P.; Muller, S. Antigenic mimicry of natural L-peptides with retro-inverso-peptidomimetics. *Proc. Natl. Acad. Sci. USA* **1994**, *91*, 9765–9769. [CrossRef]
11. Nanda, V.; Andrianarijaona, A.; Narayanan, C. The role of protein homochirality in shaping the energy landscape of folding. *Protein Sci.* **2007**, *16*, 1667–1675. [CrossRef] [PubMed]
12. Nagy-Smith, K.; Beltramo, P.; Moore, E.; Tycko, R.; Furst, E.M.; Schneider, J.P. Molecular, Local, and Network-Level Basis for the Enhanced Stiffness of Hydrogel Networks Formed from Coassembled Racemic Peptides: Predictions from Pauling and Corey. *ACS Cent. Sci.* **2017**, *3*, 586–597. [CrossRef]
13. McAulay, K.; Dietrich, B.; Su, H.; Scott, M.T.; Rogers, S.; Al-Hilaly, Y.K.; Cui, H.; Serpell, L.C.; Seddon, A.; Draper, E.R.; et al. Using chirality to influence supramolecular gelation. *Chem. Sci.* **2019**, *10*, 7801–7806. [CrossRef]
14. Bera, S.; Xue, B.; Rehak, P.; Jacoby, G.; Ji, W.; Shimon, L.J.W.; Beck, R.; Král, P.; Cao, Y.; Gazit, E. Self-Assembly of Aromatic Amino Acid Enantiomers into Supramolecular Materials of High Rigidity. *ACS Nano* **2020**, *14*, 1694–1706. [CrossRef]
15. Chen, K.; Sheng, Y.; Wang, J.; Wang, W. Chirality-Dependent Adsorption between Amphipathic Peptide and POPC Membrane. *Int. J. Mol. Sci.* **2019**, *20*, 4760. [CrossRef] [PubMed]
16. Qin, M.; Zhang, Y.; Xing, C.; Yang, L.; Zhao, C.; Dou, X.; Feng, C.L. Effect of Stereochemistry on Chirality and Gelation Properties of Supramolecular Self-Assemblies. *Chemistry* **2021**, *27*, 3119–3129. [CrossRef]
17. Qing, G.; Zhao, S.; Xiong, Y.; Lv, Z.; Jiang, F.; Liu, Y.; Chen, H.; Zhang, M.; Sun, T. Chiral Effect at Protein/Graphene Interface: A Bioinspired Perspective to Understand Amyloid Formation. *J. Am. Chem. Soc.* **2014**, *136*, 10736–10742. [CrossRef] [PubMed]
18. Hou, K.; Zhao, J.; Wang, H.; Li, B.; Li, K.; Shi, X.; Wan, K.; Ai, J.; Lv, J.; Wang, D.; et al. Chiral gold nanoparticles enantioselectively rescue memory deficits in a mouse model of Alzheimer's disease. *Nat. Commun.* **2020**, *11*, 4790. [CrossRef]
19. Zheng, Y.; Mao, K.; Chen, S.; Zhu, H. Chirality Effects in Peptide Assembly Structures. *Front. Bioeng. Biotechnol.* **2021**, *9*, 703004. [CrossRef] [PubMed]
20. Yao, X.; Hu, Y.; Cao, B.; Peng, R.; Ding, J. Effects of surface molecular chirality on adhesion and differentiation of stem cells. *Biomaterials* **2013**, *34*, 9001–9009. [CrossRef] [PubMed]
21. Ma, Y.; Shi, L.; Yue, H.; Gao, X. Recognition at chiral interfaces: From molecules to cells. *Colloids Surf. B Biointerfaces* **2020**, *195*, 111268. [CrossRef]
22. Krause, E.; Bienert, M.; Schmieder, P.; Wenschuh, H. The Helix-Destabilizing Propensity Scale of d-Amino Acids: The Influence of Side Chain Steric Effects. *J. Am. Chem. Soc.* **2000**, *122*, 4865–4870. [CrossRef]
23. Punitha, V.; Raman, S.S.; Parthasarathi, R.; Subramanian, V.; Rao, J.R.; Nair, B.U.; Ramasami, T. Molecular Dynamics Investigations on the Effect of d Amino Acid Substitution in a Triple-Helix Structure and the Stability of Collagen. *J. Phys. Chem. B* **2009**, *113*, 8983–8992. [CrossRef] [PubMed]
24. Zheng, Y.; Yu, L.; Zou, Y.; Yang, Y.; Wang, C. Steric Dependence of Chirality Effect in Surface-Mediated Peptide Assemblies Identified with Scanning Tunneling Microscopy. *Nano Lett.* **2019**, *19*, 5403–5409. [CrossRef]
25. Fairman, R.; Anthony-Cahill, S.J.; DeGrado, W.F. The helix-forming propensity of D-alanine in a right-handed. alpha.-helix. *J. Am. Chem. Soc.* **1992**, *114*, 5458–5459. [CrossRef]
26. Hu, K.; Jiang, Y.; Xiong, W.; Li, H.; Zhang, P.-Y.; Yin, F.; Zhang, Q.; Geng, H.; Jiang, F.; Li, Z.; et al. Tuning peptide self-assembly by an in-tether chiral center. *Sci. Adv.* **2018**, *4*, eaar5907. [CrossRef]
27. Hu, K.; Geng, H.; Zhang, Q.; Liu, Q.; Xie, M.; Sun, C.; Li, W.; Lin, H.; Jiang, F.; Wang, T.; et al. An In-tether Chiral Center Modulates the Helicity, Cell Permeability, and Target Binding Affinity of a Peptide. *Angew. Chem.* **2016**, *128*, 8145–8149. [CrossRef]
28. Gil, A.M.; Casanovas, J.; Mayans, E.; Jiménez, A.I.; Puiggalí, J.; Alemán, C. Heterochirality Restricts the Self-Assembly of Phenylalanine Dipeptides Capped with Highly Aromatic Groups. *J. Phys. Chem. B* **2020**, *124*, 5913–5918. [CrossRef]
29. Kralj, S.; Bellotto, O.; Parisi, E.; Garcia, A.M.; Iglesias, D.; Semeraro, S.; Deganutti, C.; D'Andrea, P.; Vargiu, A.V.; Geremia, S.; et al. Heterochirality and Halogenation Control Phe-Phe Hierarchical Assembly. *ACS Nano* **2020**, *14*, 16951–16961. [CrossRef]
30. Zhang, G.; Zhang, L.; Rao, H.; Wang, Y.; Li, Q.; Qi, W.; Yang, X.; Su, R.; He, Z. Role of molecular chirality and solvents in directing the self-assembly of peptide into an ultra-pH-sensitive hydrogel. *J. Colloid Interface Sci.* **2020**, *577*, 388–396. [CrossRef]
31. Adzhubei, A.A.; Sternberg, M.; Makarov, A.A. Polyproline-II Helix in Proteins: Structure and Function. *J. Mol. Biol.* **2013**, *425*, 2100–2132. [CrossRef] [PubMed]
32. Zarrinpar, A.; Bhattacharyya, R.P.; Lim, W.A. The structure and function of proline recognition domains. *Sci. STKE* **2003**, *179*, re8. [CrossRef] [PubMed]
33. Platé, N.A.; Shibaev, V.P. *Comb-Shaped Polymers and Liquid Crystals*; Springer: Boston, MA, USA, 2012. [CrossRef]
34. Livolant, F.; Leforestier, A. Condensed phases of DNA: Structures and phase transitions. *Prog. Polym. Sci.* **1996**, *21*, 1115–1164. [CrossRef]
35. Chakraborty, D.; Mugnai, M.; Thirumalai, D. On the Emergence of Orientational Order in Folded Proteins with Implications for Allostery. *Symmetry* **2021**, *13*, 770. [CrossRef]
36. Sidorova, A.E.; Levashova, N.T.; Malyshko, E.; Tverdislov, V. Autowave Self-Organization in the Folding of Proteins. *Mosc. Univ. Phys. Bull.* **2019**, *74*, 213–226. [CrossRef]

37. Abrusán, G.; Marsh, J.A. Alpha Helices Are More Robust to Mutations than Beta Strands. *PLoS Comput. Biol.* **2016**, *12*, e1005242. [CrossRef]
38. Pauling, L.; Corey, R.B. The Pleated Sheet, A New Layer Configuration of Polypeptide Chains. *Proc. Natl. Acad. Sci. USA* **1951**, *37*, 2451–2456. [CrossRef]
39. Tonlolo, C.; Benedetti, E. The polypeptide 310-helix. *Trends Biochem. Sci.* **1991**, *16*, 350–353. [CrossRef]
40. Kendrew, J.C.; Bodo, G.; Dintzis, H.M.; Parrish, R.G.; Wyckoff, H.; Phillips, D.C. A Three-Dimensional Model of the Myoglobin Molecule Obtained by X-Ray Analysis. *Nature* **1958**, *181*, 662–666. [CrossRef]
41. Armen, R.; Alonso, D.O.; Daggett, V. The role of α-, 310-, and π-helix in helix→coil transitions. *Protein Sci.* **2003**, *12*, 1145–1157. [CrossRef]
42. Cooley, R.B.; Arp, D.J.; Karplus, P.A. Evolutionary Origin of a Secondary Structure: π-Helices as Cryptic but Widespread Insertional Variations of α-Helices That Enhance Protein Functionality. *J. Mol. Biol.* **2010**, *404*, 232–246. [CrossRef]
43. Dasgupta, B.; Chakrabarti, P. pi-Turns: Types, systematics and the context of their occurrence in protein structures. *BMC Struct. Biol.* **2008**, *8*, 39. [CrossRef]
44. Fodje, M.; Al-Karadaghi, S. Occurrence, conformational features and amino acid propensities for the π-helix. *Protein Eng. Des. Sel.* **2002**, *15*, 353–358. [CrossRef] [PubMed]
45. Religa, T.L.; Johnson, C.M.; Vu, D.; Brewer, S.H.; Dyer, R.B.; Fersht, A.R. The helix turn helix motif as an ultrafast independently folding domain: The pathway of folding of Engrailed homeodomain. *Proc. Natl. Acad. Sci. USA* **2007**, *104*, 9272–9277. [CrossRef] [PubMed]
46. Pavone, V.; Gaeta, G.; Lombardi, A.; Nastri, F.; Maglio, O.; Isernia, C.; Saviano, M. Discovering protein secondary structures: Classification and description of isolated α-turns. *Biopolymers* **1996**, *38*, 705–721. [CrossRef]
47. Hutchinson, E.G.; Thornton, J. A revised set of potentials for beta-turn formation in proteins. *Protein Sci.* **1994**, *3*, 2207–2216. [CrossRef]
48. Donate, L.E.; Rufino, S.D.; Canard, L.H.; Blundell, T.L. Conformational analysis and clustering of short and medium size loops connecting regular secondary structures: A database for modeling and prediction. *Protein Sci.* **1996**, *5*, 2600–2616. [CrossRef] [PubMed]
49. Joo, H.; Chavan, A.G.; Fraga, K.J.; Tsai, J. An amino acid code for irregular and mixed protein packing. *Proteins* **2015**, *83*, 2147–2161. [CrossRef]
50. Rose, G.D.; Glerasch, L.M.; Smith, J.A. Turns in Peptides and Proteins. *Adv. Protein Chem.* **1985**, *37*, 1–109. [CrossRef]
51. Marcelino, A.M.C.; Gierasch, L.M. Roles of β-turns in protein folding: From peptide models to protein engineering. *Biopolymers* **2008**, *89*, 380–391. [CrossRef] [PubMed]
52. Koch, O.; Klebe, G. Turns revisited: A uniform and comprehensive classification of normal, open, and reverse turn families minimizing unassigned random chain portions. *Proteins Struct. Funct. Bioinform.* **2009**, *74*, 353–367. [CrossRef] [PubMed]
53. Kabsch, W.; Sander, C. Dictionary of protein secondary structure: Pattern recognition of hydrogen-bonded and geometrical features. *Biopolymers* **1983**, *22*, 2577–2637. [CrossRef] [PubMed]
54. Munoz, V.; Henry, E.; Hofrichter, J.; Eaton, W.A. A statistical mechanical model for β-hairpin kinetics. *Proc. Natl. Acad. Sci. USA* **1998**, *95*, 5872–5879. [CrossRef]
55. Dasgupta, B.; Pal, L.; Basu, G.; Chakrabarti, P. Expanded turn conformations: Characterization and sequence-structure correspondence in α-turns with implications in helix folding. *Proteins Struct. Funct. Bioinform.* **2004**, *55*, 305–315. [CrossRef] [PubMed]
56. Skipper, L. Proteins. Overview. In *Encyclopedia of Analytical Science*; Elsevier: Amsterdam, The Netherlands, 2005; pp. 344–352. [CrossRef]
57. Leszczynski, J.F.; Rose, G.D. Loops in Globular Proteins: A Novel Category of Secondary Structure. *Science* **1986**, *234*, 849–855. [CrossRef]
58. Wang, X.; Wang, M.; Tong, Y.; Shan, L.; Wang, J. Probing the folding capacity and residual structures in 1–79 residues fragment of staphylococcal nuclease by biophysical and NMR methods. *Biochimie* **2006**, *88*, 1343–1355. [CrossRef]
59. Neuhaus, F.C. Role of the omega loop in specificity determination in subsite 2 of the D-alanine:D-alanine (D-lactate) ligase from Leuconostoc mesenteroides: A molecular docking study. *J. Mol. Graph. Model.* **2011**, *30*, 31–37. [CrossRef] [PubMed]
60. Likhachev, I.; Bystrov, V. Assembly of a phenylalanine nanotube with a molecular dynamic manipulator. *Math. Biol. Bioinform.* **2021**, *16*, 244–255. [CrossRef]
61. Zelenovskiy, P.S.; Nuraeva, A.; Kopyl, S.; Arkhipov, S.G.; Vasilev, S.G.; Bystrov, V.S.; Gruzdev, D.A.; Waliczek, M.; Svitlyk, V.; Shur, V.Y.; et al. Chirality-Dependent Growth of Self-Assembled Diphenylalanine Microtubes. *Cryst. Growth Des.* **2019**, *19*, 6414–6421. [CrossRef]
62. Bystrov, V.S.; Zelenovskiy, P.; Nuraeva, A.; Kopyl, S.; Zhulyabina, O.A.; Tverdislov, V. Molecular modeling and computational study of the chiral-dependent structures and properties of self-assembling diphenylalanine peptide nanotubes. *J. Mol. Model.* **2019**, *25*, 199. [CrossRef] [PubMed]
63. Bystrov, V.; Coutinho, J.; Zelenovskiy, P.; Nuraeva, A.; Kopyl, S.; Zhulyabina, O.; Tverdislov, V. Structures and Properties of the Self-Assembling Diphenylalanine Peptide Nanotubes Containing Water Molecules: Modeling and Data Analysis. *Nanomaterials* **2020**, *10*, 1999. [CrossRef]

64. German, H.W.; Uyaver, S.; Hansmann, A.U.H.E. Self-Assembly of Phenylalanine-Based Molecules. *J. Phys. Chem. A* **2015**, *119*, 1609–1615. [CrossRef] [PubMed]
65. Adler-Abramovich, L.; Vaks, L.; Carny, O.; Trudler, D.; Magno, A.; Caflisch, A.; Frenkel, D.; Gazit, E. Phenylalanine assembly into toxic fibrils suggests amyloid etiology in phenylketonuria. *Nat. Chem. Biol.* **2012**, *8*, 701–706. [CrossRef] [PubMed]
66. Bystrov, V.; Sidorova, A.; Lutsenko, A.; Shpigun, D.; Malyshko, E.; Nuraeva, A.; Zelenovskiy, P.; Kopyl, S.; Kholkin, A. Modeling of Self-Assembled Peptide Nanotubes and Determination of Their Chirality Sign Based on Dipole Moment Calculations. *Nanomaterials* **2021**, *11*, 2415. [CrossRef]
67. Petitjean, M. On the root mean square quantitative chirality and quantitative symmetry measures. *J. Math. Phys.* **1999**, *40*, 4587–4595. [CrossRef]
68. Peng, X.-L.; Fang, K.-T.; Hu, Q.-N.; Liang, Y.-Z. Impersonality of the Connectivity Index and Recomposition of Topological Indices According to Different Properties. *Molecules* **2004**, *9*, 1089–1099. [CrossRef]
69. Yaffe, D.; Cohen, Y. Neural Network Based Temperature-Dependent Quantitative Structure Property Relations (QSPRs) for Predicting Vapor Pressure of Hydrocarbons. *J. Chem. Inf. Comput. Sci.* **2001**, *41*, 463–477. [CrossRef]
70. McClelland, H.E.; Jurs, P.C. Quantitative Structure−Property Relationships for the Prediction of Vapor Pressures of Organic Compounds from Molecular Structures. *J. Chem. Inf. Comput. Sci.* **2000**, *40*, 967–975. [CrossRef]
71. Zhao, T.; Zhang, Q.; Long, H.; Xu, L. Graph Theoretical Representation of Atomic Asymmetry and Molecular Chirality of Benzenoids in Two-Dimensional Space. *PLoS ONE* **2014**, *9*, e102043. [CrossRef] [PubMed]
72. Mezey, P.G. The proof of the metric properties of a fuzzy chirality measure of molecular electron density clouds. *J. Mol. Struct. Theochem* **1998**, *455*, 183–190. [CrossRef]
73. Gilat, G.; Schulman, L. Chiral interaction, magnitude of the effects and application to natural selection of L-enantiomer. *Chem. Phys. Lett.* **1985**, *121*, 13–16. [CrossRef]
74. Zabrodsky, H.; Peleg, S.; Avnir, D. Continuous symmetry measures. *J. Am. Chem. Soc.* **1992**, *114*, 7843–7851. [CrossRef]
75. Pinsky, M.; Dryzun, C.; Casanova, D.; Alemany, P.; Avnir, D. Analytical methods for calculating Continuous Symmetry Measures and the Chirality Measure. *J. Comput. Chem.* **2008**, *29*, 2712–2721. [CrossRef] [PubMed]
76. Luzanov, A.V.; Nerukh, D. Simple One-electron Invariants of Molecular Chirality. *J. Math. Chem.* **2006**, *41*, 417–435. [CrossRef]
77. Raos, G. Degrees of Chirality in Helical Structures. *Macromol. Theory Simul.* **2002**, *11*, 739–750. [CrossRef]
78. Ramachandran, G.; Ramakrishnan, C.; Sasisekharan, V. Stereochemistry of polypeptide chain configurations. *J. Mol. Biol.* **1963**, *7*, 95–99. [CrossRef]
79. Baruch-Shpigler, Y.; Wang, H.; Tuvi-Arad, I.; Avnir, D. Chiral Ramachandran Plots I: Glycine. *Biochemistry* **2017**, *56*, 5635–5643. [CrossRef]
80. Wang, H.; Avnir, D.; Tuvi-Arad, I. Chiral Ramachandran Plots II: General Trends and Protein Chirality Spectra. *Biochemistry* **2018**, *57*, 6395–6403. [CrossRef] [PubMed]
81. Mannige, R.V. An exhaustive survey of regular peptide conformations using a new metric for backbone handedness (h). *PeerJ* **2017**, *5*, 3327. [CrossRef]
82. Robinson, S.W.; Afzal, A.M.; Leader, D.P. Bioinformatics: Concepts, Methods, and Data. In *Handbook of Pharmacogenomics and Stratified Medicine*, 1st ed.; Padmanabhan, S., Ed.; Academic Press: New York, NY, USA, 2014; pp. 259–287.
83. Johansson, M.U.; Zoete, V.; Michielin, O.; Guex, N. Defining and searching for structural motifs using DeepView/Swiss-PdbViewer. *BMC Bioinform.* **2012**, *13*, 173. [CrossRef]
84. Richardson, J.S. The anatomy and taxonomy of protein structure. *Adv. Protein Chem.* **1981**, *34*, 167–339. [CrossRef]
85. Lewis, P.N.; Momany, F.A.; Scheraga, H.A. Energy Parameters in Polypeptides. VI. Conformational Energy Analysis of the N-Acetyl N′-Methyl Amides of the Twenty Naturally Occurring Amino Acids. *Isr. J. Chem.* **1973**, *11*, 121–152. [CrossRef]
86. Wilmot, C.; Thornton, J. Analysis and prediction of the different types of β-turn in proteins. *J. Mol. Biol.* **1988**, *203*, 221–232. [CrossRef]
87. De Brevern, A.G. Extension of the classical classification of β-turns. *Sci. Rep.* **2016**, *6*, 33191. [CrossRef] [PubMed]
88. Ting, D.; Wang, G.; Shapovalov, M.; Mitra, R.; Jordan, M.; Dunbrack, R.L. Neighbor-Dependent Ramachandran Probability Distributions of Amino Acids Developed from a Hierarchical Dirichlet Process Model. *PLoS Comput. Biol.* **2010**, *6*, e1000763. [CrossRef]
89. Shapovalov, M.; Slobodan, V.; Dunbrack, R.L. A new clustering and nomenclature for beta turns derived from high-resolution protein structures. *PLoS Comput. Biol.* **2019**, *15*, e1006844. [CrossRef]
90. Fang, C.; Shang, Y.; Xu, N. Improving Protein Gamma-Turn Prediction Using Inception Capsule Networks. *Sci. Rep.* **2018**, *8*, 15741. [CrossRef]
91. Fiser, A.; Do, R.K.G.; Šali, A. Modeling of loops in protein structures. *Protein Sci.* **2000**, *9*, 1753–1773. [CrossRef] [PubMed]
92. Duddy, W.J.; Nissink, J.W.M.; Allen, F.H.; Milner-White, E.J. Mimicry by asx- and ST-turns of the four main types of β-turn in proteins. *Protein Sci.* **2008**, *13*, 3051–3055. [CrossRef]
93. Eswar, N.; Ramakrishnan, C. Secondary structures without backbone: An analysis of backbone mimicry by polar side chains in protein structures. *Protein Eng. Des. Sel.* **1999**, *12*, 447–455. [CrossRef]
94. Sidorova, A.E.; Lutsenko, A.O.; Shpigun, D.K.; Malyshko, E.V.; Tverdislov, V.A. Methods to Determine the Chirality Sign for Helical and Superhelical Protein Structures. *Biophysics* **2021**, *66*, 357–363. [CrossRef]

95. Sidorova, A.; Malyshko, E.; Lutsenko, A.; Shpigun, D.; Bagrova, O. Protein Helical Structures: Defining Handedness and Localization Features. *Symmetry* **2021**, *13*, 879. [CrossRef]
96. Nicholson, H.; Anderson, D.E.; Pin, S.D.; Matthews, B.W. Analysis of the interaction between charged side chains and the.alpha.-helix dipole using designed thermostable mutants of phage T4 lysozyme. *Biochemistry* **1991**, *30*, 9816–9828. [CrossRef]
97. Ludwig, M.L.; Pattridge, K.A.; Metzger, A.L.; Dixon, M.M.; Eren, M.; Feng, Y.; Swenson, R.P. Control of Oxidation−Reduction Potentials in Flavodoxin from Clostridium beijerinckii: The Role of Conformation Changes. *Biochemistry* **1997**, *36*, 1259–1280. [CrossRef] [PubMed]
98. Nataraj, D.V.; Srinivasan, N.; Sowdhamini, R.; Ramakrishnan, C. α-Turns in protein structure. *Curr. Sci.* **1995**, *69*, 434–447.
99. Taylor, N.R.; Cleasby, A.; Singh, O.; Skarzynski, T.; Wonacott, A.J.; Smith, P.W.; Sollis, S.L.; Howes, P.D.; Cherry, P.C.; Bethell, R.; et al. Dihydropyrancarboxamides Related to Zanamivir: A New Series of Inhibitors of Influenza Virus Sialidases. 2. Crystallographic and Molecular Modeling Study of Complexes of 4-Amino-4H-pyran-6-carboxamides and Sialidase from Influenza Virus Types A and B. *J. Med. Chem.* **1998**, *41*, 798–807. [CrossRef] [PubMed]
100. Baker, E.; Dodson, E.J. Crystallographic refinement of the structure of actinidin at 1.7 Å resolution by fast Fourier least-squares methods. *Acta Crystallogr.* **1980**, *A36*, 559–572. [CrossRef]
101. Likhachev, I.V.; Balabaev, N.K.; Galzitskaya, O.V. Elastic and Non-elastic Properties of Cadherin Ectodomain: Comparison with Mechanical System. *Adv. Intell. Syst. Comput.* **2019**, *938*, 555–566. [CrossRef]
102. Glyakina, A.V.; Likhachev, I.V.; Balabaev, N.K.; Galzitskaya, O.V. Comparative mechanical unfolding studies of spectrin domains R15, R16 and R17. *J. Struct. Biol.* **2018**, *201*, 162–170. [CrossRef]
103. Lemak, A.S.; Balabaev, N.K. A Comparison Between Collisional Dynamics and Brownian Dynamics. *Mol. Simul.* **1995**, *15*, 223–231. [CrossRef]
104. Lemak, A.S.; Balabaev, N.K. Molecular dynamics simulation of a polymer chain in solution by collisional dynamics method. *J. Comput. Chem.* **1996**, *17*, 1685–1695. [CrossRef]
105. HyperChem Download—Sophisticated Molecular Modeling Environment. Available online: https://hyperchem.software.informer.com/ (accessed on 21 October 2021).
106. Rocha, G.B.; Freire, R.O.; Simas, A.M.; Stewart, J.J.P. RM1: A reparameterization of AM1 for H, C, N, O, P, S, F, Cl, Br, and I. *J. Comput. Chem.* **2006**, *27*, 1101–1111. [CrossRef] [PubMed]
107. The Protein Data Bank. Available online: http://www.rcsb.org/ (accessed on 1 October 2020).
108. Bystrov, V.S.; Kopyl, S.A.; Zelenovskiy, P.; Zhulyabina, O.A.; Tverdislov, V.; Salehli, F.; Ghermani, N.E.; Shur, V.; Kholkin, A.L. Investigation of physical properties of diphenylalanine peptide nanotubes having different chiralities and embedded water molecules. *Ferroelectrics* **2018**, *525*, 168–177. [CrossRef]

nanomaterials

MDPI

Article

Numerical Study on Broadband Antireflection of Moth-Eye Nanostructured Polymer Film with Flexible Polyethylene Terephthalate Substrate

Jun Lan [1,2], Yong Yang [1,*] and Song Hu [1]

1 State Key Laboratory of Optical Technologies on Nano-Fabrication and Micro-Engineering, Institute of Optics and Electronics, Chinese Academy of Sciences, Chengdu 610209, China; lanjun19@mails.ucas.ac.cn (J.L.); husong@ioe.ac.cn (S.H.)

2 University of Chinese Academy of Sciences, Beijing 100049, China

* Correspondence: yangyong@ioe.ac.cn

Abstract: The application of moth-eye nanostructured polymer film on the flexible polyethylene terephthalate (PET) substrate is an effective way to improve its antireflection (AR) performance. However, many factors affect the AR properties of the moth-eye structure in the actual manufacturing process. Moreover, the antireflection research based on PET substrate has been relatively lacking compared with the silicon substrate. In this paper, we simulate and analyze the AR performance of the moth-eye nanostructured polymer film on PET substrate by using the finite-difference time-domain method within the wavelength range of 400–1100 nm. Simulation results show that the parabola-shaped moth-eye structure (PSMS) can suppress the Fresnel reflection significantly. Moreover, the height and filling ratios are the dominant factors that affect the AR performance of PSMS. Additionally, the base diameter, residual layer thickness, and the refractive index of PSMS polymer film also affect the reflectivity of PET slightly. As a result, an optimal PSMS with base diameter of 400 nm, height of 300 nm, and the hexagonal close-packed arrangement is appropriate, and the solar-weighted reflectivity of PET can be suppressed to 0.21%, which shows a prominent advantage over the bare PET (≈6%). Therefore, this research has promising potential for improving the optical performance of optoelectronic devices by using nanostructured polymer materials.

Keywords: optoelectronic devices; nanostructured polymer film; antireflection coating; finite-difference time-domain method

Citation: Lan, J.; Yang, Y.; Hu, S. Numerical Study on Broadband Antireflection of Moth-Eye Nanostructured Polymer Film with Flexible Polyethylene Terephthalate Substrate. *Nanomaterials* **2021**, *11*, 3313. https://doi.org/10.3390/nano11123313

Academic Editor: Vladimir S. Bystrov

Received: 15 October 2021
Accepted: 30 November 2021
Published: 6 December 2021

1. Introduction

It is well-known that suppression of optical reflection is essential for many optoelectronic devices [1–4]. Multilayer antireflection coatings and biomimetic moth-eye structure (MS) are recognized as two effective methods for reducing reflection in the wide wavelength range. However, the application of multilayer antireflection coatings has been accompanied by some problems, such as material selection, thermal stability, thickness mismatch, etc. [5]. Compared with multilayer antireflection coatings, MS can effectively reduce the Fresnel reflection by forming a gradient refractive index (GRIN) profile between the air and the medium of MS [6]. At present, there has been much research carried out to investigate the antireflection (AR) performance of MS based on silicon or glass [7–10]. However, with the development of technology, optoelectronic devices have gradually transformed to become flexible, ultra-thin, and lightweight. PET is widely employed as device substrate or protection cover in various optoelectronic devices such as solar cells, display panels, transistors, and photodetectors due to its excellent flexibility, light weight, low cost, and high transparency [11–14]. At normal incidence, the reflectivity is about 6% for a single-side surface of bare PET, which is similar to glass substrates and will degrade the performance of optoelectronic devices [15]. Therefore, in order to guarantee the superior performance of optoelectronic devices, it is meant to reduce the reflectivity of PET by coating the moth-eye structured polymer film. So far, there are few studies of AR based on PET. In general, although there has been some experimental research on the fabrication of AR moth-eye structured polymer film on PET substrate [16,17], there has been very little work on the optimization of preliminary structural design, such as shape, size, material properties, etc. However, it is more complicated, high-cost, and time-consuming to fabricate MS with different geometry sizes for structural optimization. It is worth mentioning that the finite-difference time-domain (FDTD) method has been proved to be able to accurately predict the AR performance of MS by solving the time-dependent Maxwell's equations [18]. In order to suppress reflectance effectively, it is significant to optimize the shape and key structure parameters of moth-eye structured polymer film by using the FDTD method within the spectral range of 400–1100 nm. In this study, we investigate the influences of both the PET substrate and the thickness of the residual layer on the AR performance of MS. Furthermore, at the end of this paper, we explore the effect of the refractive index of moth-eye nanostructured polymer film on its AR performance.

2. Materials and Method

The FDTD method was first developed by Kane Yee in 1966 and is an accurate numerical method to calculate problems in electromagnetics. In this research, all optical simulations were performed using 3D FDTD methods (FDTD solution 2021-v212, Lumerical, Vancouver, British Columbia, Canada). To obtain broadband AR performance for moth-eye structured polymer film, in the simulation, we focused on the solar-weighted reflectance (R_{SW}) as the height (H), base diameter (D), filling ratio (R_{BDP}), residual layer thickness, and polymer refractive index variation. It is noted that the R_{SW} considers the influence of solar spectral irradiance, which is defined as the ratio between the reflected photons and the total incident photons and is calculated by using the following equation:

$$R_{SW} = \frac{\int_{400\ \text{nm}}^{1100\ \text{nm}} I_s(\lambda)R(\lambda)d\lambda}{\int_{400\ \text{nm}}^{1100\ \text{nm}} I_s(\lambda)d\lambda} \tag{1}$$

where $I_S(\lambda)$ is the solar spectral irradiance (i.e., AM 1.5 G) and $R(\lambda)$ is the total reflectance within the range of wavelength from 400 nm to 1100 nm, respectively.

It is well-known that Fresnel reflections can be effectively suppressed by forming a gradual refractive index change between air and materials. Further, both the cone-shaped moth-eye structure (CSMS) and the parabola-shaped moth-eye structure (PSMS) are the most popular AR structures which can suppress the Fresnel reflection effectively [19]. The schematic of structure array models is shown in Figure 1a,b, respectively. Both of them adopted conventional hexagonal arrangement, which is similar to the arrangement of the corneal array structure of natural moth eyes [20,21]. Although many polymers have been used to fabricate antireflection structured polymer film, the UV-curable NOA63 (Norland Products Inc, Cranbury, NJ, USA) stands out for its superior optical properties, appropriate refractive index, environmental adaptability, and suitability for large-area preparation. Specifically, the UV-curable NOA63 has almost no absorption at near-infrared wavelength [22], and its refractive index is close to that of PET. Compared with most polymers, it has a very high Young's modulus (1.65 GPa) and thermal stability, ensuring that the optoelectronic devices are not vulnerable to the external environment with excellent performance. Moreover, NOA63 polymer films with nanostructures can be rapidly prepared in large areas on PET surfaces by roll-to-roll nanoimprint lithography. It is also worth noting that the high viscosity and fast curing properties of NOA63 ensure antireflection films' processing speed and durability. Therefore, in this research, the UV-curable NOA63 was used as the structural material of MS, and the influence of different polymers on the AR properties of MS will be analyzed in detail in the last chapter of this paper.

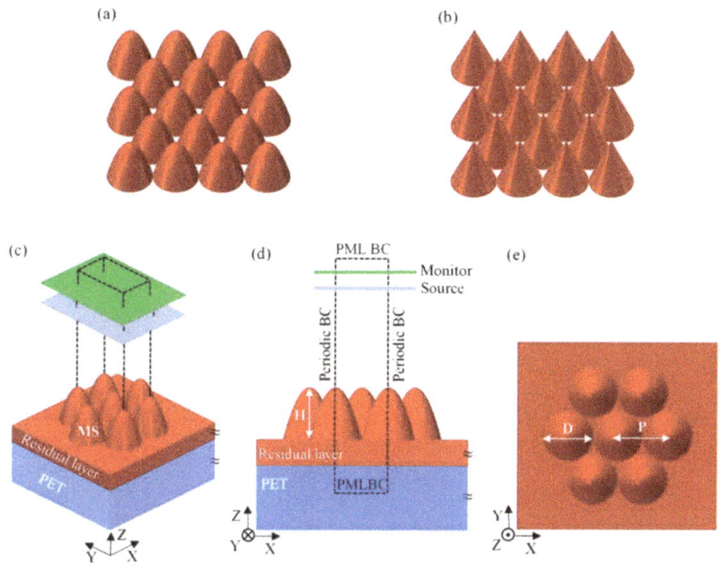

Figure 1. Schematic diagram of 3D physical model and simulation model: (**a**) parabola-shaped moth-eye structure (PSMS) array, (**b**) cone-shaped moth-eye structure (CSMS) array, and (**c**–**e**) parameter setting of simulation unit with the hexagonal close-packed arrangement.

In this simulation, classical and theoretical physical models of PSMS and CSMS are established based on the functional relation of structural parameters as follows [23]:

$$\mathrm{PSMS}: r(h) = \frac{D}{2}\sqrt{1-\frac{h}{H}} \tag{2}$$

$$\mathrm{CSMS}: r(h) = \frac{D}{2}(1-\frac{h}{H}) \tag{3}$$

where $r(h)$ is the radius of the cross-section of MS at a height of h.

The sketch of the simulation model is shown in Figure 1c–e. Where the dashed frame represents the unit of the simulation region, the optical calculation is considered a one-sided surface of PET. A plane wave light source with normalized intensity and X-axis polarization was normally incident from the air to MS. The perfectly matched layer boundary conditions (PML BC) were set in the direction of propagation of light (Z direction), and the boundary conditions of the X and Y directions were set as periodic boundary conditions (Periodic BC). In the simulation, the refractive index of PET was acquired from the refractive index website [24]. The refractive index of NOA63 polymer was set at 1.56 and the extinction coefficient was not taken into account because it could be ignored [25].

3. Result and Discussion

3.1. Shape Optimization of the Moth-Eye Structure Array

In order to determine the most suitable shape of MS for reducing the reflectivity of PET, the AR properties of CSMS and PSMS were compared. Figure 2a,b show the contour plots of solar-weighted reflectance (R_{SW}) of NOA63 CSMS polymer film coated on the PET (i.e., NOA63 CSMS/PET) and NOA63 PSMS/PET in the wavelength range of 400–1100 nm as the diameter and height of MS varied from 100 nm to 1000 nm, where the thickness of the NOA63 polymer residual layer was fixed at 2 μm and MS with the hexagonal close-packed (D = P) arrangement.

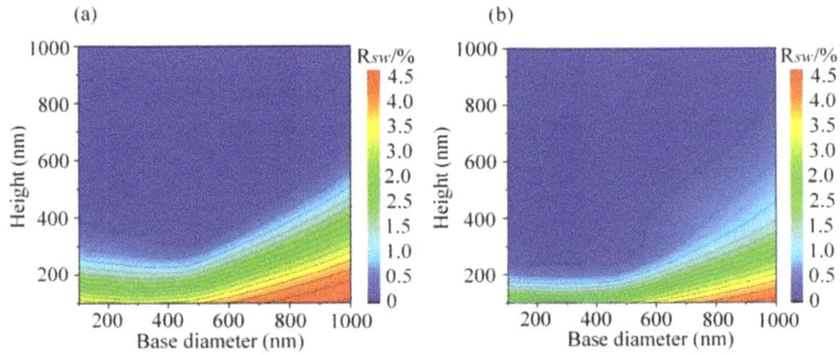

Figure 2. Contour plots of solar-weighted reflectance of (**a**) NOA63 CSMS/PET and (**b**) NOA63 PSMS/PET.

It is noted that the R_{SW} decreases gradually with the increase in the height when the base diameter is fixed. Further observation of the density of the contour line shows that it mainly becomes sparser and sparser with the increase in the height, which indicates that when MS is high enough, the decreasing rate of reflectivity will be slow. Therefore, there will exist a balance between height and its difficulty of fabrication. On the other hand, when the height of MS is fixed, the R_{SW} shows no obvious changes as the base diameter ranges from 100 nm to 400 nm, which means that the base diameter is not the main factor in affecting the AR properties of subwavelength MS. However, the R_{SW} increases significantly when the base diameter is larger than 400 nm; this phenomenon can be explained by the grating equation under normal incidence lighting [26]:

$$\sin\theta_d = m\lambda/n\Lambda, \tag{4}$$

where n is the refractive index of the incident medium, Λ is the grating period, m is the diffraction order, λ is the wavelength of the incident light, and θ_d is the diffraction angle. The incident medium is air and $n_{air} = 1$. In addition, MS is distributed with the close-packed arrangement, indicating that $\Lambda = D$. Based on the grating equation, it can be known that when the period of MS is larger than the shortest wavelength of incident light, the high-order diffraction will be generated and cause the diffraction loss [27]. Therefore, in this study, we mainly focus on the design and optimization of MS on submicron scale.

Moreover, it follows from the comparison between Figure 2a,b that PSMS exhibits better AR properties because it can achieve excellent AR performance at a smaller height, which reduces the cost and difficulty of the manufacturing process. Specifically, the effect of shape on AR performance can be explained by using the effective medium theory (EMT). Based on the EMT, subwavelength MS can be regarded as a homogeneous medium with a GRIN profile between air and the NOA63 film. The effective refractive index of subwavelength MS at normal incident can be approximately represented as [23]:

$$n_{\text{eff}}(h) = \left\{ f(h)n_a^{2/3} + [1 - f(h)]n_b^{2/3} \right\}^{3/2} \tag{5}$$

where $f(h)$ is the filling factor, which is equal to the ratio of the cross-sectional area of MS at height of h to the area taken up by MS. n_a is the refractive index of NOA63 $n_a = 1.56$, and n_b is the refractive index of air ($n_b = 1$). MS is distributed with the hexagonal close-packed arrangement ($D = P$), and thus the area taken up by MS is equal to $A = \sqrt{3}D^2/2$. Further, the cross-section area of MS at height of h is:

$$S = \pi r(h)^2 \tag{6}$$

Therefore, combined with the Equations (2) and (3), the $f(h)$ can be expressed by:

$$f_{PSMS}(h) = \frac{S_{PSMS}}{A} = \frac{\sqrt{3}\Pi}{6}(1 - \frac{h}{H}) \tag{7}$$

$$f_{CSMS}(h) = \frac{S_{CSMS}}{A}\frac{\sqrt{3}\Pi}{6}(1 - \frac{h}{H})^2 \tag{8}$$

Figure 3 shows the effective refractive index changes between air and the NOA63 film at normal incident. Compared with CSMS, PSMS can provide a nearly linear graded refractive index profile between air and the NOA63 film, which is more efficient to suppress the Fresnel reflection in wide ranges of incident wavelengths [28]. Therefore, the following work is to optimize the parameters of the simulation model of NOA63 PSMS/PET.

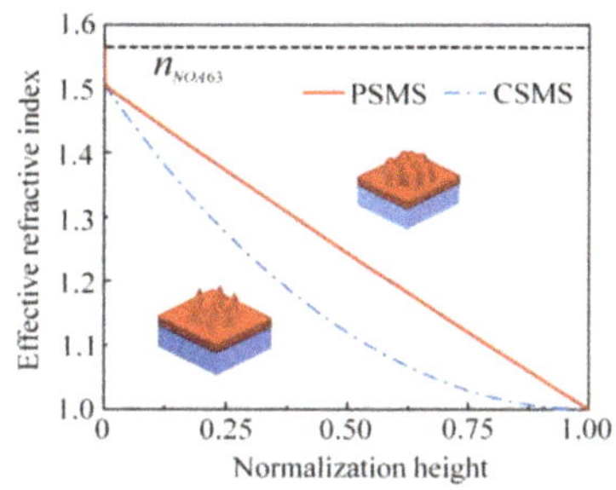

Figure 3. The effective refractive index change between air and the NOA63 film at normal incident. Normalized height 1 represents the peak of the moth-eye structure, and 0 represents the base of the surface of the NOA63 film.

3.2. Parameter Optimization of NOA63 PSMS/PET

The parameter optimization of the simulation model is helpful to fabricate MS with excellent AR properties efficiently and quickly in practical application. Figure 4a shows the R_{SW} of PSMS with the base diameter D varying between 100 nm, 200 nm, 300 nm, and 400 nm. According to the simulation results, the base diameter has little effect on the reflectivity of NOA63 PSMS/PET. As the height of PSMS increased from 100 nm to 300 nm, the R_{SW} values rapidly decreased from around 3% to 0.21%. The reason is that the effective refractive index changes more slowly as height increases, and consequently the Fresnel reflection has been suppressed more efficiently. In addition, when the reflectance is suppressed to a small value, the variation of reflectance with the increase in height is no longer significant. Thus, we can consider that 300 nm is the height with the best benefit for NOA63 PSMS/PET. Moreover, considering the difficulty of MS manufacturing, it is more appropriate to choose subwavelength PSMS with a base diameter of 400 nm.

Further, the effect of the NOA63 polymer residual layer thickness (1–10 μm) on AR performance of NOA63 PSMS/PET is shown in Figure 4b, where the base diameter and height are fixed at 400 nm and 300 nm, respectively, which are also the best parameters as proved above. It can be found that the R_{SW} varies slightly, and the fluctuation range is not more than 0.007% in the residual layer thickness varying from 1 to 10 μm. Therefore, it is almost certain that the residual layer thickness has little influence on the AR performance of NOA63 PSMS/PET under the ideal conditions, which is also consistent with the experimental results of previous relevant articles [29]. This result can be explained by the refractive index of the UV-curable NOA63 being very close to that of PET. Therefore, the Fresnel reflection due to the refractive index difference of the material is effectively suppressed. In addition, the extinction coefficient of the UV-curable NOA63 is negligibly small, so it is not considered in the simulation process. However, in the actual fabrication process, the residual layer thickness of NOA63 film should be as small as possible under the suitable technology. In this simulation, the residual layer thickness is fixed at 2 μm.

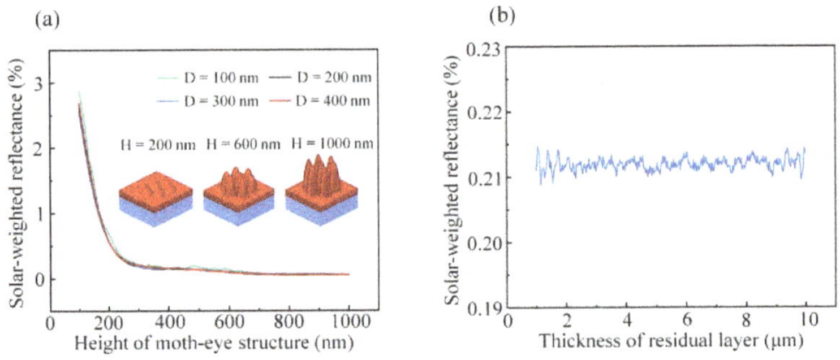

Figure 4. The solar-weighted reflectance for different parameters of (**a**) the height of moth-eye structure and (**b**) the residual layer thickness of NOA63 polymer.

What is worth mentioning is that the filling ratio is considered to be another critical factor besides base diameter and height. For the hexagonal arrangement of the PSMS array, the filling ratio is defined as D/P (i.e., the ratio of the base diameter of MS to the period between the MS array; R_{BDP}). Thus, the filling ratio is directly related to the effective refractive index value of the bottom of the PSMS and ultimately affects the AR performance of the PSMS. Figure 5 shows the R_{SW} of NOA63 PSMS/PET with the filling factor ranging from 0.1 to 1, where the base diameter and the height are fixed at 400 nm and 300 nm, respectively. As shown in Figure 5, the R_{SW} decreases with the filling ratio varying from 0.1 to 1. Specifically, it can be seen that the R_{SW} reduces rapidly with the R_{BDP} varying from 0.1 to 0.7, and then decreases slowly after the R_{BDP} is greater than 0.7. In principle, the R_{BDP} has a significant effect on the reflectivity of NOA63 PSMS/PET. Furthermore, the R_{BDP} should be at least over 0.7, and the best AR properties of PSMS are obtained when the filling ratio is 1. In this simulation, the filling ratio of PSMS with the hexagonal close-packed arrangement is 1.

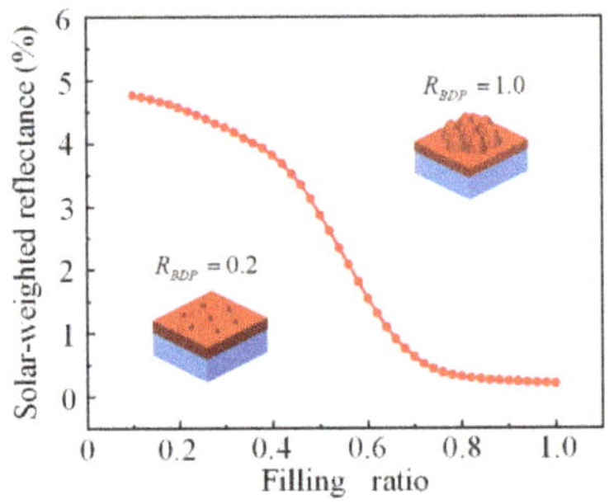

Figure 5. Curve of the solar-weighted reflectance with the filling ratio from 0.1 to 1.

In practice, many polymers show the property of reducing the reflectance of PET substrate in experiments. Most of them are low refractive index polymers with refractive index ranging from 1.40 to 1.60. In this study, we further explore the influence of polymer refractive index on the AR performance of PSMS. The R_{SW} of PSMS/PET with different refractive indices of polymer and structural parameters are shown in Figure 6. By analyzing the density and trend of the contour lines in Figure 6a, it can be seen that when the height of PSMS is less than 300 nm, the height is the dominant factor affecting the reflectivity because the R_{SW} changes rapidly with the varying of the height. On the contrary, when the height of PSMS is larger than 300 nm, the contour lines trend changes from horizontal to vertical, which indicates that the refractive index of the polymer becomes the main factor affecting the reflectivity. However, it is noted that the R_{SW} changes slowly with the varying of refractive index. As shown in Figure 6a, the reduction of reflectivity is less than

0.004 (isoline spacing = 0.002) as the refractive index of polymer increases from 1.40 to 1.60. Considering the fabrication difficulty factors, PSMS with a height of 300 nm is appropriate for most polymers to exhibit excellent AR performance.

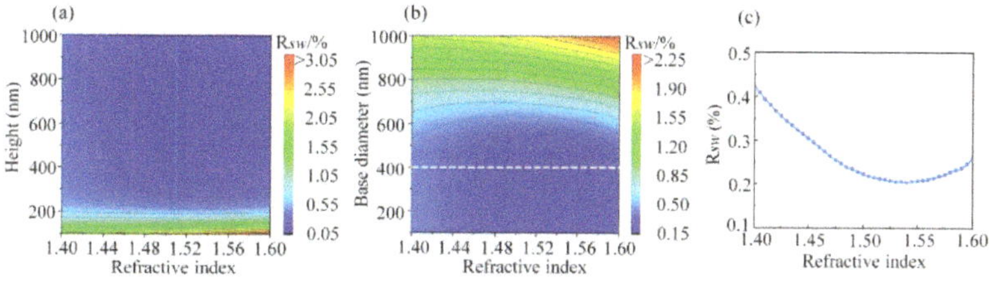

Figure 6. Contour plots of solar-weighted reflectance of PSMS/PET at different polymer refractive indices when (**a**) base diameter is fixed at 400 nm and (**b**) height is fixed at 300 nm. (**c**) The curve of solar-weighted reflectance changing with different polymer refractive indices when D = 400 nm and H = 300 nm.

When the height is fixed at 300 nm, Figure 6b shows the contour plots of the R_{SW} of PSMS/PET as the base diameter increases from 100 nm to 1000 nm and the refractive index of polymer varies from 1.40 to 1.60. It can be seen that when the base diameter is less than 400 nm, the reflectivity is mainly affected by the refractive index of polymer, and the reduction of reflectivity is less than 0.003 (isoline spacing = 0.0015) as the polymer refractive index increases from 1.40 to 1.60. Further analysis shows that the contour line approaches being completely vertical, which indicates that the reflectivity of PSMS is almost not influenced by a base diameter below 400 nm. It is also consistent with the previous simulation result. At the same time, it further proves the correction of selecting the best height with the base diameter of PSMS fixed at 400 nm in our simulation. On the other hand, when the base diameter of PSMS is larger than 400 nm, the R_{SW} of PSMS/PET is mainly affected by the base diameter, and the influence is relatively strong. Based on the above analysis, we can determine that compared with the structural parameters of PSMS, the refractive index of the polymer is not the dominant factor that affects the AR performance of PSMS. In order to further accurately reflect the variation of R_{SW} caused by the polymer refractive index, the curve of solar-weighted reflectance changing with the polymer refractive index when the height is fixed at 300 nm and the base diameter is fixed at 400 nm is shown in Figure 6c. With the increase in the refractive index, the R_{SW} of PSMS/PET decreases at first and then increases. Furthermore, it is more appropriate to choose a polymer with a refractive index around 1.55 to fabricate PSMS for reducing the reflectivity of PET.

4. Conclusions

This study investigated the AR effects of moth-eye nanostructured polymer films on top of PET substrate in the wide wavelength range of 400–1100 nm. The shape of the MS array and the design parameters of NOA63 PSMS/PET were optimized by using FDTD simulation. In general, a close-packed hexagonal arrangement (R_{BDP} = 1) of subwavelength PSMS with height of 300 nm and base diameter of 400 nm exhibits excellent AR performance, which can effectively inhibit the reflectivity to about 0.21%. Furthermore, the base diameter, residual layer thickness, and refractive index of polymers also slightly affect the reflectivity of PET; the determination of these parameters needs to consider the fabrication costs. In principle, the thickness of the residual layer should be as small as possible, and the refractive index of the polymer near 1.55 is more appropriate. This study provides theoretical guidance for the design of broadband antireflection using the moth-eye nanostructured polymer film on PET substrate, which is beneficial to further improving the performance of optoelectronic devices and flexible displays.

Author Contributions: Conceptualization, J.L. and Y.Y.; methodology, J.L.; software, J.L.; validation, S.H., Y.Y. and J.L.; formal analysis, Y.Y.; investigation, J.L.; resources, J.L.; data curation, J.L.; writing—original draft preparation, J.L.; writing—review and editing, J.L. and Y.Y.; visualization, J.L.; supervision, S.H.; project administration, S.H.; funding acquisition, Y.Y. All authors have read and agreed to the published version of the manuscript.

Funding: The research is funded by Special-Funded Program on Key Scientific Instruments and Equipment Development of Sichuan Province (Grant No.20ZDZX0004); the Frontier Research Fund of Institute of Optics and Electronics, China Academy of Sciences (Grant No. C21K004).

Institutional Review Board Statement: Not applicable.

Informed Consent Statement: Not applicable.

Data Availability Statement: The data is available on reasonable request from the corresponding author.

Conflicts of Interest: The authors declare that they have no known competing financial interest or personal relationships that could have appeared to influence the work reported in this paper.

References

1. Fang, C.; Zhang, Z.J.; Zhuang, Y.; Liu, S.; He, X.; Zhang, Y. Biomimetic diodon-skin nanothorn polymer antireflection film for solar cell applications. *Sol. Energy Mater. Sol. Cells* **2020**, *206*, 110305. [CrossRef]
2. Xu, Z.P.; Qiao, H.L.; Huangfu, H.C.; Li, X.W.; Guo, J.W.; Wang, H.Y. Optical absorption of several nanostructures arrays for silicon solar cells. *Opt. Commun.* **2015**, *356*, 526–529. [CrossRef]
3. Tan, G.; Lee, J.H.; Lan, Y.H.; Wei, M.K.; Peng, L.H.; Cheng, I.C.; Wu, S.T. Broadband antireflection film with moth-eye-like structure for flexible display applications. *Optica* **2017**, *4*, 678–683. [CrossRef]
4. Liu, W.; Wang, Y.; Guo, X.; Song, J.; Wang, X.; Yi, Y. Light Trapping in Single Elliptical Silicon Nanowires. *Nanomaterials* **2020**, *10*, 1740. [CrossRef] [PubMed]
5. Ko, Y.H.; Yu, J.S. Highly transparent sapphire micro-grating structures with large diffuse light scattering. *Opt. Express* **2011**, *19*, 15574–15583. [CrossRef] [PubMed]
6. Walheim, S.; Schaffer, E.; Mlynek, J.; Steiner, U. Nanophase-separated polymer films as high-performance antireflection coatings. *Science* **1999**, *283*, 520–522. [CrossRef] [PubMed]
7. Haldar, A.; Reddy, M.S.; Vijaya, R. Enhancement of light collection through flexible polymeric films patterned using self-assembled photonic crystals. *J. Phys. D Appl. Phys.* **2015**, *48*, 265103. [CrossRef]
8. Haldar, A.; Reddy, M.S.; Vijaya, R. Inexpensive graded-index antireflective surfaces for silicon-based optoelectronic devices. *J. Opt. Soc. Am. B* **2016**, *33*, 2331–2338. [CrossRef]
9. Kuo, W.K.; Hsu, J.J.; Nien, C.K.; Yu, H.H. Moth-Eye-Inspired Biophotonic Surfaces with Antireflective and Hydrophobic Characteristics. *ACS Appl. Mater. Interfaces* **2016**, *8*, 32021–32030. [CrossRef]
10. Dudem, B.; Jung, J.W.; Yu, J.S. Improved light harvesting efficiency of semitransparent organic solar cells enabled by broadband/omnidirectional subwavelength antireflective architectures. *J. Mater. Chem. A* **2018**, *6*, 14769–14779. [CrossRef]
11. Yoon, J.; Jo, S.; Chun, I.S.; Jung, I.; Kim, H.S.; Meitl, M.; Menard, E.; Li, X.L.; Coleman, J.J.; Paik, U.; et al. GaAs photovoltaics and optoelectronics using releasable multilayer epitaxial assemblies. *Nature* **2010**, *465*, 329–333. [CrossRef] [PubMed]
12. White, M.S.; Kaltenbrunner, M.; Glowacki, E.D.; Gutnichenko, K.; Kettlgruber, G.; Graz, I.; Aazou, S.; Ulbricht, C.; Egbe, D.A.M.; Miron, M.C.; et al. Ultrathin, highly flexible and stretchable PLEDs. *Nat. Photonics* **2013**, *7*, 811–816. [CrossRef]
13. Stutzmann, N.; Friend, R.H.; Sirringhaus, H. Self-aligned, Vertical-Channel, Polymer Field-Effect Transistors. *Science* **2003**, *299*, 1881–1884. [CrossRef] [PubMed]
14. Fan, F.R.; Tian, Z.Q.; Wang, Z.L. Flexible triboelectric generator. *Nano Energy* **2012**, *1*, 328–334. [CrossRef]
15. Leem, J.W.; Choi, M.; Yu, J.S. Multifunctional Microstructured Polymer Films for Boosting Solar Power Generation of Silicon-Based Photovoltaic Modules. *ACS Appl. Mater. Interfaces* **2015**, *7*, 2349–2358. [CrossRef]
16. Lee, S.H.; Han, K.S.; Shin, J.H.; Hwang, S.Y.; Lee, H. Fabrication of highly transparent self-cleaning protection films for photovoltaic systems. *Prog. Photovolt Res. Appl.* **2012**, *21*, 1056–1062. [CrossRef]
17. Chuang, C.H.; Lu, D.M.; Wang, P.H.; Lee, W.Y.; Shaikh, M.O. Antireflective polymer films via roll to roll UV nanoimprint lithography using an AAO mold. *Microsyst. Technol.* **2018**, *24*, 389–395. [CrossRef]
18. Yang, L.M.; Pan, C.Y.; Lu, F.P.; Chang, C.W.; Feng, S.W.; Tu, L.W. Anti-reflection sub-wavelength structures design for InGaN-based solar cells performed by the finite-difference-time-domain (FDTD) simulation method. *Opt. Laser Technol.* **2015**, *67*, 72–77. [CrossRef]
19. Song, Y.M.; Choi, H.J.; Yu, J.S.; Lee, Y.T. Design of highly transparent glasses with broadband antireflective subwavelength structures. *Opt. Express* **2010**, *18*, 13063–13071. [CrossRef] [PubMed]
20. Han, Z.W.; Jiao, Z.B.; Niu, S.C.; Ren, L.Q. Ascendant bioinspired antireflective materials: Opportunities and challenges coexist. *Prog. Mater. Sci.* **2019**, *103*, 1–68. [CrossRef]

21. Sun, J.Y.; Wang, X.B.; Wu, J.H.; Shen, J.J.; Cooper, M.A.; Zheng, X.T.; Liu, Y.; Yang, Z.G.; Wu, D. Biomimetic Moth-eye Nanofabrication: Enhanced Antireflection with Superior Self-cleaning Characteristic. *Sci. Rep.* **2018**, *8*, 5348. [CrossRef] [PubMed]
22. Leem, J.W.; Guan, X.Y.; Choi, M.; Yu, J.S. Broadband and omnidirectional highly-transparent coverglasses coated with biomimetic moth-eye nanopatterned polymer films for solar photovoltaic system applications. *Sol. Energy Mater. Sol. Cells* **2015**, *134*, 45–53. [CrossRef]
23. Stavenga, D.G.; Foletti, S.; Palasantzas, G.; Arikawa, K. Light on the moth-eye corneal nipple array of butterflies. *Proc. R. Soc. B* **2006**, *273*, 661–667. [CrossRef] [PubMed]
24. Refractive Index. INFO—Refractive Index Database. Available online: https://refractiveindex.info/ (accessed on 18 June 2021).
25. Myers, J.D.; Cao, W.; Cassidy, V.; Eom, S.H.; Zhou, R.; Yang, L.; You, W.; Xue, J. A universal optical approach to enhancing efficiency of organic-based photovoltaic devices. *Energy Environ. Sci.* **2012**, *5*, 6900–6904. [CrossRef]
26. Kim, M.C.; Jang, S.; Choi, J.; Kang, S.M.; Choi, M. Moth-eye Structured Polydimethylsiloxane Films for High-Efficiency Perovskite Solar Cells. *Nano-Micro Lett.* **2019**, *11*, 1–10. [CrossRef] [PubMed]
27. Leem, J.W.; Kim, S.; Lee, S.H.; Rogers, J.A.; Kim, E.; Yu, J.S. Efficiency Enhancement of Organic Solar Cells Using Hydrophobic Antireflective Inverted Moth-Eye Nanopatterned PDMS Films. *Adv. Energy Mater.* **2014**, *4*, 1301315. [CrossRef]
28. Qarony, W.; Hossain, M.I.; Dewan, R.; Fischer, S.; Meyer-Rochow, V.B.; Salleo, A.; Knipp, D.; Tsang, Y.H. Approaching Perfect Light Incoupling in Perovskite and Silicon Thin Film Solar Cells by Moth Eye Surface Textures. *Adv. Theory Simul.* **2018**, *1*, 1800030. [CrossRef]
29. Heo, S.Y.; Koh, J.K.; Kang, G.; Ahn, S.H.; Chi, W.S.; Kim, K.; Kim, J.H. Bifunctional Moth-Eye Nanopatterned Dye-Sensitized Solar Cells: Light-Harvesting and Self-Cleaning Effects. *Adv. Energy Mater.* **2013**, *4*, 1300632. [CrossRef]

nanomaterials

MDPI

Article

Temperature Dependence of Dielectric Properties of Ferroelectric Heterostructures with Domain-Provided Negative Capacitance

Maksim A. Pavlenko [1], Yuri A. Tikhonov [1,2], Anna G. Razumnaya [1,3,*], Valerii M. Vinokur [3] and Igor A. Lukyanchuk [1,2]

[1] Faculty of Physics, Southern Federal University, 344090 Rostov-on-Don, Russia; makpavlenko@sfedu.ru (M.A.P.); ytihonov@sfedu.ru (Y.A.T.); lukyanc@ferroix.net (I.A.L.)
[2] Laboratoire de Physique de la Matière Condensée, Université de Picardie Jules Verne, 80080 Amiens, France
[3] Terra Quantum AG, CH-9400 Rorschach, Switzerland; vmvinokour@gmail.com
* Correspondence: agrazumnaya@sfedu.ru; Tel.: +7-918-5518552

Abstract: It is well known that the ferroelectric layers in dielectric/ferroelectric/dielectric heterostructures harbor polarization domains resulting in the negative capacitance crucial for manufacturing energy-efficient field-effect transistors. However, the temperature behavior of the characteristic dielectric properties, and, hence, the corresponding behavior of the negative capacitance, are still poorly understood, restraining the technological progress thereof. Here we investigate the temperature-dependent properties of domain structures in the $SrTiO_3/PbTiO_3/SrTiO_3$ heterostructures and demonstrate that the temperature–thickness phase diagram of the system includes the ferroelectric and paraelectric regions, which exhibit different responses to the applied electric field. Using phase-field modeling and analytical calculations we find the temperature dependence of the dielectric constant of ferroelectric layers and identify the regions of the phase diagram wherein the system demonstrates negative capacitance. We further discuss the optimal routes for implementing negative capacitance in energy-efficient ferroelectric field-effect transistors.

Keywords: ferroelectrics; heterostructures; domains; negative capacitance

Citation: Pavlenko, M.A.; Tikhonov, Y.A.; Razumnaya, A.G.; Vinokur, V.M.; Lukyanchuk, I.A. Temperature Dependence of Dielectric Properties of Ferroelectric Heterostructures with Domain-Provided Negative Capacitance. *Nanomaterials* **2022**, *12*, 75. https://doi.org/10.3390/nano12010075

Academic Editor: Alain Pignolet

Received: 1 November 2021
Accepted: 21 December 2021
Published: 28 December 2021

1. Introduction

Modern microelectronics is turning to the extensive use of ultrathin ferroelectric films in a broad scope of devices, ranging from ferroelectric random access memory to logic elements and circuits for neuromorphic computers [1,2]. Numerous studies have shown [3] that ferroelectricity, in thin films, persists to nanometer thicknesses, provided that the depolarizing fields arising due to termination of the polarization lines at the film edges are properly reduced. An efficient mechanism for reducing depolarization fields was suggested by [4] and consists of forming a regular array of domains carrying alternating and opposite polarizations. This arrangement leads to the macroscopic neutrality of the depolarizing charges on the surface. However, it has been long believed that such a regular domain structure in ferroelectric films can hardly emerge, due to the screening of the depolarizing field by free charged particles or semiconductor electrons arising from impurities or vacancies. Only recently were domains in thin ferroelectric films discovered and they have become the subject of intense research [5–9]. The domains, predicted in early theoretical studies [5] are studied using the Landau–Kittel approximation, in which the polarization is assumed to be constant in the volume of domains. Subsequently, it has been demonstrated that they have a spatially dependent soft structures [6,10], resembling that of alternating vortices [9].

Superlattices, consisting of ferroelectric and dielectric layers, provide a unique opportunity for studying the behavior of nanodomains in a ferroelectric layer [11] as functions on external conditions, such as the applied field, temperature, and deformation. Due to the dielectric layers protecting the electric neutrality of the ferroelectric layer, the screening of the field by external, trapped charges does not occur. The response of the domain structure to an external applied electric field is of particular interest. As has been shown theoretically and confirmed experimentally, Refs. [12–15], the average electric field inside the ferroelectric layer, $\langle E_f \rangle$, is directed against the average induced polarization of the domain structure, $\langle P_f \rangle$. Thus, the effective average dielectric constant of the ferroelectric layer, $\varepsilon_f = 1 + \langle P_f \rangle / \varepsilon_0 \langle E_f \rangle$, emergesnegative, and this constitutes the negative capacitance effect viewed as an irreplaceable mean for reducing the power consumption of field-effect transistors [16].

As has been suggested [17,18], the use of negative capacitance in the gate structure of a field-effect transistor may enhance the voltage, controlling the semiconducting channel. Yet, the fact that the negative capacitance owes to the emergence of the domain structure is not adequately appreciated. In addition, the temperature dependence of the susceptibility of the domain structure has not been studied; and, methods for integrating a ferroelectric layer hosting domains into the field effect transistors have not been fully explored.

In this paper, we utilize the phase-field modeling of ferroelectric domains in heterostructures consisting of alternating ultrathin layers of the ferroelectric $PbTiO_3$ (PTO) and dielectric $SrTiO_3$ (STO) to investigate polarization distribution as it depends on temperature and the applied field. Numerical simulations are combined with analytical calculations. The obtained results enable us to explore the behavior of the effective dielectric constant of ferroelectric layers over a wide temperature range. Building on these results, we discuss the possibility of utilizing the domain structures, ensuring negative capacitance for designing energy-efficient ferroelectric transistors and highlighting considerations requiring special care.

2. Materials and Methods

Phase-field simulations of PTO/STO heterostructures are carried out using the FEniCS software package, version 2019.2.0.dev0 (https://bitbucket.org/fenics-project, accessed on 8 April 2021, USA) [19]. The appearance of the domain structure is described by the overdamped equation, defining the relaxation of polarization, **P**, to the state minimizing the system's free energy, $\mathcal{F}$, to its ground state with time, t.

$$-\gamma \frac{\partial \mathbf{P}}{\partial t} = \frac{\delta \mathcal{F}}{\delta \mathbf{P}}. \tag{1}$$

Equation (1) is closed by the Poisson equation, taking into account the electrostatic effects

$$\varepsilon_0 \varepsilon_i \nabla^2 \varphi = \nabla \cdot \mathbf{P}. \tag{2}$$

Here, φ is the electric potential and $\varepsilon_i \simeq 10$ [20] is the background dielectric constant due to nonpolar ions. The free energy functional describing the ferroelectric phase transition in a strained PTO has the form of the Ginzburg–Landau functional for a ferroelectric layer subjected to an electric field [21], and is written in the form [22] (the gradient and electrostatic terms are included):

$$\mathcal{F} = \int F dV, \qquad F = \left[a_i^*(u_m, T) P_i^2 + a_{ij}^* P_i^2 P_j^2 + a_{ijk} P_i^2 P_j^2 P_k^2 \right]_{i \le j \le k} + \frac{1}{2} G_{ijkl} (\partial_i P_j)(\partial_k P_l) + (\partial_i \varphi) P_i - \frac{1}{2} \varepsilon_0 \varepsilon_i (\nabla \varphi)^2, \tag{3}$$

where F is the free energy density; each of the indices i, j, k, l cyclically takes values of 1, 2 or 3 (or x, y, z). The numerical values of the Landau expansion coefficients, a_i^*, a_{ij}^*, a_{ijk}, and gradient energy coefficients, G_{ijkl}, for the PTO layers, compressively strained by the STO substrate with lattice mismatch $u_m = -0.013$ in the layer plane, are taken from [21,23] and given in Table 1.

Table 1. Numerical values of coefficients in the free energy functional $\mathcal{F}$.

Coefficient	Value	Units
a_1^*, a_2^*	3.8×10^5 (T–479 °C) $- 11 \times 10^9 u_m$	C^{-2} m^2 N^{-1}
a_3^*	3.8×10^5 (T–479 °C) $+ 9.5 \times 10^9 u_m$	C^{-2} m^2 N^{-1}
a_{11}^*, a_{22}^*	0.42×10^9	C^{-4} m^6 N
a_{33}^*	0.05×10^9	C^{-4} m^6 N
a_{13}^*, a_{23}^*	0.45×10^9	C^{-4} m^6 N
a_{12}^*	0.73×10^9	C^{-4} m^6 N
$a_{111}, a_{222}, a_{333}$	0.26×10^9	C^{-6} m^{10} N
$a_{112}, a_{113}, a_{223}$	0.61×10^9	C^{-6} m^{10} N
a_{123}	-3.7×10^9	C^{-6} m^{10} N
G_{1111}	2.77×10^{-10}	C^{-2} m^4 N
G_{1122}	0.0	C^{-2} m^4 N
G_{1212}	1.38×10^{-10}	C^{-2} m^4 N

The value of the parameter γ, which determines the time scale of relaxation, is not essential for the static case and is assumed be equal to one. In our calculations the coordinate axis, z, is directed across the ferroelectric layer, and the x and y axes lie within the plane of the ferroelectric layer. The coordinate origin is chosen in the middle of the ferroelectric layer. It is assumed that the domains have the structure of stripes, elongated along the y axis. For the numerical solution of Equations (1), Newton's method is used. At each step of calculation, a linear system is solved using the iterative generalized minimum residual method (GMRES) [24,25]. The implementation of the method is described in detail in [26].

3. Results

Figure 1 shows the calculations for different layer thicknesses' temperature dependencies on the average polarization, $P = \langle \mathbf{P}^2 \rangle^{1/2}$, the inverse of average dielectric constant of the ferroelectric layer, ε_f^{-1}, and the total capacitance of the heterostructures, c, taken per unit area, respectively.

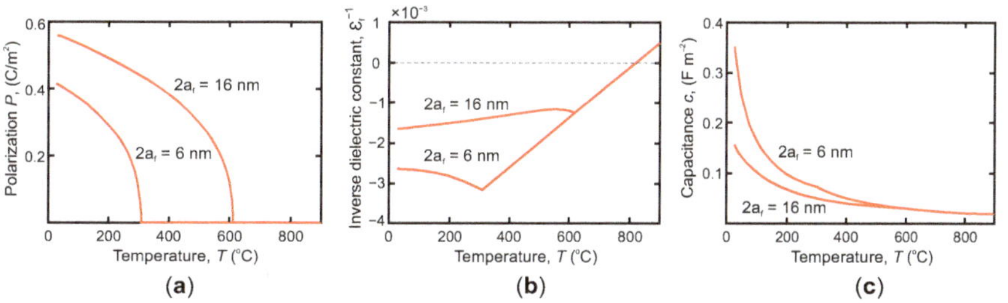

Figure 1. (**a**) Temperature dependence of the average polarization magnitude in the domain structure of ferroelectric PTO layers with thicknesses $a_f = 6$ nm and $a_f = 16$ nm; (**b**) temperature dependence of the inverse effective dielectric constant ε_f^{-1} for PTO layers; (**c**) temperature dependence of capacitance (calculated per unit area) for STO/PTO/STO heterostructures with layer thicknesses $a_f/2a_f/a_f$, where $2a_f = 6$ nm and $2a_f = 16$ nm.

The temperature dependence of the average polarization in the domain state, $P(T)$, shown in panel (a), is typical of second-order phase transitions, where the transition temperature T_{cd} for the 16-nm-thick film being higher than that of the film with 6-nm thickness. This transition is clearly seen as a kink in the dependencies of the inverse dielectric constant upon the temperature, $\varepsilon_f^{-1}(T)$, presented in panel (b). At the same time, the manifestation of the ferroelectric phase transition in the temperature dependence of the total capacitance of the heterostructure, $c(T)$, presented in panel (c), is minor. In the Discussion, we will deliberate these features in detail.

The temperature-thickness phase diagram of the ferroelectric layer "T-$2a_f$" [6], reconstructed on base dielectric and polarization calculations is shown in the Figure 2. It demonstrates four characteristic states, I–IV, revealing the different dielectric response to the applied field. There appear three important characteristic temperatures: T_{c0}, the temperature that corresponds to the critical temperature of the bulk short-circuited sample in the absence of depolarization effects (Curie temperature); T_{cd}, is the temperature of the transition to the multi-domain state; and T_d^*, the temperature at which the soft (vortex) domains acquire the Landau-Kittel domain profile. At T_{c0}, the paraelectric transforms into a so-called incipient ferroelectric state [15,27,28] where the dielectric constant becomes negative, but the depolarization field still suppresses establishing of the global ferroelectricity. The temperature T_{cd} marks the transition from the incipient state to the state where soft ferroelectric domains emerge [11]. Finally, at T_d^*, the blurred domain walls of the soft domains become rigid and narrow, and, as mentioned above, Landau-Kittel domain structure sets in [10]. Respectively, states I and II correspond to the paraelectric phases, while states III and IV correspond to the ferroelectric phases in which domains are formed.

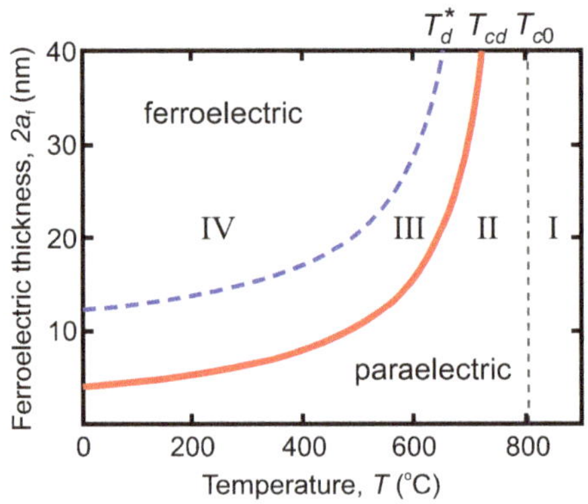

Figure 2. The temperature, T—layer thickness, $2a_f$, phase diagram of the response of polarization states in a ferroelectric layer to an external field. The roman numbers stand for the states having the different dielectric properties; T_{c0} is the temperature of the transition to the single-domain state when depolarization effects are absent; T_{cd} is the temperature of the transition to the multi-domain state; T_d^* is the temperature at which soft (vortex) domains acquire the Landau-Kittel domain profile.

The Figure 3 depicts polarization textures in a ferroelectric layer with thickness $2a_f = 16$ nm, obtained for different temperatures T below T_{cd}. The Figure 3a shows the structure of soft domains (state III) arising slightly below T_{cd}, and Figure 3b shows the structure of classical Landau-Kittel domains arising below the crossover temperature T_d^* (state IV).

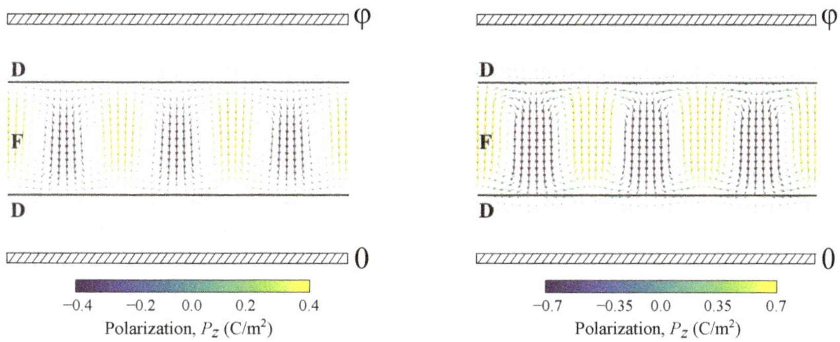

(a) (b)

Figure 3. The polarization texture of the dielectric(D)/ferroelectric(F)/dielectric(D) heterostructure STO/PTO/STO with a ferroelectric layer thickness of $2a_f = 16\,\text{nm}$ (**a**) at temperatures slightly below T_{cd} (soft domains), (**b**) at temperatures below T_d^* (Landau-Kittel domains).

4. Discussion

To describe the dielectric response of the paraelectric and ferroelectric states I–IV, we note that the transition temperature to the multidomain state [6],

$$T_{cd} = \left(1 - 2\pi\left(\frac{C}{T_{c0}\varepsilon_\perp}\right)^{1/2}\frac{\xi_0}{2a_f}\right)T_{c0}, \tag{4}$$

is lower than the transition temperature $T_{c0} \approx 800\,^\circ\text{C}$ for a short-circuited (i.e., depolarization field-free) single-domain ferroelectric PTO film, compressively-strained by the STO substrate. Here $C \approx 4.1 \times 10^5\,^\circ\text{C}$ is the Curie constant of the PTO in the paraelectric phase, $\xi_0 \simeq 1\,\text{nm}$ is the coherence length, and $\varepsilon_\perp \simeq 30$—noncritical dielectric constant transverse to spontaneous polarization in the film plane.

This reduction of T_{cd} is associated with the additional gradient energy required for the domain creation. Therefore, when the temperature drops below T_{c0}, the system remains in a supercooled paraelectric state II, and its Curie-type dielectric constant (positive in a conventional paraelectric state I at $T > T_{c0}$)

$$\varepsilon_p(T) = \frac{C}{T - T_{c0}} + \varepsilon_i, \tag{5}$$

becomes negative below T_{c0}. This paraelectric state II with the negative Curie-type dielectric constant exists in the temperature range $T_{cd} < T < T_{c0}$, and then transforms into the so-called soft domain state, denoted as state III in Figure 2. The dielectric constant at the transition point is determined by the relations (4) and (5) as

$$\varepsilon_p(T_{cd}) = -\frac{1}{2\pi}\left(\varepsilon_\perp\frac{C}{T_{c0}}\right)^{1/2}\frac{2a_f}{\xi_0} + \varepsilon_i. \tag{6}$$

The soft domains structure, shown in Figure 3a, has the profile of a smoothly varying polarization with the period of $2w$, which was found from the functional (3) minimization. This profile consists of the spontaneous z-oriented polarization, $P_z \propto \cos\frac{\pi z}{2a_f}\cos\frac{\pi x}{w}$, and the depolarization-field-induced polarization with the predominantly x-orientated component $P_x = -\varepsilon_0(\varepsilon_\perp - 1)\partial_x\varphi \propto \sin\frac{\pi z}{2a_f}\sin\frac{\pi x}{w}$. Here the depolarization potential φ is determined by the Poisson equation $\varepsilon_0\varepsilon_{f\perp}\partial_x^2\varphi + \varepsilon_0\varepsilon_i\partial_z^2\varphi = \partial_z P_z$.

The shown in Figure 3a complete soft polarization profile includes both the spontaneous and induced components, $\mathbf{P} = (P_x, P_z)$. This texture, also observed experimentally [9], resembles a periodic vortex-antivortex texture and called as vortex phase as well. Negative capacitance in the soft domain phase is due to the locally negative dielectric response of the wide domain wall regions, containing the paraelectric-like phase with $P \approx 0$ [16]. Its absolute value decreases slowly with decreasing of the temperature.

Note that the transition to the soft domain phase manifests itself as an insignificant kink in the temperature dependence of the capacitance of the entire heterostructure (Figure 1c), consisting of capacities of the ferroelectric and dielectric layers connected in series, $c^{-1} = c_f^{-1} + c_d^{-1}$. The constitutive capacitancies, calculated per unit area, are expressed through the dielectric constants of the corresponding layers, ε_f (or ε_p for $T > T_{cd}$) and ε_d as $c_f = \varepsilon_0\varepsilon_f/2a_f$ and $c_d = \varepsilon_0\varepsilon_d/2a_d$, respectively. Herewith the capacitance of the heterostructure, c, is a positive value, which ensures the stability of the system. The plot in Figure 1c presumes that the total layer thicknesses of ferroelectric and dielectric materials are the same, $2a_f = 2a_d$, and that the dielectric constant STO follows the Curie-Weiss law, $\varepsilon_d = C_{STO}/(T - T_{STO})$ where $C_{STO} \approx 7.83 \times 10^4\,°C$, $T_{STO} \approx -245\,°C$.

Upon the significant decrease in temperature, the domain polarization profiles acquire the form of the classical Landau-Kittel domains, in which the regions of the oppositely-oriented almost uniform polarizations are separated by thin domain walls. The negative sign of the effective dielectric constant arises mostly due to the motion of the domain walls under the influence of the field. Although such a regime is not fully achievable in the nanoscale films (see the phase diagram in Figure 2), the corresponding dielectric constant value calculated in [13],

$$\varepsilon_f = \varepsilon_\parallel \left[1 - \frac{\pi\varsigma}{4\ln 2}\left(\frac{\varepsilon_\perp}{\varepsilon_\parallel}\right)^{1/2}\frac{2a_f}{w}\right], \tag{7}$$

can be considered as the higher limit of the negative dielectric constant, attained at low temperatures. Here $\varsigma = 1 + \varepsilon_d/\left(\varepsilon_\parallel\varepsilon_\perp\right)^{1/2}$ and $\varepsilon_\parallel \approx C/2T_{c0}$ is the dielectric constants of uniform ferroelectric at low temperatures in the longitudinal to polarization direction.

The slight decrease in the numerically calculated inverse dielectric constant ε_f^{-1} from -10^{-3} to -1.7×10^{-3} (i.e., increasing of ε_f) with decreasing the temperature below $T_{cd} \simeq 600\,°C$, shown in Figure 1b for ferroelectric layer with $2a_f = 16$ nm, is in line with the analytical estimations (6) and (7) in which ε_p at T_{cd} is lower then ε_f at $T < T_{cd}$. Surprisingly, however, for thinner layer with $2a_f = 8$ nm the calculated temperature behaviour is the opposite, ε_f^{-1} increases from -3.2×10^{-3} to -2.7×10^{-3} with decreasing temperature. This difference is attributed to the fact that for such thin films the Landau-Kittel regime is not realized, see phase diagram in Figure 2, and the corresponding approximation (7) does not apply. Note that for the very thin ferroelectric layers the experimentally measured dielectric constant, negative just below T_{cd}, becomes again positive at lower temperatures [15]. This reversal behaviour can be explained either by the pinning of domain walls on the structural inhomogeneities, that prevents the domain wall motion, or by specific nonlinear interaction of the STO_3 dielectric layer with fringing depolarization field of domains, emerging outside the PTO layer [29].

Figure 4 depicts the z-dependence of the electric potential across the heterostructure at the applied voltage $V = 2$ V for states I–IV. In paraelectric states I and II, the induced electric fields inside paraelectric and dielectric layers are uniform, and the corresponding piecewise linear dependence of the potential along z is shown in Figure 4a,b by the red line. In ferroelectric states III and IV with domains, the distribution of the induced depolarization electric field is highly non-uniform due to the inhomogeneity of polarization in the domain structure. The spread of the potential dependencies along different lines parallel to z in these states around its averaged effective value (piecewise red lines) is shown in Figure 4c,d in blue gamma. Increasing the effective potential in the ferroelectric layer in states II–IV, against the background of the general drop of the potential between the electrodes, corresponds to the described effect of the negative dielectric constant.

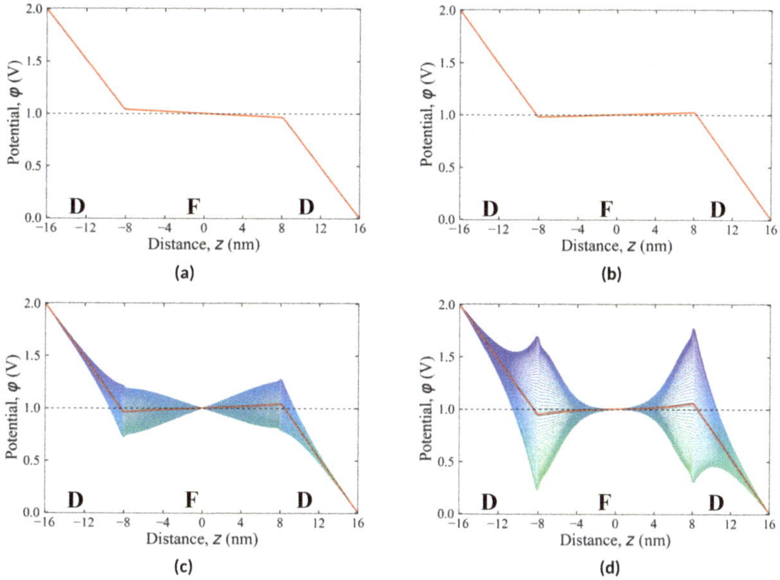

Figure 4. Distribution of the potential φ in the STO/PTO/STO heterostructure depending on its thickness for various states of the ferroelectric PTO layer described in Figure 2: (**a**) state I, (**b**) state II, (**c**) state III, (**d**) state IV.

Utilizing this effect is expected to help in devising an energy-saving field-effect transistor in which the voltage controlling the conducting channel, formed near the ferroelectric layer, is amplified due to the increase in the effective potential in the ferroelectric layer [17]. However, the domain-induced inhomogeneous distributions of the potential in the ferroelectric layer, also penetrating the dielectric layer over the penetration depth of order w, is highly nonuniform in the region of the conducting channel and varies as $\varphi \propto \cos\frac{\pi x}{w}e^{-\pi\delta/w}$, where δ is the distance from the paraelectric–dielectric interface to the channel. As shown in Figure 4c,d, the characteristic spread of the potential near the ferroelectric–dielectric interface, hence, in the channel region, is surprisingly large and is of the order of 1.2 V. Not only does it exceed the amplification of the operating voltage near the conducting channel of order of 0.3 V, arising due to the effect of the ferroelectric negative capacitance, but it is also comparable to the total voltage applied to the transistor.

Such a strong field inhomogeneity can make the conducting channel uncontrollable. Note further that the use of state II with the negative dielectric constant for creating an energy-saving transistor is a technologically challenging task [28]. Although this state does not have an aforementioned inhomogeneity of the field distribution, it has a small gain parameter that is difficult to integrate into the existing semiconductor technologies. Therefore the new routs and architectures should be devised to implement the negative capacitance effect. In particular, the integration of the intermediate floating-gate electrode, compensating the domain-induced field inhomogeneity and allowing for the control of the domain wall displacement [14] holds high technological promise.

5. Conclusions

In this work we obtained the temperature dependence of the characteristic dielectric constant of a ferroelectric PTO layer with domains integrated in the STO/PTO/STO heterostructure. We have shown that the inverse dielectric constant of a ferroelectric layer of thickness $2a_f = 16$ nm is negative and decreases from -10^{-3} to -1.7×10^{-3} with decreasing temperature in the temperature range of the phase transition temperature $T_{cd} \simeq 600\,^{\circ}$C to room temperature. For the ferroelectric layer with the thickness of 6 nm, the phase transition temperature decreases to 300 °C, and the inverse dielectric constant increases from -3.2×10^{-3} to -2.7×10^{-3} with decreasing temperature. We demonstrated that the negative capacitance effect associated with a multidomain texture can be used to create energy-efficient field-effect transistors, provided that the domain-induced inhomogeneities of the field in the channel region will be smoothed by the clever architecture of the transistor.

Note that the effect of the negative capacitance takes place in a variety of the nanosized systems including nanocylinders, nanodots and nanoparticles containing topological structures of polarization, such as vortices [30], skyrmions [26] and Hopfions [31]. The possibility of controlling such formations by an external electric field enables ferroelectric nanostructures to become a basic constituent element of next-generation transistors.

Author Contributions: Conceptualization, I.A.L.; methodology, M.A.P. and Y.A.T.; software, M.A.P. and Y.A.T.; validation, I.A.L., Y.A.T. and V.M.V.; formal analysis, I.A.L., A.G.R. and V.M.V.; investigation, M.A.P., Y.A.T. and A.G.R.; writing—original draft preparation, M.A.P., Y.A.T. and A.G.R.; writing—review and editing, I.A.L. and V.M.V.; project administration, A.G.R. All authors have read and agreed to the published version of the manuscript.

Funding: This research was financially supported by the Ministry of Science and Higher Education of the Russian Federation, grant agreement number 075-15-2021-953 from 5.10.2021, and by the European Union H2020-MSCA-RISE-MELON action, project number 872631.

Data Availability Statement: Data underlying the results presented in this paper are not publicly available at this time but may be obtained from the authors upon reasonable request.

Conflicts of Interest: The authors declare no conflict of interest.

References

1. Mikolajick, T.; Schroeder, U.; Slesazeck, S. The Past, the Present, and the Future of Ferroelectric Memories. *IEEE Trans. Electron Devices* **2020**, *67*, 1434–1443. [CrossRef]
2. Oh, S.; Hwang, H.; Yoo, I.K. Ferroelectric materials for neuromorphic computing. *APL Mater.* **2019**, *7*, 091109. [CrossRef]
3. Das, S.; Hong, Z.; McCarter, M.; Shafer, P.; Shao, Y.T.; Muller, D.A.; Martin, L.W.; Ramesh, R. A new era in ferroelectrics. *APL Mater.* **2020**, *8*, 120902. [CrossRef]
4. Landau, L.; Lifshitz, E. On the theory of the dispersion of magnetic permeability in ferromagnetic bodies. *Phys. Z. Sowjetunion* **1935**, *8*, 153.
5. Bratkovsky, A.M.; Levanyuk, A.P. Abrupt Appearance of the Domain Pattern and Fatigue of Thin Ferroelectric Films. *Phys. Rev. Lett.* **2000**, *84*, 3177–3180. [CrossRef]
6. De Guerville, F.; Luk'yanchuk, I.; Lahoche, L.; El Marssi, M. Modeling of ferroelectric domains in thin films and superlattices. *Mater. Sci. Eng. B* **2005**, *120*, 16–20.
7. Zubko, P.; Stucki, N.; Lichtensteiger, C.; Triscone, J.M. X-ray Diffraction Studies of 180° Ferroelectric Domains in $PbTiO_3/SrTiO_3$ Superlattices under an Applied Electric Field. *Phys. Rev. Lett.* **2010**, *104*, 187601. [CrossRef]
8. Baudry, L.; Luk'yanchuk, I.A.; Razumnaya, A. Dynamics of field-induced polarization reversal in thin strained perovskite ferroelectric films with *c*-oriented polarization. *Phys. Rev. B* **2015**, *91*, 144110. [CrossRef]
9. Yadav, A.; Nelson, C.; Hsu, S.; Hong, Z.; Clarkson, J.; Schleputz, C.; Damodaran, A.; Shafer, P.; Arenholz, E.; Dedon, L.; et al. Observation of polar vortices in oxide superlattices. *Nature* **2016**, *530*, 198–201. [CrossRef]
10. Luk'yanchuk, I.A.; Lahoche, L.; Sené, A. Universal Properties of Ferroelectric Domains. *Phys. Rev. Lett.* **2009**, *102*, 147601. [CrossRef]
11. Stephanovich, V.A.; Luk'yanchuk, I.A.; Karkut, M.G. Domain-Enhanced Interlayer Coupling in Ferroelectric/Paraelectric Superlattices. *Phys. Rev. Lett.* **2005**, *94*, 047601. [CrossRef] [PubMed]
12. Bratkovsky, A.M.; Levanyuk, A.P. Very large dielectric response of thin ferroelectric films with the dead layers. *Phys. Rev. B* **2001**, *63*, 132103. [CrossRef]
13. Luk'yanchuk, I.; Sené, A.; Vinokur, V.M. Electrodynamics of ferroelectric films with negative capacitance. *Phys. Rev. B* **2018**, *98*, 024107. [CrossRef]
14. Luk'yanchuk, I.; Tikhonov, Y.; Sené, A.; Razumnaya, A.; Vinokur, V.M. Harnessing ferroelectric domains for negative capacitance. *Commun. Phys.* **2019**, *2*, 22. [CrossRef]
15. Zubko, P.; Wojdeł, J.C.; Hadjimichael, M.; Fernandez-Pena, S.; Sené, A.; Luk'yanchuk, I.; Triscone, J.M.; Íñiguez, J. Negative capacitance in multidomain ferroelectric superlattices. *Nature* **2016**, *534*, 524–528. [CrossRef]
16. Íñiguez, J.; Zubko, P.; Luk'yanchuk, I.; Cano, A. Ferroelectric negative capacitance. *Nat. Rev. Mater.* **2019**, *4*, 243–256. [CrossRef]

17. Salahuddin, S.; Datta, S. Use of Negative Capacitance to Provide Voltage Amplification for Low Power Nanoscale Devices. *Nano Lett.* **2007**, *8*, 405–410. [CrossRef]
18. Salvatore, G.A.; Rusu, A.; Ionescu, A.M. Experimental confirmation of temperature dependent negative capacitance in ferroelectric field effect transistor. *Appl. Phys. Lett.* **2012**, *100*, 163504. [CrossRef]
19. Logg, A.; Mardal, K.A.; Wells, G.N. (Eds.) *Automated Solution of Differential Equations by the Finite Element Method;* Lecture Notes in Computational Science and Engineering; Springer: Berlin/Heidelberg, Germany, 2012; Volume 84. [CrossRef]
20. Mokrý, P.; Sluka, T. Identification of defect distribution at ferroelectric domain walls from evolution of nonlinear dielectric response during the aging process. *Phys. Rev. B* **2016**, *93*, 064114. [CrossRef]
21. Pertsev, N.A.; Zembilgotov, A.G.; Tagantsev, A.K. Effect of Mechanical Boundary Conditions on Phase Diagrams of Epitaxial Ferroelectric Thin Films. *Phys. Rev. Lett.* **1998**, *80*, 1988–1991. [CrossRef]
22. Baudry, L.; Lukyanchuk, I.; Vinokur, V.M. Ferroelectric symmetry-protected multibit memory cell. *Sci. Rep.* **2017**, *7*, 42196. [CrossRef]
23. Wang, J.; Shi, S.Q.; Chen, L.Q.; Li, Y.; Zhang, T.Y. Phase-field simulations of ferroelectric/ferroelastic polarization switching. *Acta Mater.* **2004**, *52*, 749–764. [CrossRef]
24. Balay, S.; Abhyankar, S.; Adams, M.F.; Benson, S.; Brown, J.; Brune, P.; Buschelman, K.; Constantinescu, E.M.; Dalcin, L.; Dener, A.; et al. PETSc Web Page. 2021. Available online: https://petsc.org/ (accessed on 28 September 2021).
25. Balay, S.; Abhyankar, S.; Adams, M.F.; Benson, S.; Brown, J.; Brune, P.; Buschelman, K.; Constantinescu, E.; Dalcin, L.; Dener, A.; et al. PETSc/TAO Users Manual. *Technical Report ANL-21/39—Revision 3.16;* Argonne National Laboratory: DuPage County, IL, USA, 2021.
26. Tikhonov, Y.; Kondovych, S.; Mangeri, J.; Pavlenko, M.; Baudry, L.; Sené, A.; Galda, A.; Nakhmanson, S.; Heinonen, O.; Razumnaya, A.; et al. Controllable skyrmion chirality in ferroelectrics. *Sci. Rep.* **2020**, *10*, 8657. [CrossRef]
27. Sené, A. Théorie des Domaines et des Textures non Uniformes dans les Ferroélectriques. Ph.D. Thesis, Université de Picardie Jules Verne, Amiens, France, 2010.
28. Cano, A.; Jiménez, D. Multidomain ferroelectricity as a limiting factor for voltage amplification in ferroelectric field-effect transistors. *Appl. Phys. Lett.* **2010**, *97*, 133509. [CrossRef]
29. Zubko, P. (University College London, London, UK). Private Communication, 2017.
30. Lachoche, L.; Lukyanchuk, I.; Pascoli, G. Stability of vortex phases in ferroelectric easy-plane nano-cylinders. *Integr. Ferroelectr.* **2008**, *99*, 60–66. [CrossRef]
31. Luk'yanchuk, I.; Tikhonov, Y.; Razumnaya, A.; Vinokur, V.M. Hopfions emerge in ferroelectrics. *Nat. Commun.* **2020**, *11*, 2433. [CrossRef] [PubMed]

MDPI
St. Alban-Anlage 66
4052 Basel
Switzerland
Tel. +41 61 683 77 34
Fax +41 61 302 89 18
www.mdpi.com

Nanomaterials Editorial Office
E-mail: nanomaterials@mdpi.com
www.mdpi.com/journal/nanomaterials

www.ingramcontent.com/pod-product-compliance
Ingram Content Group UK Ltd.
Pitfield, Milton Keynes, MK11 3LW, UK
UKHW060123010826
14089UKWH00006B/127

9 783036 547398